# OUTLINES

## OF THE

# GEOGRAPHY OF PLANTS

# OUTLINES

## OF THE

# GEOGRAPHY OF PLANTS:

WITH PARTICULAR ENQUIRIES CONCERNING

THE NATIVE COUNTRY, THE CULTURE, AND THE USES

OF THE

## PRINCIPAL CULTIVATED PLANTS ON WHICH THE PROSPERITY OF NATIONS IS BASED.

By

**F. J. F. MEYEN, PH.D. M.D.**

*Translated by*

MARGARET JOHNSTON

GYAN

*Published by*

Gyan Publishing House
5, Ansari Road
Daryaganj, New Delhi-110002
Phone: 011-47034999, 9811692060
E-mail: books@gyanbooks.com

*Distribution Network*
gyanbooks.com
India, USA, Canada, UK, Australia, France



ISBN : 978-81-212-6332-0 (PB)
First Published, 1846

2nd Impression 2022

*Printed at:* Gyan Press, Delhi.



**OUTLINES OF THE GEOGRAPHY OF PLANTS**
Author: F. J. F. MEYEN

# OUTLINES

OF THE

# GEOGRAPHY OF PLANTS:

WITH PARTICULAR ENQUIRIES CONCERNING

THE NATIVE COUNTRY, THE CULTURE, AND THE USES

OF THE

PRINCIPAL CULTIVATED PLANTS ON WHICH THE PROSPERITY OF NATIONS IS BASED.

BY

F. J. F. MEYEN, PH.D. M.D.

TRANSLATED BY

MARGARET JOHNSTON.

1846

# TABLE OF CONTENTS.


| | Page. |
|---|---|
| INTRODUCTION TO BOTANICAL GEOGRAPHY, ... ... | 1 |
| CHIEF WORKS ON BOTANICAL GEOGRAPHY, ... ... ... ... | 6 |

## PART I.

| | |
|---|---|
| ON THE CONDITIONS OF CLIMATE WHICH DETERMINE THE PRESENCE AND DISTRIBUTION OF PLANTS, ... ... | 8 |
| Influence of the Winds and the Moisture of the Air on the regular Distribution of Heat, and of the Vegetation dependent on it, ... ... | 9 |
| On the Daily Progress of Heat, ... ... ... ... | 11 |
| Explanation of the Cause of the great Difference between Coast and Continental Climate, and the consequent Difference in the Vegetation, | 13 |
| On the Mean Heat of a Place, and its Influence on the Vegetation, ... | 15 |
| Importance of the Isothermal, Isotheral, and Isochimenal Lines to Botanical Geography, ... ... ... ... ... ... | 20 |
| Parallelism between the Decrease of Heat and the Change in Vegetation from the Equator to the Poles, compared with that from the Plain in the Tropics to the Tops of the Mountains, ... ... ... | 25 |
| On the Height of the Final Limit of Vegetation in the different Latitudes, which, on the whole, coincides with the Height of the Limit of Perpetual Snow, ... ... ... ... ... ... | 30 |
| The Heat of the Ground has also been regarded as Influential on Vegetation, ... ... ... ... ... ... ... | 34 |
| On the Influence of Warmth on the Development of the Leaves and Flowers in Spring, ... ... ... ... ... ... | *ib.* |
| Influence of the Moisture of the Atmosphere, and of the Ground on Vegetation, ... ... ... ... ... ... ... | 38 |
| On the Operation of Currents in the Air and in Water, in extending the Station of Plants by Migration, ... ... ... ... | 39 |

## PART II.

Page.

ON THE CONDITIONS BY WHICH THE SOIL INFLUENCES THE STATION AND DISTRIBUTION OF PLANTS, ... 46

The Difficulty of Explaining the ultimate Causes of the Influence of the Conditions of Soil on the Stations of Plants, ... ... ... *ib.*

Consideration of the Various Local Conditions of the Stations of Plants, 50

1. Aquatic Plants:— ... ... ... ... ... *ib.*
   - Marine Plants, ... ... ... ... ... ... 51
   - Fresh-water Plants, ... ... ... ... ... 54
   - Submerged Plants, ... ... ... ... ... 56
   - Floating Plants, ... ... ... ... ... ... *ib.*
   - Lake Plants, ... ... ... ... ... ... 57
   - Ditch Plants, ... ... ... ... ... ... 58
   - River Plants, ... ... ... ... ... ... *ib.*
   - Fountain Plants, ... ... ... ... ... ... *ib.*
   - Saline Plants, ... ... ... ... ... ... 59
   - Amphibious and Inundated Plants, ... ... ... ... 60
   - Maritime and Shore Plants, ... ... ... ... *ib.*
   - Mangrove Forests, ... ... ... ... ... 61
2. Land Plants:— ... ... ... ... ... ... 62
   - Influence of the Geognostic Nature of the Soil, ... ... *ib.*
   - Sand and Flint Plants—Chalk Plants, ... ... ... *ib.*
   - Gypsum Plants—Turf Plants, ... ... ... ... 63
   - Bog Plants, ... ... ... ... ... ... 64
   - Marsh Plants, ... ... ... ... ... ... 65
   - Influence of the Soil in respect of its Aggregate Condition—Rock Plants—Stone Plants, ... ... ... ... ... *ib.*
   - Sand Plants—Gravel Plants, ... ... ... ... 66
   - Influence of the Soil on the Station of Plants in respect of its Nature—
     1. Station of Plants on other Living Plants, ... ... 67
        - True Parasites, ... ... ... ... ... *ib.*
        - Parasites, ... ... ... ... ... ... 69
        - False Parasites, ... ... ... ... ... *ib.*
        - Leaf Fungi or the Exanthema of Plants, ... ... ... 72
     2. Station of Plants on Dead Organic Matter, ... ... 73
     3. Station of Plants on Artificial Productions—
        - Wall Plants, ... .. ... ... ... ... *ib.*
        - Roof Plants—Plank Plants—Rubbish Plants, ... ... 74
   - Influence of the Soil in respect of its State of Culture—
     - I. Plants which Grow in Cultivated Soil, ... ... ... 75
        - Field Plants, ... ... ... ... ... ... *ib.*
        - Fallow Plants—Garden Weeds—Garden Plants, ... ... 76
        - Fence Plants—Hedge Plants, ... ... ... ... 77
     - II. Plants which Grow on Uncultivated Ground, ... ... *ib.*
        - Field Plants—Desert Plants—Meadow Plants, ... ... *ib.*

| | Page. |
|---|---|
| Pasture Plants, ... ... ... ... ... | 78 |
| Heath Plants, ... ... ... ... ... | ib. |
| Mountain Plants, ... ... ... ... ... | 79 |
| Bush Plants, ... ... ... ... ... | ib. |
| Forest Plants, ... ... ... ... ... | ib. |
| On the Social Growth of Plants, ... ... ... ... | 80 |
| Definitions of the Station and Distribution of Plants, ... ... | 88 |
| Station of Plants, ... ... ... ... ... | ib. |
| Distribution of Plants, ... ... ... ... ... | ib. |
| Zone of Latitude of Plants, ... ... ... ... | 89 |
| The Vertical Range or Region of Plants, .. ... ... | ib. |
| Polar and Equatorial Limits of the Stations of Plants, ... | ib. |
| Longitudinal Zone of Plants, ... ... ... ... | ib. |
| Interrupted and Uninterrupted Areas of Plants, ... ... | ib. |
| Natural and Artificial Areas of Plants, ... ... ... | 90 |
| Extent of the Areas of Plants, ... ... ... ... | 92 |
| Recapitulation of the Uses of Botanical Geography, ... ... | 96 |

## PART III.

| | |
|---|---|
| ON THE DISTRIBUTION OF PLANTS OVER THE SURFACE OF THE EARTH, WITH PARTICULAR RESPECT TO THE PHYSIOGNOMY OF NATURE, ... ... ... ... | 98 |
| General considerations of the Distribution of Plants, ... ... | ib. |
| I. THE PHYSIOGNOMICS OF VEGETATION, ... ... ... | 105 |
| A. Special consideration of the Physiognomy of the Principal Forms of Plants, ... ... ... ... ... ... | 106 |
| 1. The Grasses or Gramineous Plants, ... ... ... | ib. |
| 2. The Scitamineæ, ... ... ... ... ... | 111 |
| a. The Banana Form, ... ... ... .. | ib. |
| 3. The Form of Pandanus, ... ... ... ... | 113 |
| 4. The Form of the Ananas Plants, ... ... ... | 116 |
| 5. The Agave Form, ... ... ... ... | 118 |
| 6. The Palms, ... ... ... ... ... | 121 |
| b. The Cycadeæ, ... ... ... ... ... | 124 |
| 7. The Fern Form, ... ... ... ... ... | 125 |
| 8. The Mimosa Form, ... .. ... ... | 127 |
| 9. The Abietinæ, ... ... ... ... ... | 130 |
| 10. The Forms of Protea, Epacris, and Erica, ... ... | 134 |
| 11. The Myrtle Form, ... ... ... ... | 135 |
| 12. The Form of Dicotyledonous Trees, ... ... ... | 138 |
| a. Those with tender leaves, ... ... ... ... | 139 |
| b. Those with leathery and glossy leaves, ... ... | ib. |
| c. The Willow Form, ... ... ... ... | 140 |
| d. The Dicotyledonous Trees with large and finely shaped leaves, | 141 |

| | Page. |
|---|---|
| 13. The Cactus Form, ... ... ... ... | ib. |
| 14. The Succulent Plants, ... ... ... ... | 148 |
| 15. The Lily Form, ... ... ... ... ... | 149 |
| 16. The Lianas or Climbing Plants, ... ... ... | 151 |
| 17. The Pothos Plants, ... ... ... ... | 153 |
| 18. The Orchideæ, ... ... ... ... ... | 155 |
| 19. The Mosses, ... ... ... ... ... | 156 |
| 20. The Lichens, ... ... ... ... ... | ib. |
| *B.* General Phytogeographical Division of the Surface of the Globe, according to the Physiognomy of Vegetation, ... ... | 157 |
| General Remarks, ... ... ... ... ... | ib. |
| *a.* Division of the Horizontal Range of Vegetation into Zones, | 161 |
| 1. The Equatorial Zone, ... ... ... ... | ib. |
| 2. The Tropical Zone, ... ... ... .. ... | 171 |
| 3. The Sub-Tropical Zone, ... ... ... ... | 176 |
| 4. The Warmer Temperate Zone, ... ... ... | 190 |
| 5. The Colder Temperate Zone, ... ... ... | 199 |
| 6. The Sub-Arctic Zone, ... ... ... ... | 209 |
| 7. The Arctic Zone, ... ... ... ... ... | 216 |
| 8. The Polar Zone, ... ... ... ... ... | 220 |
| *b.* Division of the Vertical Range of Vegetation into Regions, | 223 |
| 1. The Region of Palms and Bananas, ... ... .. | 229 |
| 2. The Region of Tree Ferns and Figs, ... ... ... | 231 |
| 3. The Region of the Myrtaceæ and Laurineæ, ... ... | 234 |
| 4. The Region of the Evergreen Dicotyledonous Trees, ... | 237 |
| 5. The Region of Oaks and European Dicotyledonous Trees, | 240 |
| 6. The Region of Abietinæ, ... ... ... ... | 243 |
| 7. The Region of Rhododendrons, ... .. ... | 247 |
| 8. The Region of Alpine Plants, ... ... ... | 250 |
| II. The Statistics of Plants, ... ... ... ... | 259 |
| On the number of Existing Species, ... ... ... | 260 |
| The Vegetation of Islands seems not to be poorer in Species, than Continental Tracts of Land under equal Conditions, ... | 261 |
| Vegetation gradually becomes richer towards the Equator, not only in number of Species, but in number of individuals, ... | 263 |
| Nature under similar Conditions always produces similar or perfectly the same creatures, ... ... ... ... ... | 265 |
| Nature still produces the lower Plants as well as the lower Animals without Germs, ... ... ... ... ... | 266 |
| General Rules for the Method of proceeding in Statistical Reckonings of the floras of single Countries, ... ... .. | 271 |
| On the proportion of Cryptogamia to Phanerogamæ no Laws can be at present laid down, as the Materials are still too imperfect, | 274 |
| Statistical Proportions of Ferns, ... ... ... ... | 275 |
| Proportional Numbers of Monocotyledones to Dicotyledones in different Zones and different Regions, ... ... ... | 278 |

Page.

Considerations of the Statistical Proportions of Various Families of Plants, ... 281

In Statistical Reckonings of the Flora of a Country, each Region into which it can be divided must be reckoned singly, ... 286

## SUPPLEMENT.

HISTORY OF CULTIVATED PLANTS, CONTAINING INQUIRIES CONCERNING THE NATIVE COUNTRY, THE DISTRIBUTION, THE CULTURE, AND THE USES OF THE PRINCIPAL CULTIVATED PLANTS, WHICH SERVE AS WELL FOR THE FOOD, AS FOR THE COMFORT, THE LUXURY, AND THE TRADE OF NATIONS, AND ARE THE FOUNDATIONS OF THEIR PROSPERITY.

I. THE CULTURE OF THE CEREALS, ... 290
- 1. Wheat, ... 292
- 2. Spelt, Barley, Rye, and Oats, ... 295
- 3. Rice, ... 297
- 4. Maize, ... 302
- 5. Millet, ... 306
- 6. Quinoa, ... 308
- 7. Buck-wheat, ... 309

II. THE CULTURE OF THE PRINCIPAL TUBEROUS ROOTS—
- 1. The Potato, ... 310
- 2. The Arum or Aron, ... 314
- 3. The Manioc or Mandiocca, ... 316
- 4. The Batata or Camota, ... 318
- 5. The Igname or Yam, ... 320
- 6. The Oca, the Tacca, the Tubers of Sagittaria, and Dracontium, *ib.*

III. THE CULTURE OF THE PRINCIPAL TREES WHOSE FRUITS ARE THE GENERAL FOOD OF NATIONS—
- 1. The Bread-Fruit, ... 321
- 2. The Plantain or Banana, ... 324
- 3. The Olive, ... 328
- 4. The Cocoa Palm, ... 331
- 5. The Date, ... 336
- 6. The Chilian Palm, ... 338
- 7. The Mauritius Palm, ... *ib.*
- 8. The Sago Palm, ... 340
- 9. The Guinea Oil Palm, ... 341
- 10. The Wine Palm, ... 342
- 11. The Water Nut, ... 344
- 12. The Chesnut, ... 346
- 13. Edible Acorns and Pine Seeds, ... 347
- 14. Araucaria Almond, &c., ... 348
- 15. Brazilian Nut, ... 349

Page.

IV. The Principal Agricultural Plants which are more or less used as Luxuries—

1. The Areca Palm, ... ... ... ... ... 351
2. The Betel Pepper, ... ... ... ... ... 353
3. Catechu, ... ... ... ... ... ... 354
4. Gambir Extract, ... ... ... ... ... 355
5. The Opium Culture, ... ... ... ... 357
6. The Tobacco, ... ... ... ... ... 361
7. The Coca, ... ... ... ... ... ... 362
8. The Vine, ... ... ... ... ... 367
9. The Maguey Plant, ... ... ... ... 376
10. The Sugar Cane, ... ... ... ... ... 380
11. The Coffee Tree, ... ... ... ... ... 383
12. The Chinese Tea, ... ... ... ... ... 385
13. The Pepper, ... ... ... .. ... 395

V. On some of the Principal Plants, the Fibres and Wool of which are used in the Manufacture of Stuffs, and other Materials indispensable to Man—

1. The Cotton Plant, ... ... ... ... ... 400
2. The various Plants which furnish Hemp, ... ... 404

The Culture of the Indigo Plant, ... ... ... 407

THE

# GEOGRAPHY OF PLANTS.

---

Nature in each zone of the earth has peculiar beauties, whether it be in the sunny isles of the South Sea, under the cool shade of the northern oaks, in the lovely well-watered mountain valleys, on the picturesque glaciers of the higher ranges, or in the midst of the Lybian deserts. But nature is silent to the unobservant man, and that rich spring of enjoyment escapes him, which has power to delight and cheer us, even when suffering from the severest blows of fate.

If we ask ourselves, what it peculiarly is, by which Nature speaks to us, we shall find that it is indeed by the whole aspect of the earth's surface; but that it is chiefly its living covering, the presence or absence of vegetation, that makes so deep an impression on our minds. Where vegetation is wanting, Nature is dead; and though she may impose by gigantic masses, by deserts which excite our horror, or by the stunning noise of waterfalls, there is nothing in these which speaks to the mind or cheers the spirit. It is vegetation which fixes the natural character of a region, and determines the conditions according to which men gather into various societies, at one time leading a nomade life, at another enjoying more or less the beneficent influences of agriculture. Where vegetation is scanty, and man is more or less confined to animal food, as in the case of the Samojedes and Esquimaux on the coasts of the Northern Ocean, civilization is impossible. In those regions, man lives like the beasts, and does not even think of raising himself above them. Some Samojedes, who were taunted with not being better than the wolves which preyed on their reindeer, dryly answered, while they devoured the raw flesh they had rescued from the wolves, that they at least were not worse, for they also left the gnawed bones behind them for their enemies! What horror does so low a state of man excite in us! But centuries may pass, and yet the inhabitants of those barren regions cannot share our culture; they must first leave their country; and yet these Samojedes sing of their dear fatherland!

Thus we may estimate the great influence which the presence of a luxuriant vegetation has on the civilization and prosperity of nations. But all who are susceptible of culture, have not an equal measure of this good fortune. While the inhabitants of the South Sea Islands enjoy, with little exertion, the splendid fruits of the banana, the bread-fruit, and the palm, the people of the north must laboriously cultivate the less fertile soil, to secure their support by the scanty fruit of the cereals. In some parts of the Philippines the earth is so exuberantly fruitful, that four crops a year are gathered in, two of rice, one of melons, and one of maize; while in the extreme north of Europe, the husbandman is content with a miserable crop of barley. On the fortunate table-lands of the East Indian mountains, as in the valley of Cashmere, at an elevation of from 5,400 to 5,500 feet, and in the valley of Nepaul, at an elevation of 4,500 feet, the people enjoy in summer the delicious fruits of the tropics, and in winter cultivate our northern cereals. The course of vegetation in the table-lands of the tropical Cordilleras is different. On the lower plains of these mountains, as in the valley of Arequipa, where the large and noble city, Arequipa, is situated at a height of 7,779 feet, the extreme dryness of the air prevents a second crop, and at the greater heights of from 12,000 to 13,000 feet, the cold of winter forbids it. The potato and the quinoa (*Chenopodium Quinoa*, L.) are there the principal objects of agriculture. Under such various conditions, which determine the food of the people of those countries, there must naturally be a corresponding variety in all their social arrangements.

I have here brought forward some of the contrasts which the cultivation of plants for food presents. They must be generally striking, and the thoughtful are naturally led on to inquire into the causes which make such a difference in the productions of the same soil. Hence the origin of a science which investigates the conditions according to which plants are distributed over the surface of the earth; a science which its founder, Baron Alexander von Humboldt, named botanical geography. This new science, which answers, in a way that had before been impossible, many of the most interesting questions on the production and distribution of organic being on the surface of the globe, has spread with singular rapidity.

The whole mass of species of plants is in a certain proportion to the different latitudes of the earth's surface. The proportion increases as we approach the equator, and decreases as we remove from it. Lapland has 500 phænogamous, and 600 cryptogamic plants; while Denmark, which is smaller, but farther south, has 1,034 phænogamous, and 2,000 cryptogamic species. According to De Candolle, France possesses 3,500 phænogamous, and 2,300 cryptogamic plants; while in the herbarium of the English East India Company, there are 6,000 of phænogamous from the East Indies alone, and it is probable that more than double that number belongs to that country. The whole of Europe, on the contrary, though considerably larger than the East Indies, has only about 7,000 phænogamous plants.

It would be extremely interesting, and highly important, for botanical geography, to know the whole number of species which clothe the earth. Many years ago conjectures and calculations as to the number were made, but they have been found insufficient by the discoveries of modern travellers. At the time of Linnæus' death, about 8,000 species were known, and now more than 66,000 have been described. The number of plants still undescribed in the herbaria of different nations must amount to several thousands, so that the whole number of plants now discovered probably reaches 80,000 species. But when we consider what immense tracts of country in America, Asia, Australia, and the South Sea Islands are still quite unexamined; when we think of Africa, which, excepting some barren deserts, is probably quite as rich in plants as Europe and Asia, we must at least double the number of plants already known, so that we obtain the number of 160,000 species. Besides, we know that many modern travellers, who have examined countries that have long been visited, have brought home an astonishing mass of new plants; so that we are justified in increasing the sum of 160,000 by at least a fourth, and in accepting 200,000 as a number which will probably be found near the truth.

If we now look over this immense variety of plants, we shall find that nature, under similar conditions of climate, has always produced similar, often even the same, forms. The naturalists, Banks and Solander, as well as the two Forsters, who accompanied Cook in his voyages round the world, and Sparmann, were not a little surprised when they found, in the region of Cape Horn, a vegetation similar to that of our northern zone. If we examine the vegetation of the plains from the extreme north to the torrid zone, we shall find, as we change the latitude, a continual change in the physiognomy of the vegetation; and if, in the torrid zone, we ascend from the level of the sea to the top of the highest mountains, which there often rise above the limit of perpetual snow, we shall find again the same order of changes more or less clearly defined. There in a short time we pass through climates corresponding to those of burning Africa, of the fair lands of the south of Europe, and of frozen Spitzbergen; and, as the climate changes with the increasing elevation, so also does the vegetation. The majestic palm and the fruitful banana are not found above the height of 7,000 or 8,000 feet; but near the limit of perpetual snow on these mountains we meet with grasses, Cyperaceæ, Gentianæ, and other plants similar to the forms of northern Europe.

When we inquire more closely into the causes which produce such a peculiar distribution of plants, we find that some of them lie open to observation, but that others depend on the most secret laws of nature, whose operation we may trace out, but cannot explain. When a plant from a hot country grows equally well in ours, as soon as we give it in our hot-houses a climate similar to its own, we have found out the first cause why this plant can grow only in a hot country and not also near the pole. When we take marsh plants from their natural station and plant them in our gardens, we see that they flourish only when they are planted in a marshy soil like that which nature has appointed them. Other plants, which naturally grow in the shade, grow luxuriantly in our gardens in a similar situation. But the laws of nature, by which certain plants can grow only in hot countries, others only in the shade, and others only in marshes, remain inexplicable, as well as the causes why different groups of plants predominate in different regions, and often are confined within a small and clearly marked out circle. We see, for instance, the singularly varied Cactaceæ in the warm parts of the temperate, and in the torrid zone of America, but we see these plants there also ascending the high mountains, and vegetating in a climate which corresponds to that of the alpine regions of Lapland, where not an individual of that singular form of plants appears.

From the few remarks now made, we may conclude that many different causes, besides conditions of climate, have an influence on the station and distribution of plants; and, therefore, we must divide the subject of the geographical distribution of plants into several parts, in which all these conditions, that they may be more easily understood, shall be discussed in a certain order, as the sequel of the book will show.

---

The principal works which have appeared on Botanical Geography, are the following:—

Alexandre de Humboldt, Essai sur la Géographie des plantes, accompagnè d'un tableau physique des régions equinoxiales. Paris, 1805. 4to.

Alexander von Humboldt and A. Bonpland, Ideen zu einer Geographie der Pflanzen, nebst einem Naturgemälde der Tropen-Länder. Tübingen, 1807. 4to. (A German translation of the former work, with some alterations.)

A. v. Humboldt, Ansichten der Natur. 1ter Band. Tübingen, 1808. 12mo.—A new edition in 2 vols. Tübingen, 1826.

G. Wahlenberg's Flora Lapponica. Berolini, 1812. 8vo.

Wahlenberg's Tentamen de Vegetatione et Climate in Helvetia septentrionali. Turici, 1813. 8vo.

Wahlenberg's Flora Carpathorum Principalium. Göttingæ, 1812. 8vo.

R. Brown, General Remarks on the Botany of Terra Australis. London, 1814.—Appendix to Flinder's Voyage to Terra Australis. It has appeared in German, in R. Brown's Vermischten Schriften. Herausgegeben von Nees von Esenbeck. 1ter Theil. Leipzig, 1825.

Alexander von Humboldt, De Distributione Geographica Plantarum. Lutetiæ Parisiorum, 1817. 8vo. Also in fol. as the introduction to the magnificent work, —"Nova Genera et Species Plantarum."

Fr. Schouw, Elements of Universal Botanical Geography.

Fr. Schouw, Grundzüge einer allgemeiner Pflanzengeographie. Danish and German at Copenhagen and Berlin, 1823.

C. T. Beilschmid, Pflanzen-geographie nach Alexander von Humboldt's Werke über die geographische Vertheilung der Gewächse, mit Anmerkungen, grösseren Beilagen aus anderen pflanzengeographischen Schriften, und einem Excurse über die bei pflanzengeographischen Floren-Vergleichungen nöthigen Rücksichten. Breslau, 1831. 8vo.—C. T. Beilschmid's Botanical Geography, according to Alexander von Humboldt's Works on the Geographical Distribution of Plants, with Remarks, large Additions from other writings on Botanical Geography, and a Digression on the Considerations necessary in a Geographical Comparison of Floras. Breslau, 1831. 8vo.

Other works of less extent, as well as the single papers which have appeared on various branches of botanical geography, will be named in the course of the work.

# PART I.

## ON THE CONDITIONS OF CLIMATE WHICH DETERMINE THE PRESENCE AND DISTRIBUTION OF PLANTS.

It is very easy to show that conditions of climate, particularly heat and moisture, are the chief causes which determine the station and distribution of plants; and therefore it is of the greatest importance to the science of botanical geography, to know exactly the modes in which the influence of the often extremely complicated conditions of climate becomes apparent. To arrive at this end, we must first, though as shortly as possible, employ ourselves with the observations which have been collected on the distribution of the heat and moisture of the atmosphere over the whole globe, and which are by no means of pure meteorological interest, but constantly point to the influence which individual meteorological phenomena exercise over vegetation.

It follows from the position of the sun with regard to the earth, that the distribution of heat runs through two periods, a daily and an annual one. There are, indeed, a number of causes which, by transmission, radiation, and equalisation, modify the fixed quantity of heat which a place receives in consequence of its relative position to the sun; but experience shows that the mean quantity of heat is constant. As we proceed from the tropics to the poles, the temperature of the air becomes always lower, as the height of the sun at noon diminishes; and in the same way it must become gradually colder as we ascend from the surface of the earth into a thinner atmosphere; for as the absorption of light (if

I may use the expression for the sake of brevity) is less in thin air, it is also heated to a less degree.

The progress of the periodical phenomena of the distribution of heat is most easily observed in tropical regions, as in them all the changes of nature go on with greater regularity than in the north.

If we consider, generally, the phenomenon of the heating of the atmosphere, we should find a regular distribution of heat from the maximum within the tropics to the minimum at the poles; but this is in reality by no means the case. There are two principal causes for this departure from law—the winds, and the quantity of moisture. The influence of the winds can nowhere be better observed than in the regions where the monsoons prevail, as on the southern coasts of China, just on the limits of the tropics. At Canton* and Macao, where the temperature of the air in summer, even at night, seldom falls under 22° Reaum.; in a country where the palm grows, where the sugar-cane, the Nelumbium speciosum, the orange, and all the delicious fruits of the south are cultivated, and where the enclosures of the gardens and fields on the borders of the river are formed of the banana, the orange, the pomegranate and myrtle hedges, the temperature during the north-east monsoon sometimes falls so low, that in the morning, particularly after a clear night, when the radiation of heat has been considerable, the leaves of the banana are seen hanging down brown and withered. However, this low temperature, which kills even tropical plants, fortunately lasts but a few hours; for as soon as the sun reappears, the temperature rises again to 12° or 15° Reaum., and often before mid-day the frosted banana is again in full splendour, its blighted leaves have recovered, and even their beautiful green partly returns. As the prevailing north-east wind causes an extremely

* See my Remarks on the Climate of the South of China. Nova Acta Acad. Cæs. L. C. V. xvii. P. ii., p. 854.

dry atmosphere, the sky at this season (the months of November and December) is almost cloudless, and at night, though it is the season of our greatest darkness, the stars at Canton shine with the most peaceful light, and there is no deposit of watery vapour. The new psychrometer shows there a difference of 6° and 7° Reaum., a phenomenon which is not known with us. But this dryness is so excessive, that the skin of the parts of the body exposed to the open air cracks, and the blood oozes out, as is the case in the high plains of the Cordilleras, where one cannot travel against the wind unless closely wrapped up in woollen garments.

But the vegetation of these parts of China shows still more clearly the influence of the prevailing winds. The most luxuriant vegetation renders them a paradise in the summer months, or, I should rather say, in the rainy season. An infinite variety of gorgeous flowers at that season adorns the shrubs and plants, a number of valuable grasses, often of the most beautiful and rarest forms, clothe the plains, while millions of insects animate the rich carpet. But when the north-east monsoon blows; when the temperature, which in June, July, and August is always above 22° Reaum., falls in November to 15°, in December to 13°, and in February even as low as 10°; when all the clouds have disappeared, and for months not a drop of rain falls, this paradise vanishes as by enchantment. The fields are bare, the sides of the mountains are scorched, the dried up vegetation becomes dust, and only the soil remains, without a trace of its former luxuriance.

In the north of Germany, where in winter the ground is frozen for months, which in summer is covered with the freshest verdure, the contrast is, indeed, the same; but the south of China lies within, or at least on the northern limit of the tropics, in a zone which is known to us chiefly on account of its heat. A number of similar instances, where the wind directly or indirectly has as decided an effect on vegetation, might be brought forward.

The influence of the *Hygrometry*, or general moisture of the atmosphere, on vegetation, is even greater than that of the winds; its absence or presence is the chief momentum which determines the presence of a vegetation. Almost the whole west coast of South America is remarkable for its low temperature in proportion to the latitude, as well as for the extreme dryness of the atmosphere. In the north of Chili, on the coast of Bolivia, and in the south of Peru, there are large tracts of country where rain does not fall for years; and there are there immense plains of sand, and ranges of mountains which show not a trace of animated being. But when the sky in those regions is covered by the clouds, known by the name of Garuas, which continue almost half the year in the northern districts of Peru, and are explained by the refrigeration caused by the coldness of the current, which Humboldt discovered in those seas, flowing from the south-west to the north-east; when this moister condition of the atmosphere extends as far as those countries, the bare and apparently herbless mountains on the coast are covered with a fresh verdure, and a number of the most beautiful and rarest plants spring up. But often, as for instance in Tarapaca, the most southern province of Peru, all this vanishes in two months, as the Garuas last only this short time.

I have before remarked what will presently be more clearly shown, that the union of heat and moisture is the chief condition of the production and distribution of plants; and, therefore, we must begin our observations by considering the distribution of heat, and its influence on vegetation. If we wish to learn the height or periodical progress of heat for any place, we may begin by making observations on the daily progress of heat; and we do this by hourly observations of the heat of the atmosphere by means of the thermometer. These hourly observations, according to the different latitudes and elevations, will more or less differ from each other; the differences will be greater in higher latitudes as well as at greater elevations; on the contrary, they will be smaller as we approach the equator, where often, particularly on the coast or at sea, there is a remarkable equality in the temperature of the day. Amongst the great number of observations of temperature which I have made known in my account of my voyage round the globe, there are hourly observations for a number of days, which frequently do not differ from each other more than 1° Reaum.

But to attain to more general results, it is necessary to find a common value for the whole sum of the daily observations, and we call this the height of the mean temperature of the day. There are various methods of finding the mean temperature of single days; the best is, to take the average of all the temperatures of the 24 hours; but the observer has it very seldom in his power to make such perfect observations. An easier way of obtaining the mean daily temperature is to add together the highest and lowest temperatures of the day, and halve the sum. The mean obtained in this way pretty nearly agrees with the mean of the hourly observations. For instance, the hourly observations which I made at sea, on the 26th October, 1830, between 12° and 13° N. latitude,* give the whole heat of 521.3° Reaum., which, divided by 24, the number of observations, gives 21.72° Reaum. as the mean heat of the day. The highest temperature on that day was 22.3° Reaum., and the lowest 21.1° Reaum., so that, by halving the sum of these numbers, we obtain 21.7° Reaum. as the mean heat—a result which agrees exactly with that obtained by the other way. The results of these two methods do not agree so well when applied to observations made in high latitudes, and in the interior of continents. At sea, even when near Cape Horn, I found an extraordinary equality in the progress of daily heat. On the 25th December, for instance, I observed 4.7° Reaum. and 3.6°

* See Meyen's Reise um die Erde, Berlin, 1834, vol. i., p. 156.

Reaum. as the maximum and minimum temperatures of the day, according to which the mean heat was 4.15°, while the average of the whole 24 observations of the day was 4° Reaum.*

Such an equality in the hourly progress of the daily temperature is of the greatest importance to the climate of a country, and to the corresponding vegetation; and, therefore, it is agreeable to our plan to explain the causes which, by the diminishing of the extremes, produce such a regular progress. It is a generally acknowledged fact that the climate of the sea-coast is not so cold as that of an inland district, in the same latitude; and upon this is founded the distinction between inland and coast climates. The phenomenon is shortly the following: the atmosphere near the sea is never heated during the day to the same degree as it is at a place in the same latitude distant from the sea; but it is in the same proportion less cooled through the night, and the result is not a colder, but a warmer climate, than the neighbouring place in the interior possesses. The absence of the extremes of daily heat and cold also diminishes the great difference between the annual maxima and minima, and thus is produced a climate which is suitable to many plants of warmer zones. Some examples will explain what has been said. It is well known that the myrtle thrives very well in Ireland, indeed, almost as well as in Portugal; while with us, though in the same latitude as Ireland, it does not stand the open air, and requires great attention. And again, the laurel grows in England, where the

* The equality in the progress of the temperature of the day is still more striking, if we observe the temperature of the sea. On the 1st of January, 1831 (See Meyen's Reise, i. p. 178), the observations on the sea, while doubling Cape Horn, gave a difference of not more than 0.2° Reaum. This extraordinary equality, however, is not general; it may be accounted for chiefly by the want of sunshine during the day. In regions where the sky is clear by day, and where the sunbeams beat the whole day on the surface of the water, the difference in the degrees of heat at different hours of the day is much greater, and frequently is more than a degree.

grape seldom ripens, and all kinds of fruit are very indifferent; on the contrary, while with us the noble vine succeeds, and the apples and pears are excellent, the laurel can only be grown in green-houses. These examples are sufficient to show how important the diminution of the maxima and minima of diurnal heat is to the distribution of vegetation. The explanation of the phenomenon is this: when the air is very dry, and its transparency seldom diminished by clouds, the rays of light and heat can very easily penetrate it; and, therefore, it will be very considerably heated by day, and cooled by night, as the heat can radiate towards the clear sky from the earth without any obstruction. But the difference between the extremes of daily heat is smaller, the more moisture the air contains, first, because the low clouds moderate, and partly prevent the transmission of the rays of light; and secondly, observations have shown that the loss of heat during the night by radiation is compensated in a peculiar manner. When the atmosphere is cooled in consequence of the radiation of heat from the earth, its watery particles are precipitated, and their latent heat, returning again by radiation to the cooled air, warms it again.

The small difference between the extremes of daily heat, which characterises coast and insular climates, is still further diminished in the atmosphere of the open sea, because there the air contains a greater quantity of moisture. I have already shown how the mean daily temperature may be found, either from the mean of the whole number of observations, or from the extremes of daily heat; but if the observer has not it in his power to make so great a number of observations, or has before him only such as have been made at a few periods of the day, he must choose for observation those times which give the maximum and minimum of daily heat, or approximate to them. The lowest temperature is generally observed shortly before sunrise, and the highest a few hours after the greatest height of the sun.

It has also been tried to determine the time of day when the temperature is equal to the mean daily temperature; but all attempts of this kind only in some degree approach the truth, for it is easy to see that the various lengths of the day must cause a great difference in this. And these times must also vary according to the latitude, as well as the length of the day. In northern countries, where in summer the temperature is frequently as high, and sometimes higher than in the tropics, this great degree of heat must be explained by the length of the day, for in the tropics the day lasts only twelve hours. The sun there rises about six o'clock, while with us it appears in summer by three A.M., and does not set until eight o'clock, and, therefore, shines more than five hours longer than within the tropics. On account of this greater length of the day, the times of the maxima and minima, as well as of the approximate means of daily temperature, must vary in different zones, and the greatest difference will be found in the extreme north and extreme south. Thus in Lapland, at the season when the sun is never visible above the horizon, the highest temperature of the day is just at morning.*

The mean temperature of single days being found, we may go on to determine the mean annual temperature, which is at the same time the mean heat of the place where the observations are made.

The mean temperature of the year is very easily found, either by taking the average of all the daily observations, or by dividing the sum of the daily means by the number of days. In tropical countries the observations of one year are sufficient in order to learn the mean heat of a place, for the progress of daily and monthly temperatures is there remarkably regular; but in the temperate, and especially in the arctic zone, the observations of a succession of years are necessary to ascertain the real mean temperature. In the tropics, the

* See Wahlenberg, Flora Lapponica, p. 43.

mean temperature of one year never differs from that of another so much as 1° Reaum.; on the contrary, at Stockholm, there is a difference of a whole degree between the means of two successive periods of ten years.

Formerly, one who wished to learn the temperature of a place with a view to the study of botanical geography, was content with the knowledge of its extremes of heat and cold. But it has been shown that this method is very inconclusive, for the extremes of temperature very seldom occur, and then last but a short time, so that they have not the fatal influence on such plants as require a higher or lower temperature, which would follow from their continuance. While plants are at rest in winter, they can bear a great degree of cold. In those regions of Siberia where the cold is known to freeze quicksilver, the vegetation in summer is much more luxuriant that that of the North Cape, where cold, such as is felt in Siberia, is quite unknown.*

The mean heat which is found from the average of all the mean daily observations, gives us a correct idea of the quantity of heat which a place receives, but it by no means gives us a correct scale of the vegetation of the place. When plants are at rest during winter, the sur-

* I can here cite the results of some investigations, which give the extremes of heat and cold which the seeds of our different grains can bear. No seed germinates at a temperature below the freezing point, and experiments with various grains have shown that they do not germinate under 7° Cels. (5.6° Reaum.; according to H. Goeppert's recent observations at 3° Reaum.). Other experiments to ascertain the degree of cold which would destroy the germinating power of seeds, have given the following result, that even the extreme cold at which quicksilver freezes, does not destroy the vitality of seeds. It is, indeed, probable that a continuance of such a degree of cold would kill the seeds; but it is not easy to determine this by experiment, as so low a degree of cold cannot be maintained so long. It is otherwise with the influence of heat, for seeds no longer germinate in water at the heat of 50° Cels. In vapour it requires a heat of 62° Cels. to destroy the vitality of seeds of corn, and in dry air 75° Cels. are necessary to prevent these seeds germinating. However, the influence of a high temperature is strikingly different according to its longer or shorter continuance, for a temperature of 35° Cels. for three days destroys the germinating power of grain. (See Ann. des Sci. Nat., 1834, p. 257–270.

rounding temperature has but little influence on them; but when they unfold their leaves in spring, when they blossom in summer, and form their fruit in autumn, everything depends on their receiving, during these important periods, that degree of temperature which is appointed them by nature. At Enontekis, in Lapland, the mean temperature is equal to —2.86° Cels., and at the Hospice on St. Gotthard, it is equal to —1.05° Cels. according to ten years' observations;* yet there are at Enontekis pine and birch woods, while on St. Gotthard we are far above the limit of trees.

Thus we may bring a plant from a southern to a more northern country, where the winter is severe but the summer very fine, if we protect it from the cold of winter, and do not expose it to the open air until the spring is far advanced. We shall afterwards, when we have learned more of the distribution of the vine, see that it requires, in order to produce good wine, at least five months of a mean heat of 15° Cels.; if September and October, the season when the grape fully ripens, have not this degree of heat, the wine is sour; and a country where this is the case, is, therefore, unsuitable to the culture of the vine.

From this it is evident, that, for application to the purposes of botanical geography, the mean temperatures of different seasons and of single months are chiefly to be observed, but also the extremes of heat and cold. The copiousness of these investigations of the temperature of a place must be determined by the aim which is to be attained by them. They must be special, when we wish to compare the vegetation of places near each other, but more general, when we examine the vegetation of the principal zones.

The method of drawing the curves of temperature is in this respect of great importance: if the curves of different places are laid down together, as is done in the

* See Kämtz, Meteorol. ii. p. 93.

accompanying Table, the similarity or dissimilarity of the climates is seen at the first glance, and we get also an insight into the vegetation of these countries. On the Table, for instance, there are first drawn the curves of temperature for five tropical places, which lie almost under the same parallel, Canton, Macao, Calcutta, Havanna, and Hawaii. These five places lie close on the northern tropic; therefore a comparison of the progress of temperature in these places is very important in many respects, particularly as the climate of each is strikingly modified by various causes. It may be seen by the table, that the curves for Canton, Macao, Calcutta, and Havanna, at least in summer, nearly agree, while the temperature of Hawaii, which has an insular climate, is more than two degrees lower during summer, but in compensation has the same temperature without interruption for six months. But when we consider the minimum of heat of these five designated curves, we perceive the most astonishing difference between them. The months of January, February, and December, stand extraordinarily low for Canton, while Hawaii, representing an insular climate, which showed so low a minimum temperature, has a very high mean temperature for these winter months. However, the great deviation of the curve of temperature for Canton, from those of the other places, is easily explained. Canton lies in a region, where, as I have already remarked, the most decided monsoons we are acquainted with prevail; the north-east monsoon, which blows during the winter months, brings with it such cold air, that the temperature in February is frequently lower than 4° R., and it sometimes even freezes for some hours. Let us consider what that indicates in a climate where the palm and banana grow.

But by a comparison of the mean temperature of these five tropical places, viz.:—Calcutta, 21° R.; Havanna, 20.35° R.; Hawaii, 19.2° R.; Canton, 17.56° R.; and

Macao, 17.87° R.,* we certainly do not receive a correct idea of the climate of Canton and Macao, where the heat during the summer months, June, July, and August, is often intolerable, while at Hawaii, during this season, the temperature is very pleasant.

Thus we see, that it is the mean temperature of different seasons of the year which gives us a correct idea of the climate and vegetation of a place, and, therefore, I have shown on the table the mean temperature of the different seasons at these five tropical places ; and we see such an agreement in the maxima of heat, that we can no longer wonder that a tropical vegetation prevails in all those five places, notwithstanding their differences of mean temperature. We said before that the fields of the southern parts of China, are, during the winter months, deprived of their verdure, and that not a trace remains of the luxuriant vegetation which clothed them in summer. The roots, bulbs, and seeds remain dormant in the ground during this season, and only return to life when the north-east wind ceases, and the rainy season commences with the south-west monsoon.

The curves of temperature of Berlin, Söndmör, Enontekis, and Melville's Island, which are also marked on the Table, give, at the first glance, a representation of the great differences between the maxima and minima of heat, which are peculiar to the climate of arctic regions. On examining the curve of Berlin, (I have purposely chosen this place, because the progress of its temperature is well known to us,) we find indeed a difference of 20° Cels. between the maxima and minima, but we see that this place has at least three months' summer of a very pleasant temperature of from 16° to 18° Cels., (12.8°—14.4° R.) mean heat. At Enontekis, in Lapland, 16° further north, the summer is only two months long ; and on Melville's Island there is but one

* See Meyen, Ueber das Klima im südlichen China, l. c.

month of summer, during which the mean temperature does not reach 6° Cels.; so that the curves of temperature of Berlin, Enontekis, and Melville's Island are always more pointed as we go towards the north. Even the insular climate, which Melville's Island in some degree possesses, cannot protect it from the intense cold, as the air is too dry to prevent radiation, or moderate the cold caused by it. The curve of Melville's Island shows a difference of 40° Cels. between the maximum and minimum of heat, and that of Enontekis a difference of 33° Cels.

I remarked, when speaking of the method of determining the mean daily heat by hourly observations, that certain hours show the heat, which comes nearest the mean of the whole day, so that from a single observation the mean temperature of the day might be determined. The same may be said of the annual progress of temperature, for the observation of a single day in autumn or spring might suffice to determine the mean temperature of the whole year. But unfortunately this method is as little applicable to the observation of the annual distribution of heat as of the daily; for we can only know the exact day for each particular place after the temperature of the year has been exactly determined by a great number of observations; for there are a number of causes which change the day for observation in each place.

This is all that it is necessary for us to remark on the progress of daily heat, and on its influence on the annual distribution of heat; and, therefore, we may now pass on to the application of the mean temperature to the purposes of botanical geography.

Baron Alexander von Humboldt has here also marked out the path of this science. He connected those places which possess an equal degree of heat by lines, which he called Isothermal,* that is, lines of equal heat. As the

* From ἴσος and θέρμος.

mean temperatures of different places are very various, there must also be various isothermal lines, which are always designated by the height of the mean temperature of the place; so that we have isothermal lines of from 0° R. or 0° Cels., up to 26° Cels., or down to —16° Cels. Observations have shown that these isothermal lines by no means run parallel to the parallels of latitude, but incline towards them, particularly in the higher latitudes, but less near the equator, where they pretty nearly coincide with the parallels of latitude.

In the northern hemisphere of the globe, all the east coasts of continents and isolated masses of land, are colder than the west coasts of the same latitudes; thousands of observations have confirmed this phenomenon, although its explanation is not yet quite found out,* and, therefore, the isothermal lines frequently rise and fall. For instance, Ireland, England, and Belgium are countries which have the same isothermal line; but on the east coast of Asia, it passes just above Pekin, which is in the same latitude as Naples. Canada is further south than Paris, and it shows the temperature of Drontheim. The trees which grow at New York, in the latitude of Naples, flower at the same time as they do at Upsal.

The isothermal lines do not run in straight lines, but in curves. The isothermal lines rise in their course from the eastern coast of America towards western Europe, but they sink again towards the south in the interior of the continent, and that so quickly, that Scotland lies in the same isothermal line as Poland, and England as

* Ad. v. Chamisso (Linnæa, 1829, page 59), explains the phenomenon very naturally, although the explanation does not go to the bottom, and account for the heat of the sea. "The seas," he says, "are the equalizers of temperature. As the east winds are constant between the tropics, the west winds are prevalent in the higher latitudes. As they blow over the warmer sea, they cause a milder winter on the west coasts of the continents, and on the contrary, they cause a severe winter on the east coasts, which they reach after blowing over the colder land which is covered with snow," &c.

Hungary. We find, however, that it is only near the coast that they sink so rapidly, and that it is in consequence of the great difference which was previously shown to be between coast and continental climates of the same latitudes: it is probable that it does not take place in the interior of large continents, but that there the isothermal lines run in a straight direction, though the observations necessary to determine this are still wanting.

In the interior of the New, as well as of the Old World, the isothermal lines incline towards the south. Thus if we go towards the pole in the interior of the two great continents, we find that the temperature decreases much more rapidly than in the interjacent seas. It is well known, that for many years attempts have been made to penetrate the frozen Arctic Sea. By Behring's Straits, where the voyager is constantly near the great continents, he has penetrated only a little further than 70° north latitude; by sailing along the American coasts through Baffin's Bay, he has reached the 77th degree north latitude; while in the open sea, in the meridians of Norway and Sweden, it is easy to sail to Spitzbergen, which lies above 81° north latitude. We thus learn that the pole is not the coldest point of the earth, but that there are two poles of cold, one in the interior of each continent.

But we have already seen that the mean temperature of the whole year has not so great an influence on the vegetation as the mean temperature of the different seasons; and, therefore, it is still more important to know the places which, though in different latitudes, possess the same summer or winter temperature. Baron A. von Humboldt first paid attention to this also; he named the lines which connect places of the same mean winter temperature, Isochimenal lines (from ὀχειμών, cold), and those which connect places of the same mean summer temperature, Isotheral lines (from τὸ θέρος, heat).

The isochimenal lines in the interior of continents bend considerably towards the south, which is principally obvious near the Atlantic, where the curves, when they come near the coast, make a sudden turn towards the north. For example, the isochimenal line of —5° Cels. runs north from the North Cape (—4.62° Cels.), then pretty nearly parallel to the chain of Scandinavian mountains south to Drontheim (—4.78° Cels.), afterwards south from Upsal (—4.02° Cels.) and north from Abo (—5.38°) into the interior of Russia; here it seems to turn very suddenly to the south, as Petersburg has a winter temperature of 9.03°. In the interior of North America, the isochimenal lines appear to bend still further south; for Fort Sullivan, Fort Howard, and Fort Snelling, all within the 45th degree, show the following winter temperatures, —5.17°, —7.23°, and —8.99°, thus always lower as we proceed into the interior. But here, also, the isochimenal lines turn as quickly to the north, when we approach the west coasts of North America from the interior of the continent; the winter temperature of Fort George, in the latitude of 46° 18′, is = 3.75° Cels.; while at Washington, on the east coast of the same continent, the latitude of which is 38° 53′, the mean winter temperature is 2.96° Cels. While the inhabitants of Quebec in winter complain of the piercing cold, the natives of the same latitude on the west coast go with scarcely any clothing.

The importance of the difference between the climates of the east and west coasts of North America to the distribution of vegetation was early observed, at least much earlier than it was known by thermometrical observations. Barton has remarked that the plants of North America grow on the west coasts in higher latitudes than on the east; for instance, Æsculus flava grows on the east coast as high as 36°, and west of the chain of mountains as high as 42° north latitude; Juglans nigra on the east coast to 41°, and on the west to 44°;

and Gleditschia triacanthos to 38° in the east, and to 41° in the west.

The eastern coasts of Hudson's Bay are desert and herbless, while on the western coasts there is a tolerably rich vegetation.

The Gleditschia triacanthos is now planted with us, and grows vigorously far above 52° N. lat. For instance, in the park of Oranienburg at Berlin, there are two gigantic trees of this species. This brings us to the conclusion, that the temperature of North America is much lower than in the corresponding latitudes of Europe, which we shall return to again in another place.

The course of the isotheral lines is directly opposite to that of the isochimenal; they bend considerably north near the coast, but the farther we go into the interior of continents, the nearer the isotheral lines approach the parallels of latitude. The isotheral line of 18° Cels. hardly touches the south of England, reaches Holland in 51° north latitude, passes somewhat south of Berlin, reaches Moscow, and afterwards appears to run directly south. The summers of Paris and Moscow are nearly equal, though the winter of Moscow is quite dreadful.

All that we said before on the inclination of the isochimenal lines towards the south, partly holds good with respect to the inclination of the isotheral lines to the north, particularly in regard to the new continent;—that is, a coast climate has a less degree of heat than the climate of the interior of continents, and therefore the isotheral lines in the interior ascend further north.

The difference between the climates of the east and west coasts of continents and islands, has also been observed in the southern hemisphere, but here the west coasts are colder than the east, while in the northern hemisphere the east coasts are the colder. On account of the peculiar configuration of the continents in the southern hemisphere, fewer points of comparison offer themselves than in the northern; but the greatest disadvantage is that there are few places where exact meteorological observations have been made.

South America, which extends furthest south, shows very decidedly this arrangement of a warm east coast, and a cold west coast. Various attempts have been made to explain the proportionably very cold climate of the west coasts of South America, and many causes have been assigned which would, indeed, diminish the heat there; but the chief cause seems to be the same, which makes the east coasts of the northern hemisphere proportionably colder than the west.

In the same way as the mean temperature diminishes from the equator to the poles, it decreases in the different regions of the mountains as we ascend from the plains, until at last we reach the icy region, where perpetual snow and ice prevent all vegetation. This decrease of temperature is most strikingly and regularly seen, when we ascend in a direct line by means of a balloon. Gay Lussac ascended in this way at Paris, on the 16th September, 1805, to the height of 21,480 feet, where the temperature fell as low as 7.6° Reaum., while on the surface of the earth at the same time it was 22.2°. When one ascends a high mountain a similar gradual decrease of temperature is observed, and at the same time the most striking differences in the vegetation. He will remark, that at the foot of the mountain the plants of the plain appear, that they gradually disappear, that trees then ascend to a certain height, after which bushes prevail, which, as he ascends higher, give place to herbs, and at last to a few lichens, &c.

The traveller who has visited the countries of the north, when ascending high mountains in the south, will very soon enter regions, in the vegetation of which he will recognise the northern plants. At the limit of permanent snow on these mountains, he will miss but a few of the forms of the plants of the arctic zone, and even will find the same species, which do not once appear in the plains in the whole space between the arctic regions and the summit of these mountains.

When I many years ago travelled over Switzerland, and came upon the high valleys between the lakes of Zurich and Zug, I was not a little surprised, and very much delighted, when I saw in a noble meadow all the most beautiful plants of Lithuania, which were vividly imprinted on my memory by my first botanical excursions, and which I had not seen for many years.

It is an inexpressible pleasure, which only a botanist can feel, when, coming from the north, he ascends a high mountain in a southern region, and finds one well-known plant after another; even in the Swiss mountains his pleasure is great; but how much greater is it when, far from home, he is wandering on the mountains of the southern hemisphere? The sight of a little Gentian, very similar to our Gentiana uliginosa, and G. nivalis, at a height of 14,000 or 15,000 feet, as in the Cordilleras of Southern Peru, can enchain the botanist for hours; he again and again gathers this little plant, which takes him, at least in imagination, home.

There is, therefore, a certain parallelism between the distribution of vegetation from the level of the sea to the limit of perpetual snow, and that from the equator to the poles, although the gradual change of vegetation takes place much slower towards the poles than with the increasing altitude. It is also, with our present knowledge, no longer difficult to perceive that this parallelism exactly agrees with that which we find between the gradual decrease of heat from the equator to the poles, and that from the plain to the limit of permanent snow. Here we first clearly see of what advantage botanical geography may be to agriculture and to the general culture of the country.

We have hitherto been occupied with the mean progress of the distribution of heat over the surface of the earth, and have learned that vegetation almost keeps pace with it. The invention (if I may call it so) of isothermal, isotheral, and isochimenal lines, gives us the means of easily applying those meteorological observations to the distribution of vegetation.

If the decrease of heat in proportion to the increasing altitude were quite equal in the same latitude, different places at the same height in the same latitude would belong to one isothermal line, which, as it approached the north, would descend nearer the plain, so that at last it would coincide with the isothermal line of the plain. Applying this fact to the distribution of vegetation, we shall find that a plant, which grows high in the mountains under a fixed isothermal line, can grow in the plain only when it finds there the temperature of the same isothermal line, or at least of one very near it. Alpine plants of high regions will not grow in our gardens, at least not without particular attention, and when they do grow, they assume forms quite different from those which they had on the mountains. But we may almost know beforehand whether a plant from the plain will succeed on high mountains, when we know the relative temperatures of those regions. We have already seen, while inquiring into the distribution of heat over the surface of the earth, that the distribution of vegetation is determined less by the isothermal than the isotheral lines, particularly as respects annuals and the various kinds of corn, which are cultivated as annuals. The distribution of perennials is regulated more according to the isothermal lines, and the extremes of the cold which prevail at a place in winter. Corn is cultivated in the north of Europe at inconceivably high latitudes; at 69°, or even 70° N. lat., as at Lyngen, Alten, and on the frontiers of Norway, Sweden, and Russia, even in regions where the mean temperature is below the freezing point. But when we consider the rich and beautiful vegetation of the shores of the great lake of Titicaca, at a height of 12,700 feet, and see that only barley and oats succeed there, although I have never heard that the lake freezes in winter, we find that we must seek farther for the cause of so striking a difference, and I believe it to be this, that the isotheral line of this country is far below that which lies in 69° and 70° north latitude. The mean heat at Enontekis is —2.86° Cels., but it lies in the isotheral line of 12.80° Cels., while the isochimenal is as low as —17°. On the contrary, the mean temperature of the shores of Lake Titicaca is certainly above the freezing point, while the summer heat is less than at Enontekis; for even during the summer on these table lands, which exactly corresponds to the winter of the plains at the level of the sea, I have never observed a greater heat than 15° Reaum. at mid-day, and generally only 9° and 10° Reaum.*

Some examples would set this in a clearer light, but unfortunately we have still too few thermometrical observations on the decrease of temperature at various altitudes in different latitudes.

The observations of the temperature on St. Bernard show very clearly, that, as the elevation increases, the great differences disappear which are seen between the temperatures of the hottest and coldest seasons in northern countries of the same mean heat; this has been observed not only here, but on other heights. For instance, the monks of St. Bernard envy the Laplanders their fine climate, because although they have the same mean temperature as St. Bernard, their summer is warmer. I have already remarked, that on the plateau of southern Peru, in the basin of the lake of Titicaca, neither wheat nor rye thrives, and only oats and barley ripen, although the mean temperature at this great height is not below the freezing point.

It is a pity that we have not a sufficient number of thermometrical observations from this country, to enable us to make comparisons which would be of particular value, with a point of the earth so important in respect to botanical geography. That this table land possesses

* See Meyen's Reise, i.

so high a mean temperature, depends on another cause, which Baron Alexander von Humboldt first pointed out; for he discovered, by very close investigation, that the diminution of temperature takes place much more slowly on mountain plains than on the declivity of steep mountains, which is naturally explained by the radiation of light and heat from large surfaces. Saussure had already remarked the rapid decrease of heat when ascending steep mountains, which depends on the same cause.

After a certain parallelism was discovered between the gradual decrease of heat and the increase of height, the next idea was to express the corresponding proportions by numbers; and the attempt was made to ascertain what elevation would correspond to a decrease of 1° of heat. The observations of Baron von Humboldt, as well as of M. Gay Lussac during his balloon ascent at Paris in 1805, have determined this point. An altitude of from 90 to 100 toises will pretty nearly correspond to a decrease of heat of 1° Cels. From the average of Saussure's observations in Switzerland (80 toises in summer, and 94.4 in winter), and from those of D'Aubuisson (75 toises for each degree), it follows, that in summer an altitude of 75–80 toises, and in winter of 94–110 toises, will correspond to a decrease of one degree of heat.

To prove what has been said, the curves of temperature which I have drawn on the Table may be compared. Geneva and St. Bernard lie in the same latitude, but the place of observation on St. Bernard is more than 1000 toises higher than Geneva. The mean temperature on St. Bernard is = —1.0° Cels., and that of Geneva = 9.7° Cels.; thus the decrease of temperature for the difference of 1000 toises in elevation, is about 10.7° Cels.; therefore, here more than 100 toises corresponds to each decrease of one degree of heat.

When considering the decrease of heat in proportion

to the increasing altitude, our attention was drawn to those regions where the temperature is so low that the snow and ice lie all the year, which puts a stop to vegetation. This limit is called the limit of perpetual snow, to distinguish it from that limit, above which the snow lies all the winter months, which we usually call the lower snow limit. The region of permanent snow does not, however, as one would suppose, show a mean temperature of 0° Cels.; but we find very great differences in this respect in different latitudes, which may be easily explained when we have a greater number of accurate and extensive observations. Under the equator, the snow limit is at the mean temperature of +1.5° Cels.; in the temperate zone at —3.7° Cels.; and in the arctic zone not until —6° Cels.

But as the low temperature which determines the snow limit on different mountains, is found at lower elevations the nearer they are to the poles, the points of the snow limit of different mountains from the polar regions to the equator, when connected by lines, form a curve, the plane of which will surround the globe like a dome. This dome sinks as low as the level of the sea in polar regions, where a permanent and impenetrable mass of ice opposes all progress. But on the continents of the arctic zone the limit of perpetual snow is not on the level of the sea; under the most favourable circumstances this is the case in 81° north latitude, on the northern side of Spitzbergen. This dome rises most considerably above the level of the sea under the equator; its elevation there, according to Baron von Humboldt's observations, is usually about 14,760 feet. But more recent observations, both in the south of Peru, and on the Himalayas of India, show that the limit of perpetual snow in these countries must be placed much higher; indeed, wherever very extensive tracts of country lie at such elevations, at least as high as 16,000 or 17,000 feet. After Hälstrom's* investigations on the

* De termino atmosphæreæ terræ nivalis. Aboae, 1823.

curve of perpetual snow, it was concluded that the dome was not perfectly uniform, but slightly bent in the regions of the equator; but the later observations on the greater height of the snow limit within the tropics, are quite opposed to this bending in of the dome in equatorial regions.

The decrease of heat is indeed more rapid on steep ascents than on high plateaus, and on the heights of great continuous mountain masses; yet the volcano of Arequipa, which rises like an insulated cone from a plain at an altitude of 11,000 feet, to the elevation of more than 18,000 feet, shows but a very little snow on one side of its highest point. The well-known mountain pass between Arequipa and the province Chuquito, called Los Altos de Toledo, is far above 15,000 feet; yet the vegetation there is extremely interesting, and there is even a hut, inhabited by men, almost at this enormous height. In Himalaya, in the Nutu pass, at a height of 16,840 feet, there is no permanent snow. The whole western part of the Himalayas, comprehending the whole of Kunawar, is as high as from 12,000 to 18,000 feet, and yet but a very little snow is seen there, and the vegetation ascends as high as 16,000 feet. Juniperus communis grows there at an elevation of 14,500 feet, and the birch at 14,000 feet.

By the following series of observations on the height of the snow limit on the mountains of different latitudes, its gradual descent from the equator to the poles is confirmed.

The height of the snow limit appears to be,

| | | |
|---|---|---|
| On Cotopaxi, at | 15,735 | Prussian feet,* according to Humboldt. |
| On Antisana, | 15,456 | ... Humboldt. |

* I have reduced the measurements in toises and metres to Prussian feet, according to the excellent tables by H. Dove (Ueber Mass und Messen, Berlin, 1835). A toise is = 1.949037 metres, and a metre is = 3.186199 Prussian feet, therefore a toise is = 6.2 Prussian feet.

| | | | |
|---|---|---|---|
| On Chimborazo, at | 15,320 | according to | Humboldt. |
| On Chimborazo, | 15,539 | ... | Hall. |
| On Pichincha, | 15,190 | ... | Humboldt. |
| In the south of Peru, | 16,851 | ... | Pentland. |
| In Mexico, in 19° N. lat., | 14,570 | ... | Humboldt. |
| On Ararat, | 13,441 | ... | Parrot. |
| On the Pic du Midi, | 9,337 | ... | Parrot. |
| On Mont Perdu, | 8,078 | ... | Parrot. |
| Medium for the Pyrenees, | 8,680 | ... | Humboldt. |
| On the Mountains of Caucasus, | 10,602 | ... | Parrot and Engelhardt. |
| On the Apennines, | 9,231 | ... | Schouw (in 42° and 43°). |
| On the Alps, | 8,494 | ... | Wahlenberg. |
| On the Alps, | 8,804 | ... | Various authors. |
| In Norway, in 62° lat. | 5,120 | ... | Hisinger. |
| In Norway, in 63° lat. | 5,019 | ... | Hisinger. |
| In Iceland, in 63½° lat. | 2,642 | ... | Hisinger. |
| At Hammerfest, in 70° lat. | 2,585 | ... | Buch. |
| At the North Cape, | 2,275 | ... | Buch. |

It has been attempted in more recent times to distinguish the limit of perpetual snow from that of the glaciers, and for that it has been proposed to take the limit of that snowy substance called Firn in Switzerland. Glaciers are great masses of ice, which are entirely composed, in a peculiar way, of crystals of ice of various sizes. These crystals are, according to all accounts, as it were jointed together, and one serves to wedge in another firmly.* The glaciers frequently sink to a considerable depth, and their limits cannot be reconciled with the limits of perpetual snow. The lower Grindenwald† glacier sinks below the village of Grindenwald to the elevation of 533 toises, while the upper Grindenwald glacier does not descend lower than 670

* See Hugi Naturhistorische Alpenreise. Berghaus Annalen, iii., 292.

† See Hugi. Berghaus, p. 290.

toises. The lower glacier of the Aar is, at its outlet, 921 toises high, while the upper glacier of the Aar does not come lower than 1330 toises. In Hugi's Naturhistorische Alpenreise, a very interesting book, there are a number of measurements of the depth of different glaciers; and I have only cited these, to point out how much the heights of the glaciers differ from each other, and from the limit of perpetual snow.

It is well known, that in the island of Iceland, which lies within the subarctic zone, the glaciers even descend into the sea, while the limit of perpetual snow is at the altitude of 423 toises; but the difference is still more striking in the Straits of Magellan, in 53° and 54° south latitude, where the glaciers descend into the sea, while the snow limit maintains an elevation of between 3500 and 4000 feet.*

The firn is a granulous mass of snow full of holes, the appearance of which Hugi proposed to regard as the limit of perpetual snow. When seen from a distance, the masses of firn have quite the appearance of glaciers; and in Switzerland, at a height of 1270 toises, the glaciers change rapidly into firn. When the sun shines on these masses of firn, it dissolves to the depth of several feet, and separates in the hand like hempseed; but at night it is again consolidated by the cold.

It is unknown to me, whether the firn is found on the mountains of other countries. Besides ascending the Alps, I have frequently ascended above the limit of perpetual snow in the Cordilleras of South America, and I found there the mass of permanent snow hard and firm, sometimes so hard that it was difficult to hew steps in it; and no appearance of dissolving during sunshine was observed on the Cordilleras of Chili and Peru.

The heat of the summer is what principally deter-

* See P. King's Remarks on Terra del Fuego and the Straits of Magellan.

mines the height of the snow line in different places; it will, therefore, in different years, vary according to the variation of the heat of the summer. This variation will become smaller as we approach the equator, since the differences between the maxima and minima of mean monthly temperature diminish with our distance from it. Near the equator, only a few observations are necessary to determine the height of the snow line; while in the temperate zone, a great number is required, in order to attain a certain result.

The heat of the ground has also been looked upon as a cause which may have an influence on the stations of plants. It is certainly the case, that the surface of the ground, in which plants are rooted, receives its heat from the influence of the atmosphere, and, therefore, it also depends on the heat of the sun. For the methods of measuring the heat of the ground, directly or indirectly, I must refer to works on physics, in which this subject is fully explained.*

In the same way as the places on the surface of the earth, which possess the same mean heat, were connected by Humboldt by isothermal lines, Kupffer has connected the points of equal temperature of the ground by lines, which he called isogeothermal, the course of which is similar to that of the isothermal lines. I may touch the more slightly on this subject, as I think the differences between the isothermal and isogeothermal lines are too small to have any considerable influence on the distribution of vegetation.

It is a universally known fact, that the leaves, as well as the flowers, of the same plant, are unfolded gradually later, as it recedes from the warmer regions, and approaches the colder; and we observe the same thing in ascending from the plains to the summits of mountains. Plants, which had flowered long ago in the

* See Kämtz, Lehrbuch der Meteorologie, ii. p. 176; Gehler's Wörterbuch, N. A. iii. &c.

plain, and are even bearing fruit there, may be found still in flower on the mountain at a proportional height. Those who live at the foot of the mountains enjoy the fruits of their country for a much longer period than the inhabitants of the plain, for the fruit of the plants on the mountains does not begin to ripen until that of the same plants in the plain has been over for a long time. In many tropical countries, which lie at the foot of high mountains, the most useful fruits are even enjoyed throughout the whole year; for they ripen gradually later as the elevation increases, until the time for another crop on the plain approaches. The exact reverse of this also takes place: all the periods in the life of plants are hastened as they approach to warmer regions.

M. de Saint Hilaire,* at the commencement of his travels, observed the peach tree without leaves or blossom at Brest, on the 1st of April; on the 8th, he found them in full bloom at Lisbon; on the 25th, at Madeira, the fruit had already set; and on the 29th he got ripe peaches at Teneriffe.

Numerous other instances might be given, as Schübler has done in a separate treatise. The lily of the valley (Convallaria majalis) flowered in the year 1829, at Parma, on the 26th April; at Tubingen, on the 10th of May; at Berlin, on the 17th May; and at Greifswald, not until the 10th of June. By observations of this kind, which were made on a number of plants by various botanists, Schübler, and before him Bijelow,† were led to discover the law according to which this acceleration in the time of flowering takes place in different latitudes. The result of Schübler's investigations is, that for each degree that the station of the plant is nearer the pole, the time of flowering is delayed almost

* Plantes Remarquables du Brésil.

† On the comparative forwardness of the spring in different parts of the United States in 1817. In short notes in Silliman's American Journal, i.

four days, and that it is accelerated by the same time for each degree that the plant approaches the equator. In this investigation it has been presupposed, what is very seldom the case, that summer advances with equal rapidity in all the places where the observations have been made; but various countries show the greatest differences in this respect, when the observations of a series of years are compared with each other. But this apparent exactness, which follows from the calculation of these observations, is not at all necessary; it is enough to know, that the flowering of plants stands in a certain pretty nearly fixed relation to the various latitudes. It is generally known that the progress of vegetation is frequently so unequal at different points of the same country, that the same plant will differ six or eight days in the time of flowering, and there is sometimes a still greater difference between plants growing close to each other, so that we must not expect too great an exactness in such comparisons.

It may be easily proved artificially, that it is chiefly the warmth of the atmosphere which causes this acceleration or delay of the various stages of the same plant. If, in the severest winter months, a twig of a tree standing in the open air, be brought into a heated room, it will very soon show buds and leaves, while the tree which remains in the open air does not bud until spring. It is also known that the warmer the air in spring is, the sooner the trees bud.

After Schübler had found that the time of flowering was advanced four days by approaching the equator one degree, he tried to calculate the quantity of heat required for this acceleration. As in the middle of Europe, the decrease of heat for each degree of latitude is equal to 0.516° Reaum., this is the quantity of heat which determines the acceleration or retardation of the flowering of plants by four days, and this proportion may be reduced to single days, so that vegetation is retarded on an average one day, if the mean temperature be diminished 0.135° Reaum. (1-7th to 10-8ths) or 7½ days for a decrease of 1° Reaum. These results are apparently very exact, but they are obtained by calculation only; and that the influence of the heat of 1-7th or 10-8ths Reaum. is so important, cannot be proved by observation. Adanson* has also formed a very ingenious hypothesis by which the differences in the flowering of plants may be explained. He accounts for the differences by the quantity of heat which each plant had previously received; and to ascertain this, he added together the degrees of heat from the beginning of the year. Thus the silver poplar flowers when it has received 168 degrees of heat, and the vine does not bloom until it has received 1770 degrees. But well-grounded and advantageous as this method of investigation appears, it is not exact when it is closely examined, as M. De Candolle has done very minutely in his physiology of plants. The temperature of the previous autumn has no small influence on the flowering of plants in spring; and, therefore, it is quite arbitrary to commence the reckoning of the degrees of heat on the first of January.

But the development of the flowers, as well as of the leaves, of plants, is affected by many extremely complicated causes, and not by heat only. All the internal causes, which determine the flowering of plants, are to be considered before the influence of heat and moisture can be measured. M. De Candolle† has made a very full series of observations, to inquire into the causes which determine the budding of the horse-chesnut at various periods; but from this investigation we may conclude, that it is neither a fixed degree of heat, nor a fixed quantity of moisture which determines the budding of trees. It is, I believe, a universally

* See De Candolle, Physiologie Végétale, ii. p. 476, in which Adanson's hypothesis is clearly stated.

† See Physiologie Végét. i. p. 432, &c.

known fact, that in spring the atmosphere sometimes shows a high degree of electricity, particularly after storms, which have been accompanied by rain, and plants immediately after grow with such rapidity, that we may almost observe the gradual unfolding of the leaves ; such a state of the atmosphere, which is even invigorating to man, affects vegetation so rapidly neither by its heat, nor its moisture ; but by something else, in all probability its electricity.

The important influence of the moisture of the atmosphere on vegetation is to be seen everywhere and at all seasons, for vegetation developes itself only where moisture is present in the atmosphere : and only where there is the joint influence of a great degree of heat and moisture, vegetation shows the luxuriance which is found in tropical countries. In regions where rain never falls, as in many deserts, the ground has but little moisture, and water is generally wanting, so that vegetation is suppressed either for a certain time, or throughout the whole year. I have already spoken of those parts of the Chinese coasts, which during winter, when often not a drop of rain falls, show not a trace of the splendour which their tropical vegetation presents to the eye in summer. Yet in the narrow valleys of the mountainous districts, where the water does not quite dry up, the same rich vegetation prevails as in summer, at the very time when close at hand every thing is burned up.* Valparaiso, the well-known port on the coast of Chili, has received its name from the beauty of the scenery there ; but one who should visit this place at any season, except spring or winter, would be astonished at the deadness of nature, and at the barrenness of the rocks and mountains which enclose the harbour, and rise in very imposing masses.

The science of modern physics explains very clearly

* See Meyen, Bemerkungen über das Klima im südlichen China, l. c. p. 862.

the phenomena of hygrometry, while even as late as the time of De Luc, the invisible powers were called to aid in explaining even the most simple rain. Dalton has proved that the atmosphere for each degree of heat can receive a certain maximum of vapour, and the evaporation of moisture goes on, until this maximum of saturation is reached; and the drier the air is, the more rapidly this evaporation proceeds. If the now saturated atmosphere be cooled, the portion of water, which at this lower temperature it can no longer retain, is set free, falls to the ground, and appears to us as rain, mist or clouds, snow or hail, &c. Thus the vapours return to their source, to be evaporated again, and in this way to render it possible and agreeable for the creatures of the earth to exist in the atmosphere.

Although, in our northern countries, the atmosphere very frequently receives its maximum of vapour, and then by cooling allows the water to fall again; that is, although it very frequently rains with us, the quantity of water is so small, that it can hardly be compared with the great quantity which falls in tropical countries. At Rome the mean annual quantity of rain is 33.1 Paris inches; on the contrary, at Macao it is as much as 63 inches, and in some years, as for instance 1812 and 1828, more than 100 inches of rain have fallen there, which is the regular quantity in other places, as in Grenada. But although so great a quantity of rain falls at Macao, the rainy season lasts only during summer; and hence one may imagine in what torrents the rain falls in these regions.

The currents of the air, that is, the winds, as well as the currents of the oceans, must also be considered here, for they are frequently mentioned as important causes of the migrations of plants. We shall first consider the winds. They blow sometimes regularly in the same direction, over more or less extensive tracts; and at other times this way, or that, without any order. A number of instances are recorded, in which the seeds of various plants have been carried by them to a great distance, and the circle of their distribution in this way enlarged. The seeds of the plants of certain families, for example, the Compositæ, are furnished with organs, which are peculiarly well adapted to facilitate their transport by the wind; these are the feathery appendages, which are called the pappus. It is not to be disputed that with the assistance of such feathery organs, certain seeds might be carried, particularly by violent storms, over a great extent, and that in this way the Syngenesia in particular may be distributed over a very extensive circle; especially those from the countries in which, in autumn, when the seeds are ripe, regular winds prevail, and blow towards countries to the south-east or south-west, where the temperature is not unfavourable to the growth of these plants. We may mention, as instances, some American plants, which have spread in a very short time, as weeds, over almost all Europe, Erigeron canadensis, Oenothera biennis, and even Galinsogea parviflora, which has escaped from the botanic garden at Berlin, and is now very widely diffused.

But the currents of water, as well the streams of rivers, which often carry with them for hundreds of miles the seeds of plants, as the currents of the oceans, present a more important determining cause of the distribution of plants. Several very interesting instances have been observed, in which true Alpine plants have been brought by streams from the mountains to the plains, where they now grow freely. Link has observed, that Circaea alpina has thus been brought down from the Harz mountains, and now grows in the surrounding plains. He has also remarked* that Linaria alpina, Rhododendron ferrugineum, Alnus viridis, &c., come down from the heights of the Alps to the valleys, where they evidently follow the streams. The rivers

* Die Urwelt, i. p. 263.

from the Harz mountains have in the same way brought Arabis Halleri to the plains of Hildesheim, where it is never far from the rivers. Chamisso found, when travelling on the coasts of Chili, several distinct Alpine forms of the genera Calceolaria and Calandrinia, which I afterwards found on the Cordilleras of Chili, just on the limit of perpetual snow, whence they had probably been carried to the coast by mountain streams.

If plants are distributed sometimes over wide tracts, by the influence of streams and rivers, their circle must be much more enlarged by the currents of the oceans, as these are of such extent that they connect remote parts of the world with each other. This is not the place to treat of the various currents of the ocean and their causes; but on account of the influence, which is constantly ascribed to these phenomena, it is necessary for me to give a very short account of them.

We set out from the result of observations, that all currents in the ocean are caused by prevailing or alternating winds. In the first case, where the wind blows throughout the year in the same direction, the current also flows in the direction of the wind all the year; but in the cases where the winds alternate every half-year, the current also changes; in the one half of the year it runs with the one wind, and in the other half of the year with the other.*

When we speak of the currents in the ocean as important causes of the distribution of plants, we of course mean only the great prevailing currents, which connect remote countries and islands with each other; the small currents, which are everywhere caused by

* The modes of designating the currents in the ocean, and those of the winds, differ from each other: the wind is named from the quarter whence it blows—*e. g.*, if it blows from the north, we call it the northward; if it blows from the east, the eastward. But the currents, on the contrary, are named by the quarter towards which they flow—*e. g.*, a current which comes from the north-east, is called a south-west current; and another which flows from the west, an easterly current; so that the current is always named by the direction opposite to that of the wind.

strong gales, and disappear after a short time, may be here passed over without notice.

But as all currents, as I have already remarked, are caused immediately by the winds, it is necessary to take first a short survey of the prevailing winds.

The winds, which blow over the ocean throughout the whole year in the same direction, are known under the name of trade-winds. In the northern hemisphere, the trade-wind blows from the north-east, and in the southern from the south-east, just in a direction opposite to that in which the earth revolves on its axis. But as, by the peculiar configuration of the land, there are in the northern hemisphere two seas, separated from each other, and in the southern hemisphere three separated seas, the Ethiopian Ocean, the Indian Ocean, and the South Pacific, there are three south-east trade-winds divided from each other, and in the northern hemisphere two north-east trade-winds also separated.

In the northern hemisphere the trade-winds begin between lat. 27° and 30°, but in the southern hemisphere they appear to extend much lower. In the equatorial regions, where the trade-winds of both hemispheres blend, there is a zone of 3 degrees in breadth, which is called the zone of calms; in it neither the north-east nor the south-east trade-wind blows, but calms alternate with the most violent tempests of rain.

In the latitudes where the north-east trade-wind in the northern, and south-east in the southern hemisphere reach their polar limit, a pretty regular wind blows from the west just in an opposite direction to the neighbouring trade-wind, which is sometimes called the retrograde, but more generally the west trade-wind. It blows in both hemispheres, and generally extends from about the 28th degree of latitude, beyond the 40th.

The general direction of the currents is the same as that of the winds, by which they are caused; but occasionally they are modified by various causes. The best known is the equatorial current in the northern half of the Atlantic, by which the water between Africa, the middle of America, and the south of Europe, is carried round in a circle. Following the course of the north-east trade-wind, which becomes more easterly as it approaches the equator, the current runs towards the north-east coasts of South America, and being obstructed there, it turns to the north, runs through the Gulf of Mexico, whence it flows along the southern coasts of North America, under the well-known name of the Gulf Stream, towards the east, and returns to the point whence it set out. By this decided circular current, several remarkable facts are explained; as, for instance, that of casks, which had been shipped in England for Havannah, and which, by the wreck of the ship near the Canary Isles, had got within the influence of the equatorial current, being carried back again to England, where they were recognised by the marks upon them.

It is also well known, that trunks of South American and West Indian trees, of Cedrela odorata, for example, have been thrown on the Canary Isles, and by such occurrences Columbus was evidently led to the conclusion that a great country lay in the west. It is clear, that such a current must be a very important cause of the diffusion of plants, for seeds which contain little oily and mucilaginous matter, and have a strong shell, can remain for a long time in the water without losing their vitality.

Eriocaulon septangulare, except in its native country, North America, grows only in the Isle of Skye, in the old world, and on that account Link* justly supposes that the seeds have been carried thither by the current.

A similar circular current, but not so decided a one as in the northern hemisphere, exists in the southern half of the Atlantic; it connects the west coasts of the south of Africa with the east coasts of South America,

* Die Urwelt, p. 266.

and the retrograde current runs between lat. 30° and 45°. By this current it is possible that plants may be carried from Africa to America, and from America to the old world ; but yet the plants of the northern half of the old world could scarcely be brought by it to South America. Indeed we must guard against attributing too much to currents. The cocoa palm is the tree usually chosen to show how plants may be transported by the sea from island to island, but yet the cocoa nut has not wandered from the south of Africa to Brazil, but was planted there. It frequently occurs in the West Indian Islands, and it has probably been carried there by the currents.

In the Pacific, there are two principal currents, one in the northern, the other in the southern hemisphere, following the course of the north-east and south-east trade-winds. These currents do not run from one continent to the other, like the currents of the Atlantic, but reach their western limit in the meridian of the Marian Islands. I may also remark that the distance which these currents run over is so immense, and the time necessary to cross those seas by the influence of the currents only so great, that the tropical plants of America could scarcely reach Asia in this way. It has indeed been affirmed that the well-known American grain, maize, was brought to Japan in this way as early as the 12th century ; but I am much inclined to doubt this fact, which has been reported in Chinese writings, for a seed like maize, which contains so great a quantity of fine mucilage, cannot keep for months in sea water of so high a temperature as the water of the equatorial current. This may possibly be the very reason that the maize has never been brought to Europe by the Gulf stream, which would be only a third of the distance between America and Asia.

In those parts of the Pacific where the trade-winds, and the equatorial currents depending on them, reach their western limit, there is a system of half-yearly winds called monsoons. In the northern hemisphere of these regions is the monsoon system of the Chinese seas, where the north-east monsoon blows from October till March, and the warm south-west monsoon for the other six months. The currents in those seas run in the direction of the prevailing wind; during the south-west monsoon the current is north-east, and during the north-east wind it runs to the south-west. In the southern hemisphere of those regions, viz., in Borneo, Java, and the eastern coasts of New Holland, another monsoon system prevails; there the wind blows six months from the north-west, and six months from the south-east, then as it were in connection with the south-east trade-wind of the Indian Ocean.

This is all it is necessary to say here on the currents in the seas; we shall afterwards be better able to judge where plants may have been distributed by currents, and where it is impossible that they could have been so. When considering this subject, we must also keep in mind, that the sea water of the warm regions is heated to a higher degree, and will, therefore, the sooner destroy the germinating power of seeds.

# PART II.

## ON THE CONDITIONS, BY WHICH THE SOIL INFLUENCES THE STATION AND DISTRIBUTION OF PLANTS.

We have shown, in the preceding part, that the distribution of plants over the surface of the earth is regulated chiefly by the distribution of heat and moisture, and we now go on to the consideration of the many local conditions which can promote or hinder the presence and distribution of plants, even though heat and moisture be present in sufficient quantity. These conditions are, for the most part, the relations in which plants stand to the soil in which they grow, and the consideration of them is one of the principal objects of botanical geography. We have already seen, that certain plants can vegetate only with a certain degree of heat, and have thus discovered the law, according to which plants are chiefly distributed, but we have not yet arrived at the explanation of this law.

It remains still a riddle to us, why the vine, for instance, must have a mean annual temperature of from 10° to 17° Cels., in order to yield good wine; why it cannot also grow in arctic regions; or why the culture of the maize will not succeed in our northern countries. If it is answered, that these plants have been appointed to grow in a warm country, and therefore will not live in a colder one, it is easy to see, that this is not an explanation; it is merely drawing attention to the law, according to which nature has distributed these plants. It is just the same when we treat of the local conditions of plants. Certain plants, we shall see, appear only on saline soils, others only on loose sand, and others only on calcareous rocks; but we are still far from knowing why these plants appear on such and on other soils. The cocoa-palm, which grows chiefly on the sea-coasts within the tropics, will not grow to any size in our palm-houses, notwithstanding all the care with which it is treated. The young plants of it in our palm-houses grow a few feet high, and then usually perish even before the nut has decayed, and before the roots of the young plant have penetrated through the mass of fibres, which forms the external covering of the nut. The attempt has been made in our palm-houses to imitate the influence of the sea in the natural situation of the plants by watering them with salt water; but notwithstanding this they perished.

But though botanical physiology is still far from explaining sufficiently the most obvious phenomena of the growth of plants; though even the most important objects in it, which are apparently so easy to determine, cannot yet be taught with absolute certainty, we may bring forward as proved, what follows on the relation in which plants stand to their natural locality.

The greatest number of plants grow in the ground; only parasitical plants, a number of Cryptogamia, and a few aquatic plants, are the exceptions. Almost every soil, even quartzose-sand, the most barren of all, contains more or less soluble matter, which if it be fine enough, may penetrate and ascend into the plants, with the moisture of the ground. Plants, which we cause to grow in insoluble matter, and water with distilled water, never attain to a perfect development; but water containing only carbonic acid is perfectly sufficient to support them. It is well known that a number of plants, and, very remarkably, just those which are most fleshy and juicy, grow in dry countries, which are often quite destitute of water, as the aloes in the southern part of Africa, and the numerous tribe of Cactaceæ in the dry west coasts of South America; and also the agaves, and Furcrææ, 70 or 80 feet high, which grow on the rocks of the lofty Mexican Cordilleras. All these plants have very small roots, in proportion to their size; and from that it was concluded, that they drew their nourishment chiefly from the atmosphere. But there are many things which may be justly urged against this opinion. Although those countries are in general dry, they have always a wet season, and it is during this that these plants chiefly grow, while in the dry season, they are in a state resembling the hybernation of animals. The lichens, mosses, and Jungermaniæ, which are so common with us on the bark of trees and on rocks, grow in the same way only during winter, autumn, and spring, while during the heat of summer they are dried up and apparently dead.

But the ingredients of plants are just as compound as their food is simple, and the question is, whence these come? An infinite number of observations and experiments have been made to determine whether plants, or whether organic life in general, can produce all the heterogeneous substances which are to be found in the perfect plant, out of the simple elements it usually receives, such as water, carbonic acid, and atmospheric air. The examination of this point belongs to botanical physiology; here we can only state the result of it, which is, that plants certainly receive many of the various ingredients of the soil, but that they can also produce other substances.

It has been wished to prove, that plants can receive only pure water, from the fact, that the membrane of plants lets through no coloured fluid; but this is quite illogical. A real solution, *e. g.* of a salt in water, passes really into the substance of the plant; and thus is explained the indisputable fact, that many plants contain a greater quantity of a salt or any other substance, when they grow on a soil which contains a greater proportion than another of that substance. However, we must take care not to judge too hastily from this fact. It cannot be denied, that plants have the power of selecting certain substances from the soil in which they grow. We see very frequently in our ditches and pools plants of the same kind, for example, different Charæ, growing close to each other. One of these is perfectly green, while another is incrusted with chalk both on the upper and under surface of its leaves. If the deposition of chalk on the surface of those plants were purely mechanical, it is evident, that all the neighbouring plants would be incrusted with it in the same proportion, which is not the case ; and if the plants had not, at least in some degree, the power of accepting some substances from the soil on which they grow, and rejecting others, it could not be explained why all the plants in the same water had not received equal quantities of the extraneous substance. If, therefore, such a fact is to be regarded as established, it may be explained in some measure, why some plants are always found on a certain soil, and in general can grow on it only ; although the stations may be at great distances from each other. A few examples will illustrate this. Certain plants, which we shall by and bye know under the name of saline plants, grow only on soils which contain culinary salt ; and as soils containing this in the same proportion are very frequent in all parts of the earth, so the plants which grow on them are very generally diffused. Salsola kali, for example, grows on almost all the coasts of Europe, as well as on the African and Asiatic coasts of the Mediterranean and Caspian seas. The soda of Alexandria, which in earlier times supplied all Europe, presupposes the existence of an immense quantity of this plant in that country. Samolus Valerandi extends itself still more widely than Salsola kali, for, besides in Europe, it is found in North America, in the south of Africa, and in New Holland.

Many of what are called Littoral plants grow in inland regions, when the soil is of the same nature as on the sea coast; as an example, I mention the Glaux maritima, which appears on the chalk mountains near Berlin. The beautiful orchis, Cypripedium calceolus, the only representative of the tropical Orchideæ in our zone, grows exclusively on chalk mountains in Harz; it is not found in the whole plain of northern Germany, and first appears again on the chalk mountains of Rügen. Such facts plainly lead to the conclusion, that, in the same way as the climate determines the appearance of certain forms in certain regions, the locality also determines the presence of peculiar forms, which are always repeated, under the same conditions of locality, provided no important obstacles prevent it.

As, in the first part of this treatise, we have enumerated the facts from which are deduced the laws, according to which climate exercises its influence on the distribution of plants, we must now do the same with those conditions of locality, which are causes of the presence of certain plants.

According to the nature of things, the local conditions, which influence the distribution of plants, must be infinitely various; and some will more clearly than others show their influence. We shall endeavour to state separately each of those local conditions, and at the same time to name the principal plants whose presence is determined by it.

The relations of plants to their stations must be very different, according as they take root in water, in the earth, in the earth and water, or in the air.

Plants which grow in water, are called aquatic plants (Plantæ aquaticæ, Hydrophyta). They again present many differences, which are very important to botanical geography. First, fresh-water and salt-water plants are very different from each other, and a few of the least organized only are common to both salt and fresh water. Conferva glomerata is probably the most perfect of these. But a great number of Diatomeæ appear both in fresh and salt water; these are little imperfect bodies, which at best belong to a kingdom intermediate between plants and animals, yet are far from being true animals.

Marine plants (Plantæ marinæ) are those which are found in sea-water, and consequently in every ocean The most of them belong to the immensely large family of Algæ, and the Zosteræ are the only phænogamous plants which appear in sea-water. The Fuci solely belong to sea-water, and form a very distinct group among the Algæ, separated by form, as well as by structure, from the others. There are true Fuci found in the Caspian Sea, though at the present day it is unconnected with the ocean. Almost all the marine plants are rooted to the bottom of the sea, chiefly on rocks, and near the shore, where the water is shallow; nor do the Fuci seem to go to a very great depth, though certainly as deep as a few hundred feet. It is true that some have been measured, and have been found to be more than 300 feet in length, *e. g.* Fucus pyriferus, at Cape Horn, the leaves of which are seven or eight feet long; but these plants, as I have observed in the Laminariæ on the west coasts of South America, do not grow in a straight direction from the bottom to the surface of the sea, but lie rather horizontally, and, therefore, though of so extraordinary a length, can grow in water of much less depth. The Straits of Magellan and La Maire are full of this gigantic Fucus, and there, in the cold water, where it grows to an extraordinary size, seems to be its true zone; yet it seems that the formation of fruit is prevented by this excessive development of the leaf. It is, at least, very remarkable, that of the many travellers who have sailed through this region, not one has found the plant in fruit; whilst this has been found on small specimens, which grow in the north. This Fucus is distributed in the New World throughout all the zones, from the extreme north to the extreme southern point; Baron Alexander von Humboldt first brought it from the tropical seas, where, however, it does not grow to so great a length as at Cape Horn. The same plant also appears at the Cape of Good Hope, but neither here is it so large as at Cape Horn.

The distribution of the Algæ, and of the sea-weeds generally, is less regulated by the latitude and longitude, than is the case with land plants; but this is natural, for the sea-water is almost every where of the same saltness;* and it is this salt-water, which chiefly determines the distribution of marine plants, just as heat is the chief determining cause of the distribution of land plants.

On the coasts of the ocean, where the great Fuci grow, they cover the bottom of the sea with an impenetrable vegetation, which serves to support millions of animals. When sailing over such regions, in a calm sea, one enjoys the splendid sight which those submarine meadows and forests present to the eye; the variety and splendour of which is increased by tall corallines of the genera Isis, Gorgonia, and Antipates, or by the varied colours of the masses of expanded Madrepores. Scarlet Sea-Anemones, gold-coloured Actiniæ, and corals of various colours, are seen amongst them. At ebb-tide the Fuci, in general, come close to the surface; they are often left quite uncovered, and begin to dry, until the returning tide again refreshes the flagging plants. But when the sea is agitated by storms, when the waves dash with fearful violence against the rocks, those marine plants are torn up, and float on the surface of the water, until they are thrown on the shore. In this state, drifting about the sea, the voyager generally finds them; he is very seldom permitted to seek for them in their place of growth. But the torn-up Fuci are seldom carried far from the coast; and, therefore, in earlier times, their appearance was the surest indication to the mariner of his approach to land. Yet Columbus was

* See Meyen's Reise, &c. ii. p. 412.

much deceived by them, on his first voyage of discovery, when he reached that part of the Atlantic which is now known by the name of the Sargasso Sea. But this Sargasso Sea is a very remarkable phenomenon, the origin of which is not yet explained, though it has been much examined, and though much has been written about it. In the Atlantic, just within the great equatorial current, is a space of at least 40,000 square miles, in which is always floating on the surface a great mass of sea-weed, all of one species,—viz., Fucus natans L., which is identical with Fucus Sargasso of Gmelin, and is now called Sargassum vulgare. This Fucus swims in that sea in heaps of various sizes, more or less plentifully. Sometimes the ship, sailing through the water there, is quite surrounded by it, and sometimes not a single plant is seen for hours. I have found the distribution of this floating tangle unequal everywhere in the Sargasso Sea (Mar de Zargasso of the Portuguese, Sargasso, in Spanish,) and I think that it cannot be otherwise in so mobile an element. I have scarcely found masses so thick that I could compare them, as Columbus did, to floating meadows; but yet I have seen single heaps connected with each other, from five to ten feet in length, which generally consisted of a single plant. The Sargasso Sea extends from 22° to 36° north latitude, and from 25° to 45° west longitude (from London). Beyond those limits which are formed by the equatorial current, there is seldom a plant of this species seen, and then it is very much damaged, in which state I have seen some pieces, floating between the Azores and the south-west extremity of England. Many very different explanations of these enormous heaps of floating tangle have been given. At one time they were supposed to be brought by the Gulf stream from the Gulf of Mexico; at another, to grow on shallows in the Sargasso Sea itself, from which they were torn by fishes, Mollusca, and the large Holothuriæ; but all these suppositions are now unnecessary; and indeed it is curious, that the station of this tangle should have been so long sought, since it was known that the Fucus natans of the Sargasso Sea, never appears, either with a root, or in fruit.* I have fished up thousands of specimens and examined them, but not a trace of a root, by which they could have fastened, was to be found; and in small specimens, it could be seen very well that they had grown in all directions from a free central point, which had never been fixed. We need, therefore, seek no other station for this floating tangle than the place in which we find it, viz., on the surface of the sea, and accordingly this is one of the few plants, which grow floating freely in the water. A great number of organs, which are hollow and contain air, serve to increase its buoyancy. The fact, that a sea-weed or Alga can grow freely in the open sea, is besides no longer isolated, as I have discovered in the tropical parts of the Atlantic, particularly near the equator, a small, extremely pretty, stellated Oscillatoria; this little plant is colourless, and so small, that it is not visible from the deck of a ship, and, therefore, it had hitherto been overlooked. What is called the root of the Algæ is of a nature quite peculiar to them; they have not a true root, as phænogamous plants have, but it is merely a continuation of their leafy substance; when the Alga roots itself, the fixed end of the plant swelling out.

The plants, which appear in fresh-water only, are called fresh-water plants; and they again fall into subdivisions, viz., into such as take root in the ground, and such as swim freely about. To the first class belong most of what are called, in general, water plants; the Nymphæ, which, in the north, adorn the still waters, with their large leaves and their beautiful flowers, scarcely raised above the surface of the water; all the Potomagetons, whose leaves float horizontally on

* See Agardh, Species Algarum, vol. i. p. 7.

the surface ; the Utriculariæ, with their finely divided leaves and beautiful golden yellow flowers ; the singular form of Stratiotes, with its pretty white blossoms, equally strange to our north, for this plant imitates the form of Pandanus ; all these plants take root in the earth, often at considerable depths, and the foliage floats in the water.

It is otherwise with the Lemna or duck-weed, which in summer never fails in the standing waters of our countries ; it, as well as a number of Conferveæ and Ocillatorieæ, which can completely cover pools of small extent, float loosely on the surfacee. It is indeed the case, at least according to the observations which have hitherto been made, that in the formation of these floating masses of Conferveæ and Oscillatorieæ, the first plants have been detached, by some means or other, from the ground, or from some firm floating body ; but as soon as a few of them have become unrooted, the little sporules get a point of support, and the formation of the large masses goes on rapidly. Exceedingly little is necessary to keep the little sporules of the Conferveæ on the surface of the water, that they may germinate there ; afterwards they float with ease on the surface, and develope great masses. In the Oscillatorieæ these are formed often in from 24 to 48 hours, so that ponds, which have been cleared of Conferveæ one day, are quite covered by them on the next.

It is remarkable, that in tropical countries the Conferveæ are very rare, yet they are by no means wanting in standing water there ; and, at considerable altitudes, where the climate more resembles that of the north, for instance, in the lake of Titicaca, on the plateau of southern Peru, they appear as plentifully as with us. In the South Sea Islands, the standing pools, particularly deserted taro-fields, are quite as full of Conferveæ, Charæ, and Potomagetons, as with us. The Lemnæ, which float freely about, are more or less wanting in tropical countries, though they also are found in a few places ;

*e. g.* Lemna minor, and Lemna trisulca have been observed in America, as well as New Holland. The Lemna of the north is replaced by the genus Pistia in the tropics; it is incredible in what quantities Pistia stratiotes covers the lakes there. When there have been storms on the Laguna de Bay, the great lake in the island Lucon, these plants are driven to the coasts, and thickly cover the water there far and wide, while heaps, several feet high, are thrown up on the shore, and when corrupting spread a dreadful stench. The Laguna de Bay also fills with these plants its outlet, the rapid and beautiful Rio Pasig, which at last carries them to the open sea. The Pistia stratiotes germinates in the morasses on the coast, and, after it has raised itself above the mud, it lives, floating on the surface of the water.

Fresh-water plants may be classed in three great divisions; those which grow in water and the atmosphere only; those which grow in the earth, water, and atmosphere; and those which grow in the earth and water only, and never rise above the surface. It is of importance to botanical geography, that these different sorts of aquatic plants should have fixed names; and accordingly, I understand by aquatic plants (Plantæ aquaticæ) such as take root in the earth, rise through the water, and come above the surface to expand their leaves there, or to develope their flowers and fruit in the air.

I call those which take root in the earth, and grow in the water, but never rise above the surface, submerged plants (Plantæ submersæ). The genera Chara, Najas, Ceratophyllum, and several others, belong to this class, and more particularly, almost all the Algæ.

Floating plants (Plantæ liberæ s. pl. natantes), on the other hand, are those which swim freely on the surface, and send out their roots in the water only. The genera Lemna and Pisti are of this class, and also a number of the Oscillatorieæ. Oscillatoria flos-aquæ, which is generally regarded as one of the floating plants, is, according to my observations, not a plant, but merely the sporule filaments of the Nostoc, which have separated from the gelatine of the plant, after it has been dissolved by putrefaction.

But aquatic plants differ among themselves still more widely, in other conditions of locality, the original cause of whose influence can as little be explained, as that of the influence of climate on the distribution of certain plants.

Thus, many aquatic plants grow only in lakes or still water, and are called on that account lake-plants (Plantæ lacustres). The Nympheæ, Scirpus lacustris, Scirpus palustris, and Arundo phragmites, are of this class ; the latter surround standing water in our countries with a thick forest, and they are repeated in the colder regions of tropical countries Thus the shores of the lake of Titicaca are enclosed by a thick forest of a beautiful rush,* just as the great lakes of Prussia are. The people of that country would live in great wretchedness, if nature had not bestowed on it those plants, for it lies far above the limit of trees, and only a few bushes grow in its neighbourhood. A few sticks, an oar to put the balsas or boat, woven of rushes, in motion on the lake, or a pole for a mast on which to hoist the sail, also woven of rushes, are the riches of the poor people of that country, as it is quite without wood.

Stratiotes aloides, the pretty Utriculariæ, of which there are extremely beautiful forms in the tropics, several Potamogetons, Charæ, Trapæ, several Ranunculaceæ, the Sagittariæ, Butomus umbellatus, &c., are also plants, which grow in standing water. All these genera have their representatives in the different zones of the earth. Our Sagittaria is replaced in the warm zones by the

* Malacochæte Tatora of Nees and Meyen.

beautiful genus Pontederia, and the Nympheæ by the Lotos.

Some of the lacustrine plants have received the name of ditch plants (Plantæ fossarum and Plantæ stagnariæ), because they may be found in almost all deep ditches, and stagnant pools of small extent. To this division belong Stratiotes aloides, Hydrocharis morsus-ranæ, Butomus umbellatus, Phellandrium aquaticum, Veronica anagallis, and many others. But in nature there are no such definite boundaries, as we must here mark out; the plants of the large lakes as well as of smaller pools, not only appear sometimes in ditches, but also in running water, particularly near the margin, if there are any obstacles there which prevent a rapid current. Thus we find in rivers, especially in places where there are large floats of wood, or where the course of the water is impeded by bushes, almost all the beautiful plants, which we already know as lake and ditch-plants. There are, however, some true river-plants (Plantæ fluviatiles and Plantæ rivulares), that is, such as scarcely ever appear but in brooks or rivers; as examples of this class we may mention Ranunculus fluviatilis, Conferva rivularis, &c.

With respect to other conditions of locality, we may distinguish, among aquatic plants, some other groups; for example, fountain-plants (Plantæ fontinales s. fontanæ). They are such as grow in the fresh, clear water of springs, or close to them. In the north we may regard as true fountain-plants the following, Montia fontana, Veronica beccabunga. From these true fountain-plants we must separate those, which also grow in the vicinity of springs, but only on account of the moisture, when the soil around is very dry. We can nowhere better observe this influence of springs on vegetation, than in the deserts of the tropics; the smallest spring there forms an oasis, in which, not only the most juicy Cyperaceæ and grasses, but bushes grow, and here and there even a palm rises. In the south of Peru, when journeying from the plateau of the Cordilleras to the coast, we find only the most desert, barren, and parched-up regions; but the smallest spring, which appears here and there, often at wide intervals, is the cause of a little settlement. Often it supports only a field of Alfalfa (Medicago sativa), our lucerne, a little maize field, and a few olive trees; and yet, for the sake of this scanty produce, the great roads must pass by such places, that the beasts of burden may get the refreshment necessary for them. Nothing can equal the dreariness and death-like stillness in such regions of Southern Peru; sometimes for 20 or 30 miles not a bird, not an insect nor a plant is seen; but the smallest spring calls from this dead, dusty soil a green world, and, when rich mines are near, is the source of great riches, which could not have been obtained without it. In most tropical countries, a spring is at least the spot where a few higher or more luxuriant trees grow. In the South Sea Islands I saw the springs often surrounded by the noble Pandanus and Eugenia;* and in the Philippines, beautiful palms and Barringtoniæ grew near them.

A number of the plants which grow on the sea-shore, where the soil is impregnated by salt-water, are found also in the vicinity of springs, containing culinary salt; such plants are called saline plants (Plantæ salinæ, Halophyta). A number of species of Salsola, Anabasis, Salicornia, and the Glaux maritima belong to this group; a few Charæ also grow in these salt springs, and this is the case under all zones and in all regions. It is remarkable that not only the saline plants of the sea coast agree with those of salt springs, but the flora of the Steppes chiefly consist of similar plants, from which we may conclude what has been the earlier condition of the Steppes.

Next to aquatic plants come those which grow partly in water, partly on dry ground; and have hence

* Pandanus odoratissimus, and Jambosa malaccensis, Dec.

been called amphibious plants (Plantæ amphibiæ). They have, at least very frequently, leaves, which differ in form, according as they have grown in the water or on the ground. Nasturtium palustre and N. amphibium, Cardamine pratensis, Rumex hydrolapathum, &c., are examples of this. Several species of Mentha also belong to this class. Other plants again are found most frequently in places which are inundated in winter or spring, and have been, therefore, called inundated plants (Plantæ inundatæ); Limosella aquatica, Peplis portula, Juncus bufonius, Caltha palustris, and several others, come into this division. Thus the Plantæ inundatæ, and the Plantæ amphibiæ are very different; the former grow in a soil which is inundated only at a certain time of the year; the latter grow sometimes in water, sometimes on land, and their leaves usually differ in form in the different situations.

Plants, which grow on the shores of large sheets of water, are called shore plants. Littoral plants (Plantæ littorales seu maritimæ), when they grow on the sea shore; but shore plants (Plantæ ripariæ) generally, when they grow on the margin of fresh water, as well of lakes, as of rivers and brooks. As the sea shore is impregnated with culinary salt, many of the littoral plants exactly agree with the saline plants, which grow in the vicinity of inland salt springs. Glaux maritima, Salsola kali, Samolus Valerandi, Eryngium maritimum, Chenopodium maritimum, &c., are examples of this; and in the tropics, Lythrum maritimum, several species of Heliotropium and Vitex, &c. I have already remarked that the Steppe plants agree with the littoral and saline plants. But the Steppe or saline plants are not so similar in all parts of the earth, as is usually assumed. In the deserts of Egypt they are Dactylis repens, Cynodon dactylon, Zygophyllum album, Cressa cretica, &c. In North America they are Uniola maritima, Spartina glabra, Gerardia maritima, Aster subulatus, and several others. In South America we found, in similar situations, the gregarious Poa thalassica, *Humb. et Kunth.*, Salsola corticosa *mihi*, Salsola glomerata *mihi*, &c., and in the salt Steppes of Asia, there are Salsola prostrata, Statice tartarica, Glycirrhiza hirsuta, G. lævis, &c.

The vegetation of the sea coasts of tropical countries is very peculiar; that is, wherever the sea is not encompassed by rocks or sand, but by shores consisting of mould, partly firm, partly marshy, and strongly impregnated by moisture when the tide flows. On such soils in tropical countries, there are vegetable forms quite peculiar to them, forming thick, impenetrable woods for miles along the coast. The most common of these sea-shore plants is the Mangle or Mangrove (Rhizophora Mangle L.), which most frequently grows at the mouth of great rivers; it has this peculiarity, that its seeds do not fall to the ground and take root there, but sprout from the fruit, and send down their roots until they reach the marshy ground, from which they shoot again, so that in a short time, from a single trunk, is formed a forest, in which one can wander about at ebb tide. After the Rhizophora, the Avicenniæ are the chief trees which form the groves on the sea shore (Mangrove forests in the Brazils). In Brazil, the Avicennia nitida, and A. tomentosa L. form the Mangrove forests; on the tropical coasts of Africa, Rhizophora and Avicennia tomentosa; and on the coasts of India and New Holland, appear the genera Rhizophora, Avicennia, Ægiceras and Bruguiera, and frequently on the shores of rivers near their mouth, the splendid Barringtonia.

Baron Alexander von Humboldt* has collected at the mouth of the Rio Sina several Fungi (of the genera Boletus, Hydnum, Helvella and Thelephora), which hang on the Rhizophora, and flourish, though washed at full tide with salt water.

The plants which take root in the earth and grow in

* Reise, &c. Theil 6, 2te Hälfte, p. 57.

the atmosphere, are called land plants (Plantæ terræ adfixæ). They differ widely according to the chemical composition of the soil, though, as we have already seen, plants absorb very little of its constituents. When treating of aquatic plants, we mentioned those which appear in salt springs, as well as the shore plants which agree with them. Those plants are also land plants, and grow in soil, which is impregnated with culinary and other salts.

But the geognostic nature of the soil has also a remarkable influence on the presence of certain plants. Although it is true that the limits are not so distinctly drawn by nature as we must draw them, one can scarcely overlook this influence on a general survey of the phenomena. The most important groups, in respect of the geognostic nature of the soil, are :—

1st, Sand plants, called also flint plants (Plantæ arenariæ, plantæ silicaceæ). These are of a peculiar character in all parts of the earth, and the greatest number of them are probably grasses. On our sandy plains are chiefly found Carex arenaria, Herniaria glabra, Arundo arenaria, several species of Tussilago and Potentilla, Sedum acre, and several other plants ; in shifting sand, where there is seldom a constant vegetation, Elymus arenarius is in its fitting situation, and is often used with great advantage to bind loose sand, when no other plant will grow on it. We also distinguish the Plantæ sabulosæ, that is, plants which grow in river sand ; as Elymus sabulosus, and several species of Tussilago and Salix.

2d, Chalk plants (Plantæ calcareæ) grow on calcareous rocks. There are some plants, for example, the family Orchideæ, which particularly like this soil, and some species appear only on calcareous rocks. Cypripedium calceolus is an example of this, which I have already given. Teucrium montanum, Sesleria cœrulea, and several other plants, are produced on a calcareous soil. Chalk mountains show several other peculiarities in their vegetation ; they have for the most part few woods, but in general have rather a shrubby and bushy vegetation, and, therefore, possess many little plants, which grow in the shade of those bushes.

Among the various calcareous rocks, gypsum appears to have a form of plants peculiar to it, viz., the genus Gysophila, which appears generally on gypseous soils ; and, therefore, we have also gypsum plants.

A turfy soil has also a peculiar vegetation, and in northern countries, where turf moors are so frequent, this is extremely influential on the character of the landscape. The species which grow on this soil are particularly distinguished by growing socially, and they have an excessive development of root. As an example of this, we may adduce the Sphagna, which very seldom allow any other plant to appear among them. Vaccinium oxycoccus, Andromeda polifolia, Droseræ, several species of Juncus, and Salix, are the most common turf plants (Plantæ turfosæ, plantæ cæspitosæ). When turf begins to form in pits, from which it has been dug, and in which water has collected, Charæ and Confervæ are in our regions the first plants which supply the materials for it. Afterwards Spongilla lacustris appears in the sides of the pits ; then Utriculariæ, Scirpus palustris, Myriophylla, Equiseta, Nympheæ, &c., appear, which again decay and gradually fill up the pits, while the sides at the same time gradually approach each other. When the pit has filled up, and some firm soil is formed, Comarum palustre, Alisma plantago, Vaccinium oxycoccus, Droseræ, Eriophora, &c., appear, and thus the turfy soil is renewed.

I have here considered the geognostic nature of the soil with respect to its influence on the appearance of plants, but I must not neglect to remark that those plants which peculiarly belong to a particular soil, appear very frequently on other soils; and M. De Candolle* even affirms, that he has observed in France, that

* Dictionnaire des Scienc. Nat. tom. xviii. p. 377.

every plant of that country can grow on every soil, a result which, however, cannot be admitted. I do not believe that Carex arenaria can grow on turfy ground, nor Cineraria palustris in loose sand. But it is certain that the geognostic nature of the soil has less influence on the presence of plants than its chemical composition.

But the chemical and geognostic nature of the soil is in another respect important to the distribution of plants; for it appears that plants which have a preference for a soil of a particular nature, have a much wider circle of distribution than those which grow in common mould; for the local conditions, which determine the presence of the plants belonging to them, are only too often repeated.

Mould is the soil most generally suited to the growth of plants, and upon it are frequently found all those plants, which we have already mentioned as peculiar to other soils.

According to the quantity of moisture it contains, mould forms a soil, which is more or less favourable to certain plants, so that they grow in it more frequently and luxuriantly, than on a soil of a different nature. There may here be distinguished the following groups:—

Bog plants (Plantæ uliginosæ): these grow on a very wet soil, which is on that account so soft that it yields when stepped on, and rises again. In northern countries such bogs are very frequent, particularly in meadows; and also in the higher regions of mountains, as on the Alps, on the Harz mountains, on the mountains of Silesia, and even on the plateau of the Cordilleras of Southern Peru, where they are quite as extensive as on our northern mountains. The principal bog plants are Pinguicula alpina, Primula farinosa, Caltha palustris, &c. It is natural that bog and turf plants should frequently agree, as turf can be obtained in almost all boggy places.

Bogs are distinguished from marshes only by greater

firmness, and by containing less water. A marsh is so soft, that it sinks in when stepped on, and does not rise again as a bog does. As marshes frequently contain sheets of water of greater or less extent, aquatic plants often appear amongst those which are peculiar to marshes, and are called marsh plants (Plantæ paludosæ s. palustres). To these belong Menyanthes trifoliata, Hottonia palustris, Cineraria palustris, Scheuchzeria palustris, Comarum palustre, Bidens cernua, &c. As marshes very often dry up in hot summers, many of their plants disappear early in summer, and do not appear until the following year, when the marsh is again filled with water. Such places are then extremely similar to those which are inundated at certain seasons, and to which inundated plants are peculiar. It has also been wished to separate mud plants (Plantæ limosæ) from marsh plants, but they, as well as inundated and bog plants, commingle with marsh plants in such a way, that the distinction does not seem to be founded in nature, nor can it serve any purpose to make so great a number of divisions.

The qualities of the soil have, in many other respects, a most decided influence on the presence of certain plants, and we proceed to discuss these more fully in a stated order. We shall consider the influence of the soil on the presence of plants :

I. In respect of its aggregate condition.

Rock plants (Plantæ rupestres seu rupicolæ)* are divided from stone plants (Plantæ saxatiles). The first grow on bare rocks, for example Sedum rupestre, a great number of Cacti, and other succulent plants in tropical regions, but especially the greater number of lichens, ferns, and mosses. Stone plants grow on stones, which have been detached from mountain-masses, and Thlaspi saxatile is given as an example ; but I have not

* See Schouw's Pflanzengeographie, Berlin, 1823, p. 158.

been able to find out any distinction between rock and stone-plants.

Sand-plants, which we have already noticed, must also be mentioned in this place. Gravel plants (Plantæ glareosæ) yet remain, which according to Schouw, grow chiefly on the detritus of mountains ; such as Saxifraga rivularis, Ranunculus alpestris, and R. glacialis. On the plateaux in the south of Peru, at an elevation of 14,000, 15,000, or 16,000 feet, are extensive tracts, more than a day's journey in length, where the soil entirely consists of a white, weather-worn trachyte, which in appearance is very like fine sand. In this grows some species of Sida of remarkable beauty, as Sida pedicularifolia *mihi* ;* and in the lava ashes of the volcanoes of the South American Cordilleras, I found other as pretty species of Sida, as Sida borussica *mihi*, several grasses, Baccharidæ of very peculiar forms, as Baccharis phylicæformis *nob.*, Baccharis genistelloides *Hook*, Baccharis sagittalis *Lessing*, B. quadrangularis *nob.*, and the Tulostoma Meyenii *Klotzsch*, a very distinct mushroom. In other countries we found, under similar conditions, for instance, on the cone of the volcano of Maypu, very pretty plants,—the Calycera ventosa *nob.*, and several others.† But the forms of these Alpine gravel plants are more peculiar in America, for there they always appear in little bunches, which sometimes contrast very prettily with the dark lava ashes, as in the instance of the Sida borussica, with its little bunches of leaves thickly covered with hairs, from which peep out the flowers, striped even before opening, with white and violet.

II. In respect to the nature of the soil.

We must here again separate three different groups, according as the plants appear on other living plants, on dead organic substances, and on artificial products.

* See Meyen's Reise, &c. i. p. 460.
† Ibid. i. p. 356.

We have hitherto considered dead nature as the soil of plants, but there are a great number which fasten on organic forms. We shall first consider what are called parasitical plants, which, as the name implies, take root parasitically on other plants. Our misletoe (Viscum album L.) which grows on our high trees, is universally known ; it is a parasite, its seeds send out roots which penetrate through the bark of the tree, and then draw their nourishment from the wood of the tree on which the parasite takes root. In warmer countries, especially within the tropics, and in the subtropical zone, the misletoe is replaced by the genus Loranthus, the number of whose species is as great, as the splendour of their scarlet flowers. But parasitical plants differ so widely in the relations in which they stand to their soil, that is, their supporting plant, that it is necessary to arrange them in subdivisions, which are also important in respect to botanical geography.

We separate,

1. True parasites (Plantæ parasiticæ veræ). These are plants which fasten on the roots of other plants, and are so intimately united with the substance of the supporting plants, that these form from their substance a peculiar organ, which serves as a support to the parasites. The plants belonging to this class grow always on fixed species of other plants, and they are clearly distinguished from plants not parasitical, by various characters ; for instance, by the absence of pores, &c. The embryo is wanting in the seeds of all true parasites, and after closely examining the mode of connection between the parasites and the supporting plants, I maintain also, that these plants are by no means produced from seeds, which take root in the substance of the roots of other plants, but that they are to be considered as a morbid production, which has grown out of the root of the supporting plant. I have investigated this subject in the magnificent genus Brugmansia, and in a Balanophora

which Blume has brought from Java.* All that has yet been said by some to the contrary, I consider quite foreign to the purpose, and therefore I remain in my former opinion.†

True parasites are not of a green colour, but generally more or less brown, and sometimes of remarkable bright colours. They also have no pores. To this division decidedly belong the plants of two families, the Rhizantheæ of Blume and the Balanophoreæ of Richard, which only appear within the tropics, and in the sub-tropical zones. The most remarkable and rarest plants belong to these families; the Rafflesia from East India has become famous under the name of the giant-flower. It is like a gigantic mushroom, three or four feet in diameter, and in form resembles the flower of a phaenogamous plant, which, issuing direct from the ground, has immediately fastened on the root of another plant.

In the north we possess three genera of parasitical plants, Lathræa, Orobanche, and Monotropa, which fasten on the roots of other plants, as, for instance, Lathræa squamaria on the root of the beech. It appears, however,—that is, some experiments are said to have proved,—that some of them may be raised from seed; and the Orobanche is adduced as an example. But the observations of Vaucher do not appear to be demonstrative,

* See Meyen Ueber das Herauswachsen parasitischer Gewächse aus den Wurzeln anderer Pflanzen. Flora, 1829, Nro. 4.

† Link wished to refute the above opinion by the fact, that in a section of a bough, the passage of its coloured sap to the Viscum growing upon it through the spiral vessels could be observed; but Viscum does not belong to the group of parasitical plants of which I have spoken, and also in Viscum the insertion can be easily seen and traced. M. De Candolle passes by my opinion, which is irrefutably founded on the structure of Brugmansia, with the remark that it is an odd idea; but if this learned botanist, with his great penetration, had attempted to refute the opinion, it would have been of more service to the science. The Orobanches must not be taken in order to refute my statement, as I myself have expressed some doubt about them; but much less Vaucher's observations on this subject, as they want the necessary exactness. Rafflesia, Brugmansia, and Balanophora should be taken to refute or confirm my opinion.

and I do not yet believe in them. In India, the genera Æginetia and Phelypæa are the representatives of our Orobanche and Lathræa.

According to recent observations, true parasites appear in various other plants on the bark of the stem or stalk, as the genera Apodanthes *Poit.* and Pilostyles *Guill.*; but these genera seem yet very doubtful, and appear to be only the shrivelled up flowers of the very plants on which they are found.

2. Parasites (Plantæ parasiticæ). These grow on the bark of other plants, chiefly on the tops of trees, while their seeds send their roots into the substance of the bark, and indeed so deep, that they are closely connected with the woody substance of the supporting plant, and suck their food from it. The before-mentioned genera Viscum and Loranthus belong to this division; however, many species of the latter are not parasitical, but grow in the earth, and form high bushes. The Viscum in our regions has but little influence on the physiognomy of vegetation; at most, it is remarked in winter on the leafless trees, particularly in fertile districts, where the green bunches of misletoe, among the leafless boughs, appear very strange in winter, when the ground is covered with snow. But in warmer countries, where the Loranthus grows, its scarlet flowers form the most brilliant contrasts with the dark green leaves of the plant by which it is supported. In Chili a leafless Loranthus covers with a scarlet carpet a large candelabra-like cactus, whose snow-white flowers, eight or nine inches long, project from it, and present a beautiful object to the eye.

3. False parasites (Epidendra seu Epiphytes). These grow on the surface of other plants, but they do not send roots into the substance of the plant, and therefore can draw no nourishment from it. They generally fasten in the crevices and hollows of the bark of trees, and afterwards affix themselves so firmly, that they can be torn away only with considerable force.

Plants of this kind appear in all parts of the earth; they are not appointed to certain fixed species or genera, but grow wherever they can find a point of support. In our northern countries we know what a great number of lichens, mosses, and Jungermanniæ fasten on the trunks of trees; they have these plants as their soil, but do not exclusively grow upon them, for under similar conditions, they can grow in other places. The smoother the surface of the bark, and the drier the soil in which the trees grow, the smaller is the number of these Epiphytes, but it increases in proportion to the humidity of the soil and the atmosphere. When wandering in the pine woods of our dry sandy districts, we may indeed see but few lichens and mosses on the trunks of the trees, but we can from this form no idea of the quantity of plants which appear on the trees of the damp woods of our northern mountain districts. The Usneæ, which, on the trees of our dry districts, present small, ill-developed specimens, are, in the damp woods of the Harz and Riesengebirge, more than a foot long, and, by their greenish grey colour, in many respects resemble the Tillandsiæ of the tropics. In the island of St. Helena, on the western side of which the climate is very humid, there is a reddish variety of Usnea barbata, which grows in such quantities on the trees of Conyza arborea that form the alley near Napoleon's residence, that this hanging vegetation is what first attracts the eye of the traveller.

How different are the Epiphytes which grow on the surface of other plants in the tropics! With us not even a fern, but Cryptogamiæ only appear on the trees; but there, where the atmosphere with a high degree of heat contains also an extraordinary quantity of moisture, there may be found often, on a single tree, such a number of different plants, that if planted in the ground they would cover a large space.

The Pothos plants there grow on the boughs of the loftiest trees, through whose splendid foliage the large white flower rises. Strange Orchideæ, Bromeliæ, and Pitcarniæ grow in the angles of the branches and fill up every crevice in the bark of the tree. The prettiest ferns, like our Lycopodium and ivy, twine up on the surface of the trunk, while silver-grey Tillandsiæ hang from the branches; not to mention the multitude of climbers, which once rooted in the earth, have ascended the trees and continue to flourish there, when not a trace of their root remains. The long shoots of these plants sometimes stretch from one tree to another, sometimes hang like cords, more or less obliquely, down to the ground, often not showing a single leaf for a length of twenty or thirty feet; and these stretched cords serve for the monkeys and wild cats to clamber on. In the New World, Bauhiniæ, Paulliniæ, Bignoniæ, Banisteriæ, and Passifloræ, are the chief plants which show such an excessive development of stem; in the Old World, especially in India and the neighbouring islands, different species of Rattan or Calamus, which furnish the canes of commerce. The Calami are a peculiar division among the palms, and strangely enough grow quite like the climbers of the New World; they twine up the trunks of trees to the very top, pass to the next tree, and descend its stem to the ground, from which they again run up. Many of these plants are covered with bristles, and even strong spines. Attempts have been made to measure their length, and they have already been found 500 and 600 feet long; but the longest can scarcely have been measured, and, besides, I believe, that among other climbers,—viz., the Paulliniæ, Banisteriæ, and several others,—may be found plants just as long as among the Calami.

In the forests of the tropics the trees arch together at a considerable height, and darken the sky so that not a ray of light can reach the ground; while through the thick foliage run climbing plants, covered with leaves and flowers, with hundreds of shoots running from tree to tree in all directions, and twining round each other.

There is seen a profusion of various flowers in the top of the tree, but it is difficult to decide to which stem they belong; the tree must be felled before they can be reached. A quantity of fruits and flowers often strew the ground, but it is not easy to determine on which plants they grew.

We return to the Orchideæ and Aroideæ, which are the most common parasites on the bark of trees, and we find that these plants, composed of a firm and juicy tissue, have peculiar organs to enable them to imbibe readily the moisture of the atmosphere. Their roots are covered with a white veil of cellular tissue, the cells of which either wholly consist of spiral vessels, or are furnished with these peculiarly hygroscopic elementary organs on their inner surface. Even the fine fibres of many parasitical Orchideæ, by which they attach themselves to other plants, consist entirely of a spirally wound lamella, which is either equivalent to a wide spiral vessel, or is itself composed of spiral vessels, lying close to each other and soldered together.

The parasitical vegetation of the tropics is still far from being exhausted; the leaves of the parasitical Orchideæ, Aroideæ, and Ferns, are in their turn covered with parasitical plants. The tropical Jungermanniæ, often of the prettiest forms, resembling our Dendrites, grow on the leaves of these plants, and that in such profusion that there is seldom a plant of these families, particularly in damp forests, which does not show several of these little microscopical Jungermanniæ. The bark of the trees is also rich in lichens, but these seldom belong to the foliaceous tribes, which seem as few in number within the tropics as mosses and Fungi are.

4. A fourth group of parasitical plants consists of leaf-fungi, which, as produced by disease, have been called, particularly in recent times, the Exanthema of plants. The leaf-fungi appear in the greatest numbers in the northern countries, although they are also frequent in the southern polar regions; they for the most

part belong to fixed species, though there are a few which appear on various plants. In respect of the physiognomy of vegetation these leaf-fungi have but little influence, and that merely casual, as when whole trees are covered with coloured leaf-fungi, as I have seen in Chili on the Rio Tinguiririca.

As parasites grow on other living plants, there are also several kinds of the lower forms which appear on dead organic bodies, such as dead plants, dead animals, animal excrements, &c. Even in our colder regions we see that as soon as the trunk of a tree decays, its surface is covered with various species, which are almost always the same for certain regions.

We have some very authentic observations on the appearance of certain forms of fungi on certain dead animals, or dead parts of animals, and these are extremely important with reference to the doctrine of equivocal generation. It must be generally known that in autumn, when the flies begin to die, they frequently fasten to the window-frame, and the hinder part of their bodies is more or less covered with minute white fungi. The fungus which forms this white fatty covering is an Isaria, and the little circle of dust on the pane is caused by the bursting of its spores. Another Isaria has been found on dead horse hoofs, and several kinds of mould are known which grow on old animal excrements.

The growth of plants on living animals, as of several Algæ on fishes, for instance on old carp, whales, and especially mussels, is quite accidental.

We have now considered those plants which have as their soil other living plants, as well as those which appear on dead organic bodies, and we now proceed to a third division, which contains those plants which grow on artificial productions. I follow here Schouw, who distinguishes wall, ruin, roof, plank, and rubbish-plants.

Wall-plants (Plantæ murales, seu plantæ murorum) are those which appear on the walls of buildings, and certainly are very seldom wanting on them when old; but as they chiefly appear on very old decayed buildings, ruin-plants (Plantæ ruinorum) are not properly distinct from them. As belonging to this class, I may name Lecanora muralis, Dicranum murale, Asplenium ruta-muraria, Sedum acre, Sedum telephium, Hedera helix, and many others. But it is right to remark, that all these plants which we have considered as wall and ruin-plants, can grow quite as well in other situations, on the ground or on the bark of trees and on rocks; and a particular inclination to the artificial situations can only have been ascribed to them, because in certain countries they are almost always to be found upon them.

This is also the case with roof-plants (Plantæ tectorum). Sempervivum tectorum, which has a preference for such a habitat, appears likewise in natural stations; and the numerous mosses, which in the north grow on the roofs of houses, are found on the ground, on rocks, and on the bark of trees.

Board or plank-plants (Plantæ parietinæ) are those which grow on the wooden fences with which our gardens are generally enclosed. Parmelia parietina and Lecanora muralis, are the lichens which most frequently appear in such situations, but they grow equally well on stone walls and rocks. In northern countries the Usneæ are more frequent on such wood-work than with us; and in East Prussia there is seldom wanting on barn-doors a great quantity of Ramalina fraxinea, in specimens three, four, and six inches long. In the countries west of us there are found very often on palings the Confervæ, which have the form of the old Trentepohlea, and are so frequently removed from one genus to another. The plant which furnishes the well-known fragrant violet moss, found on rocks in the Riesengebirge and other places, decidedly belongs to the same genus.

Rubbish-plants (Plantæ ruderales seu ruderatæ) are

such as are found on rubbish-heaps in the vicinity of dwellings. They differ in different countries. Chenopodium vulgare, Senecio viscosus, Borago officinalis, Xanthium strumarium, Hyoscyamus niger, &c., are recognised as such plants. These plants agree with those which grow in preference in the neighbourhood of towns and villages, and are named Plantæ urbanæ; but their station is generally on places where rubbish has at one time been thrown.

Certain plants appear in fixed, and singularly peculiar stations, as, for instance, Racodium cellare, an extremely pretty fungus, which is found on wine casks; Conferva fenestrale (Byssocladium fenestrale) which grows in window-panes, and Conferva dendrita on paper. The Racodium can grow only in its assigned stations; nor can Conferva fenestralis grow excepting on a pane of glass which is constantly exposed to a damp atmosphere.

Plants furnish distinctions very important in their geography, when their relation to the soil or to the social plants near which they grow is considered. The soil is either in its natural condition, or it is cultivated, or the cultivated soil again lies fallow. A number of plants may be given for each of these cases, which seem to prove that they delight in such and no other station.

1. The plants which grow in cultivated ground (Plantæ locorum cultorum).

All plants which are artificially sown or planted in tilled ground, are called cultivated plants (Plantæ cultæ seu plantæ sativæ), although they differ very widely from each other. Linnæus understood by *ager* a cultivated field, and by *arva* a fallow field; and in this I follow his distinction, although these words have been employed in a different sense by later authors.

Field plants (Plantæ agrestres seu sativæ) are those which grow in cultivated and sown fields. They exist in great numbers, and differ in each zone of the earth.

Several appear only with certain cultivated plants, for example, Centaurea cyanus and Lolium temulentum, almost only in rye fields, but the former also amongst oats. Field plants are almost universally called Plantæ arvenses, and, therefore, all which bear this specific name, *e. g.* Spergula arvensis, Sinapis arvensis, Serratula arvensis, Convolvulus arvensis, &c., belong to this class. Suffrenia filiformis is peculiar to the rice grounds of hot countries, as our corn-flower is to rye fields, with which branch of agriculture it has probably been introduced into the East Indies.

Fallow plants (Plantæ arvenses) are those which grow on fallow fields. They are properly but little different from those of sown fields, only appearing in greater plenty, as their growth is not checked by the field being sown. The cereals are thickly sown, so that fewer field-plants can spring, and very few, such as Lolium temulentum, Centaurea cyanus, &c., shoot up through the stalks of corn; but when the same field remains unoccupied, all the seeds come to perfection, which in the preceding year were forcibly kept down; and sometimes they even grow up amongst the stubble. As such plants, I mention Rumex acetosella, Carduus nutans, which is the sign of a good soil, Convolvulus arvensis, Androsace septentrionalis, Echium vulgare, Artemisia campestris, &c.

Garden weeds (Plantæ horticulæ) are such plants as appear in the beds of gardens, and are injurious to the growth of the cultivated plants; such are the nettle (Urtica urens), Alsine media, Lamium amplexicaule, Chenopodium viride, and Chenopodium vulgare, &c. The plants cultivated in gardens are called garden-plants (Plantæ hortenses).

The enclosures of fields and gardens have also plants peculiar to them, and possess a ranker vegetation than waste places far from cultivated ground, which is explained by the soil being enriched by the manure of the cultivated ground. The plants which appear on the borders of fields are called plantæ liminæ, while those which grow in the hedges of gardens are called hedge-plants (Plantæ sepicolæ, commonly Plantæ sepeariæ). As Plantæ liminæ, I name the following: Cichorium intybus, Tanacetum vulgare, Artemisia vulgaris, Galium verum, and several others. Among hedge-plants are, Urtica dioica, Lamium album, Borago officinalis, Bryonia dioica, Xanthium strumarium, and Datura stramonium, as well with us as in India, its native country.

2. The plants which grow on uncultivated ground.

The plants which grow on uncultivated ground, have generally received from systematists the specific names sylvestris, agrestis, campestris, &c., but for the sake of botanical geography, it is necessary to make more minute distinctions. We shall, therefore, separate:

Field-plants (Plantæ campestres) are those which commonly grow on level and open fields, as Draba verna, Veronica triphyllos and Veronica hederæfolia, Echium vulgare, &c. As we have already seen that the presence of certain plants is determined by the chemical and physical nature of the soil, it follows that the plants of a dry field will be very different from those of a wet, clay soil; and the plants of a sandy from those of a calcareous soil.

Plains (campi), the soil of which is so dry and barren, that few or no plants grow on it, are called deserts (deserta), and the few plants which sometimes appear on them, desert plants (Plantæ desertarum) which differ in each desert. Meadow-plants (Plantæ pratenses) form a fourth group of field-plants. Meadows are an ornament of northern countries, which does not appear in the tropics under the same conditions. Although in tropical countries there are, at least in the rainy season, extensive tracts of green grasses, as, for instance, the Savannahs of South America, they are without all the beautiful flowers which, at certain seasons, cover our meadows with varied colours. At one time the blue flowers of Campanula glomerata, C. patula, Myosotis scorpioides, and various gentians ; at another, the white or red clovers (Trifolium pratense, T. fragiferum, T. repens), or the yellow flowers of the Ranunculaceæ of Caltha palustris, and the Lysimachiæ, adorn their green covering. It is not so within the tropics, at least never to such a degree, and the grass plains there are either not covered with our bright verdure, or it lasts but a short time, unless they are overflowed on the margins of lakes and rivers. The grass-plains of South America, as well the Savannahs on the Orinoco, as the Pampas in the southern regions, and on the table lands of the Cordilleras, are of quite a different nature. On our meadows the grasses are uniformly distributed, but on those of tropical America, they always grow in larger or smaller patches, not to mention that they belong to species and genera quite different from ours. We shall afterwards have occasion to enter more into the minutiæ of these phenomena.

The plants which grow in our pastures we call Pasture-plants (Plantæ pascuæ) ; they differ in general but little from meadow-plants, as pastures are merely meadows that produce little hay on account of possessing a small quantity of moisture. Of pasture plants may be named Gentiana campestris, G. uliginosa, Bellis perennis, Pimpinella saxifraga, Ranunculus repens, R. bulbosus, several species of Galium, and many others.

The vegetation of heaths (ericeta), which are so frequent in the north of Europe and Asia, is remarkably striking ; the plants which appear on those heaths are called Heath-plants (Plantæ ericetinæ), and they are of peculiar form. Our heath is the well-known Erica vulgaris, the representative of the large family Ericeæ, which is so extraordinarily numerous in southern regions, that in the south of Africa it even determines the character of the vegetation. The heath covers extensive surfaces in the north of Europe, often allowing no other plant to appear amongst it ; but sometimes it grows less close, and then bushes of Juniperus communis, Ledum palustre, and Andromeda polifolia, some little plants, as Parnassia palustris, some species of Sphagnum, and especially of Polytrichum spring up.

Mountain plants (Plantæ montanæ), as the name implies, grow on mountains ; their number is very great, and we shall afterwards, when we treat of the distribution of plants, learn to know them more exactly ; here we shall make but a few general remarks. Mountain plants are remarkable for their large flowers, and for a social growth, which is more observable in some than in others. They pass into alpine plants, for in lower latitudes the very same plants appear on the high Alps, which in higher latitudes grow on low mountains, and at last, even on the plain. The arctic flora and that of the Alps, therefore, remarkably agree, though, as we shall afterwards learn, there are in each zone the greatest differences in particulars.

Bush plants (Plantæ fruticetorum et dumetorum) are well distinguished by their appearance in places which are overgrown with high bushes. As such places are rich in shade and moisture, the preference of certain plants for them is easily explained. As examples of bush plants may be mentioned Origanum vulgare, Asarum europæum, Corydalis bulbosa, Asclepias vincetoxicum, &c. Such low brush-wood, with its peculiar vegetation, is found in every zone, and in various languages there are peculiar names to distinguish it from forests which consist of higher wood.

Forest plants (Plantæ sylvaticæ et nemorosæ) are those which grow in woods, or, at least, are most frequently met with in them. Woods have been divided according as they consist of various trees, and have a variety of soil. Linnæus understood by *sylvæ*, those which have a dry sandy soil, as our pine woods ; while by *nemora*, forests of hard wood only were meant.

M. De Candolle has used these words as synonymous, which, however, cannot be admitted.

To woods of pine, fir, &c. belong Linnæa borealis, Pyrolæ, Vaccinium myrtillus, Ophrys ovata, and several other plants; in those of hard wood, Atropa belladonna, Geum rivale, Hepatica triloba, Trientalis europæa, Oxalis acetosella, &c. &c. are chiefly found.

Various names for woods have also been invented from the name of the tree, of which they chiefly consist; as Pineta, Fageta, Querceta, Palmeta, Oliveta, &c., according as they are composed of species of Pinus, Fagus, Quercus, or of palms or olives.

### *On the Social Growth of Plants.*

This is a phenomenon which determines in a peculiar manner the distribution of plants, and has a very decided influence on the physiognomy of nature. If we attend, while walking in an open field, to the manner in which plants appear, we shall soon find that a greater or smaller number of individuals of some species grow together, while others grow in single scattered specimens.

To the former, the name social is applied, and the others are called unsocial or solitary. Sphagnum palustre and Dicranum glaucum are extremely social plants; they often cover the moors of the north with so thick and uniform a covering, that seldom any other plant grows up amongst them, and the plains have thence a very dreary appearance. Cenomyce rangiferina (Lichen rangiferinus) in the same way overspreads the dry regions of the north. Among aquatic plants may be named Charæ, Acorus calamus, Scirpus lacustris, Arundo phragmites, and several others, which grow so socially, that they alone give our countries a peculiar character. If we think of our lakes, the shores of which are surrounded by a broad forest of reeds (Arundo phragmites), or a leafless one of rushes, in which thousands of birds sing their morning and evening song, and bring up their young; if we think of the social willows, between which shoot up the splendid flowers of our Epilobiums,* and of the nightingale which frequents them, and delights the dweller on the shores of the lake, we must allow the social growth of plants to be of the greatest importance to the character of natural beauty. How different would appear the margin of such a lake without these social plants!

But agreeably as the social growth of plants has influenced the physiognomy of nature in the preceding instance, there are others in which the uniformity caused by so great a number of individuals of the same form, is disagreeable and dispiriting. Who does not know our heath, which is so dreadfully decried on this account? Vast tracts of country are sometimes entirely covered with it; as an example, we may mention the heaths of Luneberg, which, on a smaller scale, are often repeated in the plains of the northern part of the temperate zone, in the Old World. But the heath is the most social plant of all, and if all other plants were to occupy the surface of the earth in the same proportion, there would not be room for more than 5000 species. We have very recently received a very interesting work on the social growth of plants, from E. Meyer,† in which the distribution of the heath is treated of with particular attention. Our species of pine are, perhaps, next to the heath, the most social plants; and indeed it may be doubted whether our pine (Pinus sylvestris) would not in earlier times, when the culture of the soil did not oppose its distribution, occupy as great a surface as the heath now does; I almost think it must have been still greater. As some of the

* Epilobium palustre and E. angustifolium in particular.

† Natur-wissenschaftliche Vortrage, gehalten in der physikalischen ökonomischen Gesellschaft, zu Konigsberg, 1834, p. 160–184.

social plants of our zone, I name Polygonum aviculare, which frequently forms very extensive patches; Poa annua, Vaccinium myrtillus, Juncus bufonius, Myriophyllum spicatum, the Charæ, &c., not to mention the vast woods of our countries, in which the beech, the oak, the alder (Alnus glutinosa), the birch (Betula alba), and many others, grow socially together for miles. Humboldt has remarked, that the phenomenon of the social growth of plants belongs chiefly to the temperate zone,* and that tropical countries are not so rich in species with this habit. The greatest difference in this respect may be even observed between the northern and southern parts of the temperate zone, and Meyer very truly remarks, that Italy, though as rich in species of grasses, has no such meadows as Germany possesses, and that though there are there a far greater number of forest-trees, there are no such extensive forests as are found here. It is probable, however, that Italy may, a few thousand years ago, have been somewhat richer in woods than it is now.

But, as we have already remarked, the torrid zone has also its social plants, although not in the same proportion, as the temperate zone.

We have formerly spoken of the forests on the seashore, in the tropical zone, where the same species covers most extensive tracts. The mangle or mangrove (Rhizophora mangle L.), and the Avicenniæ, are the best known of these social plants, which form the mangrove forests in Brazil, and, by their almost uninterrupted circle of distribution, form a belt round the torrid zone of the globe. In the South Sea Islands the ferns, with moderately tall stems, grow almost always socially; and the few true tree-ferns with tall slender trunks which I have had the opportunity of observing, appear always in a fixed, generally very limited, circle of distribution, and grow in it socially, if other plants do not

* See his Ideen zu einer Geographie der Pflanzen, p. 8.

press in between them. There are several Scitamineæ, which exclusively cover great tracts. R. Brown found in New Holland several Banksiæ, as Banksia speciosa, growing socially. Protea argentea L., and Protea mellifera also grow in this way.

The Bambusæ of the tropics grow as socially as the trees of our northern forests. Bambusa arundinacca, in the East Indies and the neighbouring islands, forms the most impenetrable forests, which are little inferior to ours in extent and beauty. Humboldt also observed on the Magdalena almost uninterrupted forests of bamboo-reeds and banana-leaved Heliconiæ; and on the Savannahs of the lower Orinoco, Kyllingiæ and elegant Mimosæ grow socially in vast numbers. But when we turn from the plains to the mountains of the tropics, we find there social plants quite as frequently as in the temperate and arctic zones. The forests of Cinchona there resemble our forests of the edible chesnut; and Escallonia and Rhododendron take the place of our Genista and Ulex. In the South American Cordilleras a great number of resinous Baccharidæ grow quite as socially as Vaccinium myrtillus, our Rhododendra, and some willows do on our lower mountains.

As the beautiful red flowers of the Thymus cover, as with a red carpet, large spaces of our sandy districts, while the soil around is destitute of vegetation, so the Calandriniæ and some Verbenaceæ grow almost like turf over the barren plains of some elevated parts of the Cordilleras of Chili and Peru.

Now that we have thus exposed the distribution of social plants and its effect, we turn to its explanation. A mutual inclination to social life, as is observed in animals and man, is of course not to be admitted. Plants are fixed to the ground, and uniformity in the physical and chemical nature of the soil can alone cause their appearance in what is called a social state. If we examine a natural pine-wood, we shall, with few exceptions, see that its extension is only limited by a change of soil. When a little river or a standing pool is in the midst of such a pine wood, it is common to see on its margin, where the soil is generally better than that of the forest itself, a few other trees; sometimes the alder, sometimes willows, or other large bushes.

On examining the distribution of the heath, we find that this social plant grows always on the same soil; it is what is called a sour soil, which places an invincible obstacle in the way of all culture, but is the natural ground of the heath. As northern Europe is so rich in this soil, which is intolerable to almost all other plants, the heath appears there in great and extensive masses. The different geognostic constitution of the earth, in the temperate zone of the southern hemisphere, prevents a greater resemblance in the physiognomy of the vegetation of that zone, yet there also social plants appear in masses; and if the continents of the southern hemisphere were more extensive in the high latitudes than they are, they would certainly offer similar appearances. I have even in Chili met with several very social plants, as Acacia caven, Lycium gracile, Bambusæ, and several species of Cactus; but on the eastern side of the Chilian Cordilleras, that is, in the Pampas, the social grasses reappear more or less as on our northern meadows. Unfortunately the interior of South America, to the south of 40°, is almost quite unknown to us; but from the descriptions which the naturalists who accompanied Cook give us, nature there seems to be quite the same as with us. The interior of New Holland seems to possess a plant, Polygonum junceum,* which grows over great tracts like our heath; and Cupressus callitris, and several Proteaceæ and Eucalyptæ, according to some recent travellers, grow quite as socially as our forest trees. Thus the opinion that the social plants of the

* See Sturt's Two Exped. into the interior of Southern Australia, &c. London, 1833.

warmer countries belong to saline and littoral plants only, must yield to later observations. The vast and impenetrable forests of Bambusæ in the tropics, require quite as good a soil as our woods of beech and oak.

External circumstances certainly, for the most part, determine the social growth of plants, but as these are also very prolific, their appearance in masses is the easier explained. In this respect we may divide social plants into such as are remarkable for the great number of individuals which grow close to each other ; and into such as, by shoots from a single stem, cover a great surface. In trees, bushes, and shrubs, this distinction may be made at the first glance. In the valley of Copiapo, in northern Chili, I observed a species of Lycium, the bushes of which, growing from a single root, formed very thick mounds, which could not be seen through, even at a height of 10 or 15 feet ; and such clumps of this bush stand more or less close to each other on the sandy plain, so that the physiognomy of nature is there very peculiar. Indeed this sort of social growth is nowhere more striking than in the Alpine flora of the Cordilleras. Several species of Azorella, Bolax, Verbena, and Lycopodium assume there a very peculiar habit, to which there is nothing similar in the Old World, except in a few Cryptogamiæ. This peculiar alpine vegetation is first seen near the limit of perpetual snow. The plants at first fasten on large projecting rocks ; then by branches from the side, which begin from the very base of the stem, or in other instances by shoots from the root, they gradually extend so considerably that they often cover surfaces of rock 12 and 20 feet square. Often great blocks of rock are entirely covered with a thick and extremely hard turf, which has been formed by a single plant ; and these turfs are so close and hard that it is extremely difficult to cut into them with the sharpest instrument. The stem of such a family of plants, which is doubtless a monument of many centuries, seldom attains the length of a foot, but is sometimes as much as five or six inches in thickness, and has from its very base an infinite number of twigs and branches. By the continual increase of the stem in thickness, it gains also something in height; and thus the sod which the plant forms gradually assumes a vaulted form. On account of the quantity of resinous matter which these little umbellated plants contain, they burn very well, and in a green plant the fire keeps in for a long time. In the desert parts of the Cordilleras, where trees are entirely wanting, these clumps are frequently seen, often only half burned, and the traveller is obliged to make use of them as fuel.

On the whole the phenomenon of the social growth of plants is nowhere more frequent than in the Alpine flora; and this, as well as all other natural phenomena, appears on a larger scale on the heights of the Cordilleras. Only a very few unsocial plants are met with there; at least they usually form little bush-like clumps, which often stand singly in the most desert, sandy, or ashy places. The interesting species of the remarkable family of the Boopideæ, intermediate between the Umbelliferæ and Syngenesiæ, form often little mounds, of which I have already spoken. Several Pereskiæ also, which ascend near the limit of perpetual snow, form similar mounds a foot or 1½ feet in height, and several feet in circumference. Nothing but the reddish yellow spines, two or three inches long, is to be seen on the surface; the succulent leaves are hidden beneath them, and send up their flowers only so far that they may be protected by the spines from the cold winds. These singular Cacti may be mistaken for couching deer at a distance.

Aquatic plants, in general, grow socially oftener than land plants, and the external conditions, which determine this, are more uniform. As aquatic plants produce a great quantity of seed, this, together with favourable external influences, causes a large increase in the number of individuals. However great the disposition of a plant to social growth may be, external circumtances must also favour it. When a stagnant pool in our country is covered with duck-weed (Lemna), its growth is checked when the masses of Conferveæ become predominant, and it entirely ceases as soon as the water is dried up. We have already mentioned the gregariousness of the turf-moss; the enormous quantity of seed this plant produces, favours it, but the soil on which it grows must always be damp moory ground. With the extirpation of woods this soil disappears, and, in consequence, the moss also.

The Fuci, the inhabitants of the sea, are extremely social. On the west coasts of South America, I have found them in forest-like masses at the bottom of the sea, which, animated by millions of the lower animals, form a submarine world. The Laminariæ there form these masses, but in the Straits of Magellan are found the gigantic forms to which Fucus pyriferus, and Fucus antarcticus belong; these Algæ have been measured to lengths of 200 and 300 feet. One of the most singular phenomena of the kind is the social growth of Fucus natans, in the Sargasso Sea, within the great equatorial current in the Atlantic, of which we have already spoken.

Many species of Fungi have been mentioned, which grow socially. Some of them grow in more or less regular circles, which enlarge from year to year, the cause of which is that their common thallus, which constantly spreads from the centre, produces the new fungi on its outer margin.

Thus we have specially considered all the conditions which, by climate as well as by peculiarities of soil, exercise an influence on the appearance of plants in a fixed place, and we can now proceed with greater advantage to general considerations on the station and distribution of plants.

## *More Exact Definitions of the Station and Distribution of Plants.*

The station (*statio*) of plants denotes the relation in which plants stand to the situation in which they always grow ; or we understand by it, the locality in which a plant grows, whether it has been placed there by nature or by art. When we make a distinction here, the expression "native country" denotes that station to which a plant has been appointed by nature.

The distribution of plants (*extensio plantarum*), on the other hand, signifies the whole range of their stations, without attending to the relations in which the species, genera, or families stand to each other. If, for example, a plant appears in most of the countries of the Old World, from one end of it to the other, we say that it is distributed over all the Old World. The range of distribution, or the area of a plant (*area plantarum*), therefore includes all its stations.

As all the places on the surface of the earth are pointed out by the parallels of latitude and longitude, so are also the station and range of a plant, so far as these extend horizontally ; but where the range is vertical, as on the sides of mountains, it is marked out by degrees of height. It hence follows, that the horizontal extent of the station of a plant, or its horizontal distribution, is regulated by the latitude and longitude of the plain; while the vertical distribution has a reference to the height at which a plant appears on the mountains.

We have already often remarked and shown by examples, that the distribution of plants is chiefly regulated by the distribution of heat over the surface of the globe ; and as this again is in a certain proportion to the parallels of latitude, it follows, that the distribution of plants is chiefly regulated by the latitude, and that the longitudinal extent of the area is much less to be taken into consideration.

The area of a plant, with reference to its extent in latitude, is also called the zone of latitude, or simply zone; while it is called the region, when its vertical range is referred to. The term, zone of longitude, is more seldom applied to the horizontal range of plants according to the longitude.

The zone of every plant has a polar limit, beyond which the plant does not appear, and an equatorial limit, at which the extension of the plant towards the equator ceases. Those plants whose polar limit ascends to the extreme latitudes, are exceptions to this, as well as those which cross the equator and enter the opposite hemisphere. The former are generally called polar or arctic plants, and the latter tropical plants. But this is not quite accurate, as an arctic plant, for instance, may appear within the arctic zone, without extending to the highest latitudes. As examples we may adduce, the many bushy plants, which pass beyond the arctic circle, but afterwards entirely disappear, and which are nevertheless classed among arctic plants. Similar examples might be given of so-called tropical plants, which do not quite reach the equator.

The longitudinal zone of a plant is marked out by an eastern and western limit.

The regions of plants, or their vertical range, are marked by an upper and an under limit, which are defined by the degrees of altitude.

The area of a plant, or its range, is either uninterrupted or interrupted. We shall learn more of these relations when we afterwards consider the distribution of plants over the globe; but we have already, when treating of the distribution of temperature and the influence which particular localities exercise on the stations of plants, defined those conditions, which determine or promote an uninterrupted or interrupted range of a plant. For example, if a plant require a certain degree of heat, and its presence chiefly depends on this, it may appear in all those places which have the requisite degree of mean heat; and in this way a plant may have a range with frequent interruptions. Numerous examples can be given. Our common plants, Prunella vulgaris, Origanum vulgare, Thymus serpyllum, grow on the mountains of Himalaya, which enclose Cashmere, at an altitude of 8200 feet;* indeed, the second region of Himalaya, that is, from 5000 to 9000 feet, which Royle describes, has a perfectly European aspect; and there are found there Ranunculus arvensis, Thlaspi arvensis, Capsella bursa-pastoris, our ivy (Hedera helix), Galium aparine, Leontodon taraxacum, Acorus calamus, Phleum alpinum, Alopecurus geniculatus, Poa annua, Samolus valerandi, and a number of other plants, which chiefly belong to the higher latitudes of Europe. Or if we keep to Europe, we find the pretty Primulæ, the beautiful Anemones and the lovely Gentians of our country, or the Dryas octopetala of the north, on the Swiss Alps, under conditions of temperature similar to those of our country. Saxifraga hirculus grows in the cold moory forests of our northern provinces, but in similar circumstances it reappears quite commonly on the Swiss Alps. Other examples serve to show the influence of local conditions on an interrupted or uninterrupted range. Salsola kali, standing in a peculiar and close relation to the sea-shore, has an extraordinarily extensive, and, on the coast at least, an uninterrupted area.

The range of a plant may be either natural or artificial; in the latter case, the station of the plant is extended beyond its natural limit by artificial planting. The range of most cultivated plants—those used as food, as well as those which minister to the comfort or luxury of man—has in this way been extended, and often in a remarkable degree. How melancholy would be the condition of man, if the most of the plants used for food had not such a power of adap-

* See Royle, Illustrat. Lond. 1833, fasc. i.

tation. We need not specify the Cereals, which have followed wherever the people of the Old World have turned; but shall mention only the Vine, the range of which has been so wonderfully extended over almost all the earth. Even in the island of Java, the Vine ripens extraordinarily large grapes, and in New Holland it seems to have found its second native country. We shall afterwards become more exactly acquainted with the distribution of the Vine.

In general the range of plants has been artificially extended in an oft interrupted manner, and only the most important plants used as food, can boast of an uninterrupted artificial distribution, such as the cereals and the principal vegetables. I shall here mention a few of the commonest plants which are remarkable examples of artificial diffusion. Our common garden herbs, sage, rosemary and balm, are grown at Surinam.* Our radishes are nowhere of better flavour than at Rio Janeiro, and in the East Indies. Our fragrant clove-pinks are quite as beautiful as with us, at St. Jago de Chile, and probably still more aromatic. In the fields near Canton, are grown for the use of Europeans almost all our vegetables. The winter season is indeed chosen for their cultivation, and they are kept from perishing from the cold by an artificial covering to prevent the radiation of heat from the ground. In general we may assume, that plants of northern countries may be extended much further towards warmer regions than the reverse, for the artificial introduction of true tropical plants into colder countries is extremely difficult, and can only be done to a small extent. In the botanic gardens of tropical countries, when it is wished to grow the plants of the north, they are easily protected from the rays of the sun by a light roof, and the ground can be kept cool to any degree by the evaporation of water.

* Sack's Reise nach Surinam, 1820, i. page 181.

We have now to consider the extent of the range of a plant; and we shall find here some indications of those plants whose extension may be attempted with most success. In general, plants with a naturally wide range, may be extended much further artificially, while plants of a very limited area are generally spread with difficulty. There are some plants whose range is so extensive that they are to be found in all parts of the earth; we are accustomed to call some of these weeds, and they are found almost everywhere as such. We instance a number which are common to our country and to New Holland: Lemna minor, Lemna trisulca, Marsilia quadrifolia, Convolvulus sepium, Festuca fluitans, Arundo phragmites, Panicum crusgalli, Scirpus lacustris, Cladium mariscus, Juncus effusus, Vallisneria spiralis, Solanum nigrum L., &c.

In general we may lay it down as a rule, that the range of plants is wider, the lower the degree of their development. Of the Phanerogamæ which are common to Europe and New Holland, the greater number are Monocotyledons.* The wide range of the Crytogamiæ, particularly of lichens and mosses, and probably of Algæ also, is sufficiently known; indeed many of them seem to be distributed uninterruptedly from one end of the earth to the other. Sticta crocata, which grows in Europe, Africa, the island of Bourbon, and New Holland, is also found in the West Indies and South America. I myself have observed it on rocks and walls, in the province of San Fernando, in the centre of Chili, at an altitude of 3000 feet. I have also found Parmelia perforata in the most distant parts of the globe, even in the remote Sandwich Islands. Lecanora subfusca is another of those lichens which are everywhere to be found.

* Among the plants which Mr. R. Brown has described in his excellent Flora of Australia, there are 167 common to Europe and Australia, and these consist of 122 Acotyledons, 30 Monocotyledons, and 15 Dicotyledons.

Some other plants have a remarkably limited range, and hence it is that many plants bear the names of countries, towns, and single mountains, but this nomenclature generally originates in imperfect knowledge of the surrounding country, where the plants also grow, but have not been observed. What a limited station was formerly ascribed to Linnæa borealis, and Braya alpina, but every day new stations of these plants are discovered!

The question has been asked, whether there might not be laid down a general rule for the extent of the natural sphere of distribution, in respect to its zone of latitude, &c.? Schouw is the first who has examined into this point. It may be investigated by comparing the plants of a northern flora with those of a southern, and observing how many plants are common to both. But, as we have seen, certain plants which have a very extensive, but interrupted range, are often common to the most remote parts of the earth, and therefore it is difficult to fix an absolute zone of latitude for them. If it be wished to determine this, such plants only must be chosen as have an uninterrupted range, and of these there are very few.

Plants vary so much in the extent of their range that general rules for it can scarcely be laid down. Schouw* thought he had found, that in the temperate zone of the northern hemisphere, a breadth of 10°–15° was the most common breadth of the area of a plant, and that this was seldom under 5 or above 30 degrees.

The longitudinal extent of the zone of a plant is generally much greater than that in latitude, as the change of temperature according to the longitude is small; and there are even some plants whose zones of distribution form a belt quite round the globe, as Cyperus polystachys, Pistia stratiotes, &c. Yet there are plants whose areas are of very small longitudinal extent, which is generally caused by great expanses of water or ranges

* L. c. page 185.

of mountains. In South America, particularly in Chili, the greater number of Calceolariæ are found on the western side of the Cordilleras ; while at most but a very few species appear on the eastern side. Schouw instances Lobelia dortmanna, which grows in Norway, Sweden, Jutland, Scotland, England and Holland, but has not yet been found in the eastern part of Europe and Siberia. The Ericeæ, at the Cape of Good Hope, have a very small longitudinal range, which is evidently caused by the great oceans which bound it. The distribution of the Camelliæ, and many other plants, furnishes similar examples of this point.

It has been attempted to fix the vertical, as well as the horizontal extent of the range of plants. M. De Candolle* has tried to determine the vertical range of the plants belonging to France ; and as this has been an extremely laborious work, it is a pity that the results for a great many of the species cannot be absolutely accepted as the rule, for the plants which, in that country, reach their polar limit, do not ascend so high as in more southern countries. It is again extremely remarkable, that others ascend much higher in France, than they do in countries farther south.

Schouw† has arranged, according to certain heights, the measurements of the vertical range of the 1500 plants belonging to France, which M. De Candolle has given, and has obtained the following result, after he has excluded from his calculation all the plants which reach their polar limit in France :

11 species show a vertical range of 3000 metres.

| | | | | | | |
|---|---|---|---|---|---|---|
| 19 | species of | from | 2500 | to | 3000 | metres. |
| 72 | ... | ... | 2000 | ... | 2500 | ... |
| 200 | ... | ... | 1500 | ... | 2000 | ... |
| 391 | ... | ... | 1000 | ... | 1500 | ... |
| 194 | ... | ... | 500 | ... | 1000 | ... |
| 31 | ... | ... | 100 | ... | 500 | ... |

* Mém. sur la Géographie des Plantes de France.—Mém. de la Soc. d'Arcueil, xiv. pp. 262–322.

† L. c. page 178.

Accordingly for France, or the middle of the temperate zone, it seems that from 1000 to 2000 feet is the most usual vertical range.

Schouw has compared the vertical extent of 293 tropical plants, according to Humboldt's observations, and it is very singular that the result is quite different from that for the temperate zone. Namely, 1000 toises is the highest vertical range of the station of a plant in these countries, and from 200 to 600 metres is the most common extent of the region in which each plant grows, while in France it varies between 1000 and 2000 toises. However, too much value must not be put on these calculations of Schouw, for Humboldt's measurements of the upper and lower limits of the station of certain plants were not made with a view to this purpose. In Southern Peru I have traced the vertical extent of the station of cultivated maize for 2000 toises, and it is there still cultivated at a height of 12,700 feet above the sea, as in the celebrated island of Titicaca, in the great Alpine lake of the same name; which is not the case with any plant in Europe. Our lucerne is cultivated in that country in the burning districts of the coast, and ascends as high as 11,000 feet; on the plateau of Chuquito, at an elevation of 12,700 feet, I have no longer seen this agricultural plant, which is so extremely important to that country. Barley and oats still ripen there, but rye is cultivated only as a green crop.

Schouw has rightly observed, that those plants which have a wide range, have also a particularly extensive zone of height. The maize and Medicago sativa, which I have already given as instances of a remarkably extensive vertical distribution, have also a very wide horizontal range. In Europe the social Erica vulgaris, which shows a considerable horizontal extension on the mountains of northern Europe, ascends as high as 9000 feet above the plain.

The natural as well as the artificial distribution of each plant might be represented on a separate map of the world, and if all the points of its appearance were painted of one colour, one might at a glance take a very instructive survey. Such pictorial representations might be given of the range of each genus, as well as of each family, and Schouw has given such maps in the Atlas to his well known work on botanical geography. He has given the range of the beech (Fagus sylvatica) as a pattern for the representation of the distribution of a wild plant, while he has chosen the vine for that of a cultivated plant. In such representations of the distribution of single genera or families, by using a darker or lighter tint of the same colour, the points where the genus or family appears in greater or smaller quantities, can be indicated; this has been done very well by Martius in the plate of the distribution of the Amaranthaceæ.*

Botanical geography, as a science, would be of great interest to the scientific and the educated man, even though its application to practical life did not give it a still higher value. Whenever a sufficient number of meteorological observations have been made at the most diverse points of the surface of the earth, so that the whole course of the isothermal, isotheral, and isochimenal lines is exactly known, we shall even beforehand be able to determine exactly whether a plant may be transplanted from its indigenous station to another, or whether this trouble would be unrewarded; an object which is evidently of the greatest importance. In particular, we are still without the knowledge of the mean temperatures of table lands at great elevations, to enable us to determine what plants might be introduced into those regions. It is easily seen of what importance this is, and I shall only give one instance. The extensive country on the plain of Chuquito, around the Alpine lake of Titicaca, is thickly peopled, and a great number of magnificent towns have been built at this immense height. But wood is wanting in this country,

* Nova Acta Acad. Cæs. Leop. Carol. Nat. Cur. Vol. xiii. P. I.

where a perpetual spring reigns, and where the fertility of the soil and an abundance of the precious metals might render the inhabitants prosperous. We are indeed still without all the mean temperatures of that country which lies at the altitude of 12,700 feet, but from the few observations which I have myself made there, as well as from some others made by Pentland and Rivero, it may be considered certain that the fir, as well as the birch and alder, would grow vigorously there. What prosperity would result from the introduction of great forests, where every stick, every pole or plank is counted riches, and where the sailor must commit himself to the stormy lake in a miserable bark woven of rushes!

The prosperity of nations has generally followed husbandry and the culture of the useful plants; and with them civilization and happiness have advanced. It is surely a wrong method of proceeding to try to force the sciences and the Gospel on rude men, who live from one day to another, and are in want of necessary food. With the presence and culture of certain useful plants the relations of social life are so intimately connected, that the study of these must be also connected with that of the distribution of plants, for men would be compelled to live in very different ways, if this or that useful plant did not encourage their sloth or industry.

By such special inquiries, we shall gradually learn more and more of the influence which the varied diffusion and distribution of plants over the globe exercises on the civilization of man; yet we must here go very carefully to work, that we may not be deceived by appearances, and thus arrive at false results. How universally do we hear tropical countries extolled! How happy, it is said, must be the land where nature is so liberal, that the most precious fruits are produced without the labour of man! Yet in reality this is not the case.

# PART III.

## ON THE DISTRIBUTION OF PLANTS OVER THE SURFACE OF THE EARTH, WITH PARTICULAR RESPECT TO THE PHYSIOGNOMY OF NATURE.

WE have hitherto considered the external causes which determine the station and diffusion of a plant, without taking into view the reasons why a certain plant appears only within the range appointed it and in no other.

It is a well known observation, and one repeated by every traveller, that vegetation, and indeed organic life in general, is developed in gradually increasing profusion from the poles to the equator, and that the forms become always more perfect, and more luxuriant and beautiful, as they recede from the colder regions. This subject will be more fully considered in this part of the book. Baron Alexander von Humboldt has here also marked out for us the path of this science. His celebrated work, "Considerations on the Physiognomy of Plants," pointed out this highly interesting side from which botany may be viewed, and how it may improve the taste of nations by increasing their sensibility to the beauties of nature, and thus have an influence on the progress of the arts.

The subject of the distribution of plants over the surface of the globe may be divided into two perfectly distinct branches, one of which, called the Physiognomics, considers vegetation according to the distribution of the forms which point out the groups of plants; it is a peculiarly natural system in which similarity of form is the principle of classification. The physiognomics of vegetation investigate the predominance of this or that form of plants by the absolute mass of its individuals, or by the impression which it makes on the mind of man by its influence on the character of nature; the other branch, viz., the Statistics of plants, on the contrary, does not concern itself about the absolute predominance of this or that group of plants, or this or that type, but considers the relative proportions, founded on real numbers, which this or that group by its number of species bears either to the whole mass of known plants, or to the number of species of other groups. A group of plants, for example, may determine the natural character of a country, without therefore being predominant by the number of its species. We possess a treatise by Humboldt, on the laws which are observed in the distribution of plants,* in which this point is expressly discussed. The celebrated author in it says: in a northern country, where the Compositæ and the Ferns stand in the proportions of 1.13 and 1.15 to the sum of all the phænogamous plants (that is, when this proportion is found, where the whole number of phænogamous plants is divided by the number of the species of these two families), a single species of fern may cover ten times more ground than all the Compositæ together. In this case the ferns predominate by the mass of their individuals over the Compositæ, but not by their number of species.

The physiognomics of vegetation teach us, that nature, at the creation of plants, has distributed them over the surface of the earth according to certain laws, which are quite unknown to us. We have now learned some of the external causes which place the more developed and nobler forms of vegetation in the hot zones; but we know no cause, why the same species of plants are not always produced in the same conditions of climate. The singular group of the Cactaceæ is properly peculiar to the torrid and subtropical zones of the New World;

* Dict. des Scienc. Nat. tom. xviii. p. 422–436. 1820.

two species only have yet been found in East India and China, and there in the interior of the country, at considerable altitudes. However, the form of the Cactus has its representative in the Old World; for on its eastern, as well as its western side, we have Euphorbiæ, which we should certainly consider Cacti, were we without the knowledge of their organs of fructification; as Euphorbia nereifolia in Southern China, which the Ipomœa quamoclit twines round and decorates with its scarlet flowers, just as Loranthus aphyllus does the Cerei of Chili. Euphorbia canariensis and Euphorbia balsamifera represent the Cactaceæ in the western part of the Old World. It is equally inexplicable, why the Old World only should possess the true Ericeæ, while the Erica cœrulea *Willd.*, not a true Erica, comes in their place in the New World.

But the statistics of plants prove in the most decided manner, that nature has distributed the various forms of the vegetable kingdom amongst the different zones, according to fixed immutable laws, which will become more and more obvious, the more perfectly the whole number of species of plants belonging to certain countries is known. When we compare small districts with each other, the proportions which the different groups of plants bear to each other do not always exactly agree; but we perceive the harmony of the laws, when we compare large tracts with each other, and we may consequently with reason affirm, that the forms of the different plants stand in mutual dependence on each other.

It is true, that certain groups of plants are confined to particular, and often very limited regions, but the greater number of families are distributed over the whole globe, individual representatives of those groups appearing wherever a fertile soil is exposed to light and air.

The snow line, or the cold, puts a stop to vegetation on our globe towards the poles and at great heights; but vegetation entirely ceases wherever the soil is composed of such substances as are quite unsuitable for the nutriment of plants, even though it should be at the equator, or at the level of the sea. I have found vegetation at the greatest altitudes of the Cordilleras of Southern Peru, wherever the mantle of snow was wanting, and where there was a moist soil; but the hard, bare rock, when swept by strong winds which prevent the deposition of organic substances, shows no vegetation, even when lying at an elevation at which it is luxuriant. On the cone of the volcano of Arequipa, and also on the volcano of Maipu, I have ascended far above the limit of vegetation, though no snow line marked out this region; but the cone of the volcano of Arequipa rises far above 18,000 feet, and more than 2500 feet of its peak is formed of black lava ashes, from which break out here and there regular columns of grey and reddish trachyte. In these ashes, above the height of 15,000 feet, no vegetation is to be found; but as far as this, there grows a singular Fungus, resembling a Lycoperdon, with a long root which strikes deep among the ashes; and little Malvaceæ, extremely pretty forms of Sida, and singular Baccharidæ end the phænogamous vegetation.

In general, we remark in the distribution of plants, that the species of genera, as well as the genera of families, proceed from a point and range themselves round it in concentric circles, or spread from it like rays in various directions; or, which is still more usual, are distributed in belts of greater or smaller breadth, that are sometimes parallel to the meridians, sometimes to the parallels. In all these modes of distribution of the smaller as well as of the larger groups of plants, their social or isolated growth appears to be a very important circumstance, which has an especial influence on the character of the vegetation. It is also important to know whether certain groups are merely placed next each other, or whether their circles of distribution do not encroach upon each other; in the latter case, the plants of different groups grow mingled together. For example, the Coniferæ and the Casuarinæ, in the Old World, are very distinctly separated families, whose circles of distribution scarcely touch each other ; yet in New Holland they encroach on each other, for Casuarinæ, Araucariæ and Cupressi, there grow all together.

The genera, as well as the families, which are the larger groups, attain their maximum in some one place of the earth ; that is, they have in that place their greatest number of species, while in another place is found their minimum, where the number of their species is very small. These meanings are expressed by saying, this genus or family predominates in this place, or is wanting in that.

When a group of plants reaches its maximum in a country, and also influences its physiognomy by the masses in which it grows, it is usual to give it a special name, which is generally formed from that of the country or zone. Thus we have tropical forms, which either appear solely within the tropics, or at least attain their maximum there. The Palmæ, Musaceæ, Piperaceæ, Scitamineæ, &c., almost exclusively belong to the torrid zone, yet single representatives of them extend even to high altitudes in the temperate zones. The family of Palmæ, for instance, shows several such exceptions ; Chamærops humilis, it is known, is found in 49° north latitude, and Cocos nucifera grows still farther south. The Cocoa-palm of Chili, as it was formerly called, grows on the west coasts of South America, as low as latitude 36°, at Conception ; it is the only species which appears on the whole west coast of South America, from the extreme south of Peru, while the east coast of this continent is extraordinarily rich in palms.

When a family of plants predominates in any zone, either by the quantity in which it grows, or by the number of its species ; and in another zone are found only single forms of it, we say, that that family is represented by these few species ; or we call these species the representatives of that family. The Ericeæ of the Old World have their maximum in the south of Africa; single forms of this family, Erica vulgaris in particular, indeed predominate by quantity in the north of Europe, but the beautiful shrubby forms which belong to the Cape of Good Hope are represented in the south of Europe solely by Erica arborea. This species is indigenous to the forests of Spain and Portugal, and also appears in as great quantities in the Canary Isles. The following instance is still more striking. The Acaciæ reach their maximum in New Holland, to which country they almost entirely belong; but Acacia heterophylla, quite peculiar from the various forms of leaves which grow on the same tree, represents this large family of the southern hemisphere, in the Sandwich Islands. The ordinary leaf of this tree is merely an expanded leaf stalk, and greatly resembles the form of the leaves of the Eucalyptæ, an extremely interesting Australian family, which has the same native country as the Acaciæ. It seems as if the Acacia heterophylla not only represented the Acaciæ in the northern hemisphere, but by the form of its leaves pointed out some relationship to the Eucalyptæ.

The large family of Laurineæ, which has its maximum in the tropical zone, is represented in the extreme south of Europe by Laurus nobilis, our common laurel.

The succulent Mesembryanthema of Southern Africa are represented in the south of Europe by a few species of this genus; but we also recognize their form and habit in the numerous species of Sempervivum belonging to the Canary Isles, and of Sedum belonging to the south of Europe. Link, to whom we owe our exact knowledge of the Lusitanian flora, recognizes Drosera lusitanica, Ixis bulbocodium, and Triglochin bulbosum, as European representatives of the South African flora.* If we keep to northern Europe, we find the beautiful

* See Link, Die Urwelt und das Alterthum, 1834, p. 259.

Cistus plants of Spain and Portugal, represented in the north by our Helianthemum annuum. The Abietinæ, which reach their maximum in the arctic and temperate zones of the northern hemisphere, and frequently preponderate over every other family in quantity, become gradually less frequent towards the south, and in the southern hemisphere are represented only by the genera Araucaria, Ephedra, Cupressus, Dammara, &c.

If we consider the general features of the vegetation spread over the globe, or I might rather say, the different impressions which it at different places makes upon us, we shall soon remark certain principal groups which are more or less clearly separated from the surrounding plants. These groups, which are distinguished by their peculiar physiognomy, sometimes agree also in artificial characters, and form certain genera and families; sometimes it is the whole vegetation of the district, which has received a peculiar character from the arrangement or grouping of the different forms of its plants. If we were to classify the whole mass of vegetation according to the peculiarities in physiognomy which it presents, we have already seen that the classification must be two-fold; that is, both geographical and purely botanical. When the geographical principle is taken as the foundation of this classification, we divide the vegetation according to the countries and larger tracts in which it is found, and call such divisions, Floras, which are designated by the names of the countries. Other writers on this subject have named such divisions, regions* and phyto-geographical kingdoms.† M. De Candolle and Schouw are the first who have mapped out the whole surface of the globe into phyto-geographical divisions, and I shall point out in the sequel how far I think it necessary to follow their arrrangement.

* See De Candolle, Dict. des Scienc. Nat. tom. xviii. p. 411.

† See Schouw's Grundzüge, u. s. w., p. 505.

## I. THE PHYSIOGNOMICS OF VEGETATION.

The operation of organic force becomes gradually more active and powerful as we recede from northern countries and approach the equator; there uniformity, too often the companion of poverty, prevails; whilst here we have the greatest variety, combined with luxuriance and profusion. The forms of the plants become more and more developed; they seem as it were more perfect the nearer they approach the hottest regions of the earth, a perfection undoubtedly to be attributed to the favourable influence of the greater light and heat.

But though the greater number of plants which have large and highly organised flowers, belong to tropical countries, though there the variety of the most beautiful forms fills the man of sensibility with amazement, yet we cannot but perceive that other countries have also their peculiar beauties. It is not the quantity of vegetation, nor its great luxuriance, which makes the most pleasing impression, but rather the blending of the different forms of plants, and the harmony between the vegetation and the configuration of the surface of the earth.

If we in this way analyse the whole impression which the survey of vegetation produces on us, we cannot fail to perceive that there are certain forms of the vegetable kingdom, which more or less predominate in certain places, and on that account particularly attract our attention. These forms which determine the character of a region, are either peculiar to it, or they reappear in other regions, where they predominate also, or become merely their representatives. The beauty or singularity of these forms, their imposing size, their splendid colouring, and whatever other peculiarities they possess, give the character of the physiognomy of the country.

Infinite as is the number of plants, there may be selected from them a smaller number of principal forms, not collected into genera or larger groups by artificial characters, but which agree in the general impression they make on the mind.

To have an exact acquaintance with these principal forms of vegetation is of the greatest importance to a phyto-geographical division of the globe, as they principally fix the natural physiognomy of different countries. Humboldt is the first who has made such a classification of vegetation, and this must be taken as the foundation of all further inquiry into the subject. It is not until we are somewhat intimately acquainted with the various characteristic forms of plants, that we shall be able to recognize the peculiarities of each flora, and to characterize the physiognomy of each country.

We shall, therefore, notice separately the principal forms of vegetation, relying partly on personal observation in the Old and New Worlds, and partly on a careful study of the descriptions of the best travellers. It is obvious that the number of principal forms must be increased as the floras of foreign countries become more complete, and as countries, yet little known, are more carefully examined by botanical travellers, who may zealously devote themselves to this science.

## A. *Special Consideration of the Physiognomy of the Principal Forms of Plants.*

### 1. *The Grasses or Gramineous Plants.*

We begin with the grasses, the growth of which in great masses, in the form of meadows and pastures, is so well known to us. The beautiful verdure of extensive fields of grass makes an agreeable and enlivening impression on us ; it is a characteristic feature of nature in northern countries. It is singular that man, to ensure a supply of food, should have chosen to cultivate the grasses, as the most of them have very small seeds, and, consequently, the raising of a sufficient quantity is extremely laborious ; but I shall enter more fully on this subject in the last part of this book, in which I shall treat of the culture of the alimentary grasses.

To cultivate the cereals, man must have a settled habitation ; and they thus were one of the most important instruments in his civilization, and, at a later period, have been the source of the prosperity of nations. Now, wherever man emigrates, he carries the cereals along with him, if the rudeness of the climate do not prevent their culture. We may estimate the influence of this branch of agriculture on the physiognomy of nature in those countries which have been, for ages, the seats of civilised nations, such as Italy, Greece, the East, China, &c. The south of Europe may be called treeless in comparison with the north, yet it has doubtless in earlier ages been as richly clothed with forests as Germany or Russia are now, though even in these latter countries great progress has been made in the culture of the soil. What a beautiful sight is a field of ripening corn, which, when stirred by a gentle breeze, undulates like the sea after a storm, and over which plays a continual change of light and shade. The rice fields of the warmer countries present a similar spectacle ; and, in tropical countries, they alone show the beautiful verdure to which an inhabitant of the north is accustomed from his youth.

But the low grasses which form meadows and pastures, are peculiar to the colder regions, and to the colder half of the temperate regions ; they are replaced by large arborescent forms in the sub-tropical zone and within the tropics ; even in the south of Europe appears a gigantic grass, Arundo Donax, which chiefly belongs to the northern part of Africa, but which has now been introduced into the New World, where it grows freely in almost every part of the Spanish colonies.

The plants which are commonly called grasses, belong

to two great families, one of which includes the true grasses; the other the semi-grasses or Cyperoideæ. The grasses, as well as the Cyperoideæ, have certain forms, which especially predominate in the different zones. In the torrid zone, the Bambusaceæ, Saccharineæ, Oryzeæ, Olyreæ, Chlorideæ, and Paniceæ, predominate over the others, and some of them are even peculiar to this zone; the Hordeaceæ, Bromeæ, and Agrostideæ, on the contrary, are extra-tropical forms. There is a corresponding distribution of the grasses according to the increase of altitude. The distribution of single forms of the Cyperaceæ is still more obvious; the genus Cyperus attains its maximum in the torrid zone, and gradually diminishes beyond the tropics. On the contrary, the genus Carex reaches its maximum near the arctic circle, and decreases to the north and south of it. The genera Scirpus and Schœnus encroach on the areas of both the former genera, and are not so distinctly limited.

Although the forms of the northern and the tropical grasses differ so essentially from each other, the phenomenon of social growth is common to them both, and equally obvious. The noble Bambusæ, which form trees from 30 to 50 feet high, and which grow in the tropical and sub-tropical zones of both continents, frequently form vast and impenetrable forests. The form of these trees is remarkably elegant; their slender trunks, with reclining branches and high grassy leaves, are something quite singular; they remind the traveller from the north of the willows of his native country. The Bambusæ, in tropical countries, are planted to adorn the landscape, as the weeping willow is with us; and a beautiful lawn, surrounded by Bambusæ, such as I have seen in India, even though the turf be composed chiefly of Cyperoideæ, is one of the most charming objects in nature.

The distribution of these arborescent grasses presents several peculiarities. Though properly belonging only to the tropics, it is remarkable that several species extend far south on the western coasts of South America, for we find them in Chili below 36°, and in New Zealand at still lower latitudes. Bambusæ grow in the South Sea Islands also ; I have observed them in Oahu, one of the Sandwich Islands, but it is still unknown to what genus the species belongs.

The small family Saccharineæ is another group of the gigantic grasses of the warmer zones, and is one of the most beautiful forms among them, often forming high impenetrable tufts of sedge-like leaves, from the midst of which rise tall, slender shafts, with large bunches of flowers, which, from their silvery colour, are seen at a great distance. These silvery bunches, of which there are generally twenty or thirty in a single tuft of these plants, wave like banners in the wind. In the north of Chili, in the province Copiapó, close to the margin of the little stream which waters that province, I have found some of the most beautiful grasses of this group, viz., Gynerium Neesii *n. sp.*, and Gynerium speciosum *n. sp.*, which grow there along with tall species of Phragmites, while the gigantic Equisetum bogotense, often ten and eighteen feet in height and covered with thousands of branches, rises up amongst them. Von Martius also extols the beauty of the grasses of this form, which are said to grow in vast numbers on the margins of the rivers of Brazil. The Indians make their arrows from their flower shafts, which are sometimes without a leaf for six, seven, eight, and even ten feet. The sugar-cane and our Arundo phragmites, which grow round the ponds and lakes of the north, bear a great resemblance to these Gyneria, but are far inferior in elegance and beauty.

There are some well-marked forms of the Cyperoideæ, which have great influence on the character of the landscape. In the lower plains of the East Indies they generally form the green turf on the margin of streams ; but some genera of this group show tall, slender forms, the principal of which are the Papyrus and some species of Cladium.

Another group of gramineous plants is formed by the genus Eriocaulon, which, for the most part, belongs to tropical countries; and by Eriophora, which represent the former genus in the temperate and arctic zones. The individuals of this group bear little white heads on slender stems, which, rising from the dark green around, form with it an extremely pleasing contrast. In the tropical parts of both Indies, species of Eriocaulon are very common; in the south of China, they are always to be found on the margin of standing water, where they frequently grow two, three, and four feet high, and strike their little heads together at the slightest wind; and the pretty flowers of Utricularia bicolor and other species of this interesting genus, are seen close by them. But who is unacquainted with the effect of the white, woolly heads of the Eriophora in northern countries, which, when ripe, ornament wide tracts of wet moor? They grow there amongst tufts of the common Juncus and dwarf willows, and wave in the wind until they have completed their course and returned to dust.

The Restiaceæ, which are confined to the most southern part of Africa, and which singularly harmonize with the monotonous forms of the other characteristic plants of that country, form another very extensive group among the grasses.

I may in conclusion notice the form of rushes chiefly represented by the large genus Scirpus, which has recently been divided into so many genera. Many Juncoideæ resemble rushes very closely in form, and thus represent this truly aquatic form on dry ground. Rushes belong to the shore plants, of which we have spoken in page 60: they are imposing from the countless numbers in which their slender leafless stalks surround, as with a forest, the margins of sheets of water, but they give the landscape an extremely monotonous appearance.

## 2. *The Form of the Scitamineæ.*

The form of the grasses must not be confounded with that of the Scitamineæ in the wide sense of the term. There is a great similarity between them, but the leaf of the Scitamineæ is generally broader and more fleshy, and the flower shows a brilliancy of colouring which is totally wanting in the former family. As the grasses afford the staple food of the greater part of mankind, so there are some species of Scitamineæ, viz., the Bananas, which yield the most common food of the less civilized natives of the tropical zone. While man is gradually making the landscape more monotonous by extending the cultivation of the cereals, the rude Indian is unconsciously beautifying it in tropical countries by planting the banana. When he erects his hut of bamboo-reeds and palm leaves, he plants around it a few shoots of the banana, which yield him a supply of food and form the chief ornament of his simple dwelling.

If we attend merely to the general impression which these plants produce by their influence on the physiognomy of the country, we may divide the Scitamineæ into two sub-divisions, one of which comprehends the proper Scitamineæ and the Canneæ ; and the other embraces the Musaceæ, which are as it were arborescent Scitamineæ. To the Musaceæ belong the Banana, the noble Urania, the splendid Heliconia and Strelitzia. No other form of plants shows such variety and brilliancy in the colours of the flowers, as the Scitamineæ do ; they even surpass the Lilies and Orchideæ. The leaf of the Banana and Urania reaches such an extraordinary size, that its parenchyma, when fully developed, does not hold together, but separates at regular distances ; and, therefore, there are on every large plantain tree some leaves which hang down and are split across. The golden flower of a

Strelitzia, in the midst of its dark bluish-green leaves, is a beautiful object. It grows in the thick forests of the tropics, in damp ground, frequently near little sheets of water, and is shaded by tall slender ferns, whose finely cut leaves are in perpetually trembling motion; or lofty trees grow near, the trunks of which are covered with a host of beautiful climbers, amongst which glance out the large shining flowers of the Aroideæ. The Urania, which resembles a Musa, with its leaves placed sideways, is one of the most beautiful and remarkable of plants. Urania speciosa, called the water-tree by the Dutch, on account of the great quantity of water which flows from its stem, or leaf-stalk, when cut across, has been introduced into Java from Madagascar, and I have even seen it planted in China. I have seen a plant of this species in flower, which every day secreted, from a single sheath of flowers, more than a quart of thin liquid honey, in which thousands of insects found their death. The magnificent Urania amazonia, which Martius has made known to us,* is a giant among these plants. On a stem thirty feet in height, it shoots up a gigantic ear of keel-shaped sheaths from amongst its immense leaves, a very few of which are sufficient to thatch a hut.

The plants of the Banana form are almost entirely confined to the tropical zone; a few only passing beyond its limits, though the plantain is cultivated far within the sub-tropical zone. Within the tropics they seldom ascend above the height of 1400 feet, yet we find some very striking exceptions to this rule, just as we do among the palms. Humboldt observed at the height of 6600 feet above the sea, a form of plantain, which was 12 feet in height and formed so close a thicket, that he had the greatest difficulty in hewing a path through it. This plant† in all probability belongs to the genus Maranta or Heliconia.

* Reise nach Brazilien, iii. p. 20.

† See A. v. Humboldt's Naturgemälde, p. 61; and his Reise, &c., ii. p. 428.

The Scitamineæ, which represent in miniature the Banana form, also belong to the torrid zone; they are low plants, and therefore their beautiful foliage of the brightest green has little influence on the physiognomy of the country. Many of these plants grow socially; on the Sandwich Islands, as well as the Philippines, I have seen wide tracts quite covered with them. Some species of Alpinia, Amomum, and Canna grow to a considerable height, and their broad and shining leaves and singularly beautiful flowers contribute not a little to the attractions of the tropical plains.

### 3. *The Form of Pandanus.*

The Pandaneæ and Dracænæ join closely to the Banana form or to the Scitamineæ generally, and have a very decided influence on the character of tropical vegetation. They have long, linear-lanceolate leaves of a shining green, which are placed in regular spiral lines on the tall straight, or winding stems, and cover them to the top.

The stems of the plants belonging to the Pandanus form are generally without branches; it is only when very old that they branch at the top. Besides the Dracænæ, the large genus Pandanus and Freycinetia belong to this form. The true Pandani are peculiar to the Old World, and are in it predominant within the tropics; the genus Freycinetia extends to lower latitudes, and is even found on Norfolk Island. The Pandaneæ of the New World, viz., the genus Phytelephas *Ruiz* et *Pav.*, are stemless and have feathery leaves;* the Phormium tenax of New Zealand and the neighbouring islands, has also been considered as a stemless form of Pandanus.

Some Dracænæ resemble the Scitamineæ in form; Dracæna terminalis, the foliage of which is sometimes

* See Kunth, Handbuch der Botanik, Berlin, 1831, p. 240.

green and sometimes red, is a remarkably beautiful tree, which is very frequent in the South Sea Islands. In the Sandwich Islands it is planted round the huts, and forms a fence of singular beauty, which in general is not more than five feet in height. The tuberous root abounds in starch and sugar, and from it is obtained a spirituous liquor; for this purpose the roots are cut from the stalks, which are again put in the ground and soon send out new roots.

The colossal Dragon-tree (Dracæna draco), which stands in the little town of Orotava in Teneriffe, belongs to this form of plants; this giant among plants is 70 feet high, and 45 Paris feet* in circumference, and its antiquity must at least be greater than that of the pyramids. The trunk of this tree is hollow, so that we can ascend by a staircase in the interior to the height at which it begins to branch. On the 21st July, 1818, a violent hurricane broke off a large branch of this colossus, by which an open space was made, where several persons can stand. At the earlier stages of its growth the true Dragon-tree is branchless, like a Banana; it is only when of great age that it branches, and then it acquires a greater resemblance to the true Pandani.

The Pandani, that is the species of the extensive genus Pandanus, are found in tropical countries, whereever the soil or the atmosphere possesses a sufficient quantity of moisture. On the peninsula of Macao, in the south of China, I have seen Pandani growing luxuriantly in loose sand, but at no great distance from the sea, and consequently where the atmosphere was very damp. In the Philippines, Java, Sumatra and the Eastern Peninsula, the form of Pandanus prevails universally from the coasts to the region of the arborescent ferns.

* See Alex. v. Humboldt's Ansichten der Natur, 1808, p. 236, and Observations sur le Dracæna Draco L., par Sabin Berthelot, in the Nova Acta Acad. C.L.C. Nat. Curiosorum, tom. xiii. pars ii. p. 773, with beautiful drawings of this tree.

The Pandani often form upright trunks of considerable thickness, at the top of which is a conical mass of foliage; other species with more slender stems grow socially in greater or smaller numbers, and their leafless stems, which have foliage only at the top, wind about in various directions. But the most singular forms of Pandanus, which are peculiar to the tropics, are those whose trunks are covered with roots, which descend to the ground like tightly drawn cords and support the main trunk. On the mountains of the Philippines in the region of the tree-ferns, I became acquainted with the most singular growths of this kind; the roots from the stems of these Pandani descend from the height of 12 or 15 feet, and, as they enter the ground at a distance of several feet from the main trunk, the road frequently passes between them, if they are not so numerous as to obstruct all progress.

But the Pandani play an important part not only in the physiognomy of nature, but also in the domestic economy of man; for from their leaves are manufactured coarse mats, which we find very often used in the dwellings of the Indians. In the South Sea Islands the most common species is Pandanus odoratissimus, which is used in various ways; its flowers are of such delightful fragrance, that it is often cultivated in Egypt and Arabia for their sake.* The Indian women of the South Sea Islands strew the pollen of this beautiful plant over their hair.† The fruit reaches the size of a child's head, and is composed of several drupes; it resembles the pine apple, and is, like it, of a beautiful golden colour, but the inner part of the drupe is of the most beautiful yellow. The native women of the South Sea Islands string the stones on long threads and ornament with them their heads, necks,

* See Bové, Relation abrégé d'un Voyage Botanique en Egypte, dans les trois Arabies, en Palestine et en Syrie, &c.

† See Forster De Plantis esculentis insularum Oceani australis, Berolini, 1786, p. 41.

and arms, and also throw long strings of them over their shoulders. In time of need, when there is a great scarcity of fruit, the fruit of Pandanus odoratissimus is also used as food, but its fibrous and woody structure can yield but little nutriment ; only the pulp of the drupe is sucked out.

### 4. *The Form of the Ananas.*

The plants of this form are closely allied to the Pandani ; the form of their leaves is almost the same, but their green is of a more glaucous tint. In the splendour of their flowers they surpass most of the tropical forms of plants ; large ears or spikes of flowers, which show the most varied colours, rise from the centre of their mass of leaves. A great number of them are stemless, and only a few have a trunk, like the Pandaneæ.

The pine apple (Bromelia ananas L.), cultivated in our hot-houses, is a representative of this form of plants, which, in its native country within the tropics, reaches the height of four or five feet. I am certain that the pine apple is native both in the Old and the New World, for Pigafetta, in Magellan's first voyage round the world, describes it as one of the most delicious fruits of the Molucca Islands, but it is probable that the fruit belongs to different species. The pine apple, says Barchwitz,* when speaking of Lethy and the other South-western Islands, grows readily beside streams and in wet places, and is an ell and a half in height. In Banda, the pine apple, after being cut, is laid in spring-water for a quarter of an hour, and then eaten with wine. In the equatorial and tropical zones of America, it grows in great quantities ; and in Surinam, Brazil and the Indies, is used for enclosing the fields, and forms thick hedges, while its

* Ost-Indische Reisebeschreibung, Erfurt, 1751, p. 239.

sharp-edged and toothed leaves effectually keep out all animals. In Singapore the pine apple is extensively cultivated, and the fibres of its leaves used in manufacturing fine stuffs. The most of the plants of the Ananas form in their general appearance resemble stemless Pandani; and they especially aid in producing the rank luxuriance which distinguishes a true tropical vegetation. The greater number of the Bromeliæ, Tillandsiæ, Pitcairniæ, Guzmanniæ and some other genera, grow parasitically on the bark and branches of other trees. In Peru I have seen solitary trees and bushes almost entirely covered with Tillandsiæ, whose splendid flower-spikes rose out of the lead-grey foliage on which grew pretty golden-yellow Ramalinæ. Tillandsia usneoides L., which is of a pale, silver grey, covers like a veil the tropical trees of America, just as the long Usneæ do the pines of the damp forests of our northern mountains; but the Tillandsia reaches an extraordinary length, and hangs down in masses, which quite veil the tree, and, when stirred by the gentlest wind, wave to and fro like gigantic silver tresses. Other plants of this form are imposing from their enormous size, for Bromelia pinguin, as Martius* tells us, spreads its mass of leaves over a space 12 feet in diameter, and, although itself a parasite, it is likewise overgrown with mosses and other little parasitical plants.—Although the Ananas form properly belongs to the torrid zone only, there are some forms of Tillandsia, which appear on tropical mountains, and grow there in the Alpine regions or even extend to the region of perpetual snow. These little Tillandsiæ, which are found in the highest parts of the Cordilleras, form there extensive patches, that have the most monotonous appearance from the lead-grey colour of their leaves. An Usnea-like Tillandsia grows even high in the temperate regions of the mountains of Mexico; it covers

* Reise nach Brasilien, iii. p. 17.

the Coniferæ, particularly the Juniperus of these regions, and also the interesting Yucca trees, for instance, Yucca filamentosa, which is sometimes rather white than green, from the quantity of parasites upon it.

From the peculiar form of the leaves of these plants, which enclose the stem like scales, little reservoirs are formed in the hollow of the mass of leaves. These fill with water by the dew and rain, and keep the plant fresh, long after the commencement of the dry season, when every trace of moisture has disappeared all around. The traveller is often obliged to make use of this water, tenanted usually by insects and tree-frogs.

### 5. *The Agave Form.*

The plants of the Agave form, next to the Palms, show the most beautiful forms amongst Monocotyledonous plants; and many of them, those which are stemless as well as those which have stems of various heights, are also of gigantic size. Their foliage grows in large tufts, and the leaves are extremely strong and rigid, but often thick and fleshy; these linear-lanceolate leaves, which are frequently of enormous extent and bulk, are placed close one above another and spread out in all directions. This form is much more interesting when such a tuft of rigid leaves is placed on a tall and slender trunk, as in the genera Furcrœa, Yucca, Vellozia, and Barbacinia; all of which belong to the New World. The genus Furcrœa *Vent.* is particularly distinguished by its enormous masses, which appear in an extremely interesting form; the Furcrœæ, as well as the Agaves, their nearest relations, belong to Mexico and the northern parts of South America; those forms, which are said to have been found in Madagascar and the south of China, probably belonging to another genus. The Agaves and Furcrœæ often attain a great age before they flower, and with this epoch of their life, at which they, as it were, over-exert themselves in order to produce the enormous masses of flowers, their course is ended and they soon after die. The magnificent Furcrœa gigantea flowered in the year 1793 in the Botanic garden at Paris, where it had grown from the beginning of the century. But we have lately been made acquainted with a giant of this genus, both on account of its size and its great age, viz., Furcrœa longæva *Karw.* et *Zuccar.* This plant, which Karwinski discovered on the mountains of the Mexican province Oaxaca at the height of 9000 and 10,000 feet, had a slender trunk from 40 to 50 feet in height, and from 12 to 18 inches in diameter; at the top of this began the great tuft of leaves five and six feet long, from the centre of which rose to the height of 30 or 40 feet, a spike of innumerable white flowers; consequently this Furcrœa had reached an absolute height of 80 and 90 feet. What a striking object must such a sedge-like plant present! A period of 300 or 400 years is probably required for the development of such a monocotyledonous tree. To this form are closely allied the Mexican Yuccæ, which are found under similar conditions of locality. Schiede* speaks of these forests of the Mexican mountains, which appear there in the region of the Coniferæ, and are formed of trees exceeding the height of 30 feet; the singular Usnea-like Tillandsiæ of a silvery grey hanging from them in great quantities, and giving them a wintery appearance; while the other species of this genus, growing close at hand in the warmer regions of this country, display a loveliness, and by reason of their beautiful flowers a brilliancy, which is surpassed by few other forms of plants. As the gigantic spike of the Furcrœa is imposing from its size,—for Furcrœa longæva probably displays more than 20,000 flowers on its spike,—so the flowers of the Yucca make an agreeable impression by their beauty, of which we may form an idea, by imagining a number of tulips of the most vivid hues, growing on a lofty tree.

* Botanische Berichte über Mexico, Linnæa of 1829, p. 223.

Some species of Yucca, as well as the Vellozieæ and Barbacenieæ, have stems with a few branches, and then bear so great a resemblance to the Pandaneæ that other botanists have placed the Dracænæ amongst the Agaves.* It is certain, that the Pandanus form by the Dracænæ, and the Agave form by the Yuccæ, pass into each other. The Mexican Agaves, which are, as it were, stemless Furcrœæ, are imposing from their singular form, as well as their size. Although their leaves are large and fleshy, they grow in countries where the ground is quite destitute of water, or consists entirely of rock, with here and there a little earth. The beautiful spikes adorned by thousands of flowers, which these plants shoot up to the height of 16 and 20 feet, contribute not a little to animate the desert regions in which the Agaves for the most part grow. Some species, as is well known, are plants of great utility, and are cultivated to a very large extent in Mexico. These plants also flourish in corresponding climates in the Old World, and in some parts of it the Agaves have been planted in such numbers, that they have an essential influence on the character of the vegetation. In the Canary Islands and St. Helena they are planted at the side of roads, and when in flower, have, even at a great distance, a beautiful effect.

The Aloe plants in the Old World correspond with the Agaves of the New World; many of them in habit closely resemble the plants of the Agave form, both with and without stems. Almost all the species of Aloe are native to the south of Africa, where they occupy the sub-tropical zone; only a few forms, as Aloe vulgaris, *Dec.*, represent this genus in the sub-tropical zone of the northern hemisphere. In Greece and Arabia, and farther east towards India, even on the Mascarene Islands, appears a species of this genus, Aloe

* See v. Martius' Reise nach Brasilien, iii. p. 16.

macra *Haw.*; yet in the New World this group of plants is entirely wanting; the few species there having been introduced and cultivated. There are amongst the Aloes also gigantic individuals; Aloe dichotoma, which in our hot-houses reaches a considerable size, is said to attain a circumference of 300 or 400 feet at the Cape of Good Hope, and yet the soil is nowhere drier, than in those regions where the Aloes are found.

The Aloes are connected by some Lachenaliæ to the Phormium of New Zealand, and by it again to the plants of the Pandanus form.

## 6. *The Palms.*

The palms are considered the noblest of all the forms of vegetation. However they may vary in other respects, their character consists in a tall slender trunk, from the top of which rises an immense tuft of gigantic leaves. Some of them attain the most extraordinary height, and shoot up far above the loftiest tropical forests. Palms 70, 80, and 100 feet high are not unfrequent; but the wax-palm of the Cordilleras of Quindiu, which Humboldt and Bonpland have described, grows to the height of 160 and 180 feet, and, what is very remarkable, at an altitude of 9000 feet, while the Palms within the tropics seldom extend above the height of 3000 feet. With these strikingly contrast the stemless palms, which belong to the genera Chamærops and Nipa; these in their natural state seldom attain the height of five or six feet, though Chamærops humilis is grown in gardens to a tall tree with a slender stem. Very different from the lofty Palm trees, are the Cane-palms with slim twining stems, to which the genus Calamus belongs; these are the thorny climbers of the forests in the torrid region of East India. Often reaching the length of 400, 500, and 600 feet, their slender and generally twining stems mount to the top of the highest trees, whence they descend again, or catching hold of the neighbouring trees they bind them firmly to each other. The beautiful feathery leaves of these Cane-palms, which are twined with the stem round the trunks of large trees, assist not a little in enlivening and adorning the forests of the tropics.

The form of the leaves, which are sometimes long and simple, sometimes feathery or fan shaped, and their dark shining verdure, or the silvery-white down which clothes them, as well as their position and size, in proportion to the whole height of the tree, give the Palms a very varied aspect. How majestic must be the Jagua Palms, which grow round the granite rocks of Atures and Maypure! Their smooth and slender trunks rise, as Humboldt tells us, to the height of seventy and eighty feet, so that they shoot above the level of the forest like ranges of columns. Their leaves, of which there are generally only seven or eight, shoot up perpendicularly for fourteen and sixteen feet, and form a light and airy capital to these columns. How different is the luxuriance and beauty of the Fan-palms! the expanded horizontal leaves of which, when raised high enough by the trunk, shade a wide space around. The leaves of Manicaria saccifera *Gaertn.*, the only Brazilian Palm with undivided leaves, are twenty feet in length and six feet in breadth. These leaves, on account of their size and strength, are generally used for thatching.* The humble Chamærops, which is less beautiful, which we may even say shows no trace of the luxuriance of a fully developed Fan-palm, attains only a small height, and attracts the eye rather by its striking form than by its beauty.

Other Palms, which droop the foliage of their lofty crowns, have an effect very different from that which the slender Palms with aspiring leaves produce. On the western coast of South America, particularly in Chili, a tall Palm tree is planted in the middle of the convent

* V. Martius' Reise, iii. 989.

gardens ; in Chili the palm chosen for this is the Cocos chilensis, now called Molinaea micrococos, the name given it by the unfortunate Bertero ; its trunk is smooth, lofty, and thick, and is imposing from its size, but the pale drooping leaves of its crown of foliage, and the lead-colour of the stem make an extremely melancholy impression, corresponding to that of conventual life.

Not less varied is the share which Palms take in forming the physiognomy of nature, according as they shoot up from the midst of other plants, or grow socially in greater or smaller numbers. It is chiefly the dwarf-palms which grow in great masses, and this social growth seems to depend on the uniformity of the soil suitable to them. In the marshy parts of the Philippines, the other large islands near them, and the Moluccas, there are wide tracts, entirely covered with the Nipa-palm (Nipa frutescens L.), the stems of which are concealed in the moor. Chamærops palmetto grows in vast quantities over the extensive marshes in the neighbourhood of New Orleans. Chamærops humilis, the representative of the Palms in Europe, also prefers the most marshy ground, and grows very socially in such places in Italy, Sicily, and Spain, though on account of its lowness it has little influence on the character of the country. But there are also lofty palms which grow socially, and form groves, the enchanting pictures of which have so often excited the fancy of the poet. The Date-palm, which grows nearest us, and which forms the chief support of whole nations, grow socially in groups, in the shade of which men fix their dwellings. The well-known Cocoa-palm (Cocos nucifera L.) which adorns the coasts of India and the shores of the South Sea Islands, not unfrequently occurs in larger or smaller groups. In still greater masses grows the Fan-palm at the mouth of the Orinoco, the far-famed Mauritia flexuosa, which secures the independence of the wild Guaraunas. This tribe construct their huts on mats firmly bound to adjacent trees of these palm forests, and when the ground is inundated, they live like apes upon the trees, and go from place to place in little canoes. A view of such a scene is given in Sack's Travels in Surinam.

But much as we have heard in praise of the palm forests, and often as poets have sung the charms of such scenes ; and though it is also certain that the noble form of palms surpasses that of any other trees, and that palm-groves are peculiarly imposing from their lofty aspiring trunks, we seek in vain in them for the gay beauty which a forest of the bright foliaged trees of northern regions presents. The cool shade of our beech woods, and the choir of joyful songsters is sought for there in vain. Yet the Palm is the noblest of all plants ; it belongs almost exclusively to the hotter regions, in which was placed the cradle of mankind, and in its praise is included that of the whole warmer zone, for where the Palm grows is placed the happy clime where nature, without the aid of man, produces abundance of the most delicious food ; where the bark of the trees yields sufficient clothing, and every spot shaded by a light canopy of leaves, offers the most delightful abode to man. Where the sweet fruit of the date ripens, and where the cocoa-palm rises majestically, is placed the seat of the happy child of nature, and the poet praises the land where man lives in such a primeval state.

The form of the Palms increases in beauty the nearer we approach the equator ; their normal form is straight and unbranched, but the beautiful Palm of Thebes, Cucifera thebaica,* has a branched trunk.

The Cycadeæ, which include the genera Cycas and Zamia, form a small subdivision in the form of the Palms. The species of Cycas in habit belong entirely to the Palms, though the structure of their fruit brings them nearer the Coniferæ ; they are peculiar to the Old World, and chiefly to the eastern part of it. The

* There is a beautiful figure of this plant in the great French work on Egypt.

Zamiæ on the contrary, differ considerably from the form of the Palms, and the African and New Holland species of this genus frequently present the most singular forms. In the dry and barren plains of the south of Africa, where the ostrich takes up its abode, grow the strange forms of the Zamiæ;* their stems are thick, low, and unshapely, and have at the top a tuft of leaves spreading far apart. There is, so to speak, something awkward and disproportioned in the shape of this palm-like plant, which quite suits the monotony of a South African landscape.

### 7. *The Fern Form.*

The form of the Ferns varies so extraordinarily, and the tall and stemmed species and genera of these plants stand so near the Palms, at least in form, that it is sometimes possible to confound the stems of these two forms of plants with each other; that this has been often already done, is shown by the accounts of the many fossil Palms found in our northern regions, which are proved by later investigations to belong almost without exception to the Ferns and Cycadeæ.

The tree-ferns unite in themselves the majestic growth of the palms with the delicacy of the lower ferns, and thus attain a beauty to which nature shows nothing similar. It is chiefly the tall, arborescent species which have especial influence on the physiognomy of nature, and they also most resemble the Palms.

The whole group of the Ferns may be classed in three divisions, viz.: the herbaceous, the shrubby, and the arborescent. The herbaceous ferns chiefly belong to the temperate and cold zones, where there are only a very few species which form a small stem, and attain a height by their fronds, that gives them a resemblance

* See the figure of these plants in the Allgemeinen Gartenzeitung von 1834, Nro. 11, Tab. i. and iv.

to shrubby plants. Struthiopteris germanica is such a fern of our zone, representing in our colder regions the shrubby species of the tropical and sub-tropical zones. Our appellation, Shrubby Fern, may probably be thought improper, as all those which belong to this division are properly little trees with short stems, just like the Palms which are called stemless; but we choose the name for this group of Ferns in order to distinguish them clearly from the tall, arborescent Ferns, and to be enabled thus to give a more special physiographical delineation of a country. There are also a few of the herbaceous Ferns which attain an extraordinary height; such as Pteris aquilina, which so frequently covers wide tracts in our climates, but their fronds grow single, and are never united in tufts, as in the shrubby Ferns, the fronds of which spring from the top of a short stem.

The number of the herbaceous Ferns in warmer regions is not less great, but their influence on the character of vegetation there is of a very different nature. In tropical countries the shrubby Ferns grow on the ground in as great quantities as the herbaceous do in the colder zones, while the latter are more frequently parasites, growing on the stems and branches of other plants, and by their extremely interesting form, by the way in which they fix themselves, or by their colour, they give a peculiar character of beauty and luxuriance to the higher vegetation. The Tree-ferns are confined to the torrid zone; their slender trunks rise to the height of from twenty to twenty-eight feet, and from the tops spring large fronds, often eight and nine feet long, tripinnate and feathery, which, from their extraordinary delicacy, are put in tremulous motion by the gentlest wind.

These slender trunks, often quite smooth, and beautifully pitted by the marks of the insertion of the leaves, grow to a height above twenty feet, and are sometimes not more than three inches thick. On some of the East

Indian islands the Tree-ferns grow in such numbers that their stems are as close to each other as the slender firs and pines are in our plantations. Sometimes the trunks of these tree-ferns attain a greater thickness ; often as much as eight inches in diameter, but in such instances they are generally cased in a thick layer of roots. Wherever the Tree-ferns appear within the tropics, from the plain to the height of 3000 or 4000 feet, the soil and atmosphere are full of moisture ; indeed they seem to prefer wet places, and grow in them along with Musaceæ and Scitamineæ. The shrubby Ferns also grow in a damp atmosphere, and, therefore, they occur in great numbers on the South Sea Islands; but I think that they prevail rather at the tropics than in the equatorial zone ; and they are also found less frequently at the foot of tropical mountains, than at an elevation between 2000 and 3000 feet.

## 8. *The Mimosa Form.*

We pass from the Tree-ferns to the plants of the Mimosa form, the foliage of which is as finely cut as that of the ferns, and often is still more beautiful. The Mimosæ and Sophoræ, with their finely feathered leaves, also belong, almost without exception, to the torrid zone, or, at least, attain their maximum there. They form bushes and trees, which spread out their branches like the Firs, and particularly like the Araucariæ of Chili. Thousands of leaflets are ranged close to each other with the greatest regularity, and form the feathery leaf of the Mimosa ; they are endowed with extreme sensitiveness, and at the slightest touch alter their direction and fold close together, while the phenomenon of periodical sleep is more clearly seen in this than in any other group of plants. There are tracts in the hot and damp zones which are entirely covered with Mimosæ ; their feathery foliage has a light and beautiful effect, and the vibration caused by a horse galloping past, is sufficient to set the whole mass of it in motion.*

The development of vegetation becomes gradually more perfect as it advances from the poles to the equator, and this is especially the case in the Leguminosæ, and among them particularly in the Mimosa form. The true Mimosæ, which are found both in the Old and New Worlds, do not pass beyond the tropics; and a large group of Ingæ, viz. those without thorns, belong to America only. The genera Acacia, Prosopis, and Gleditschia, &c., on the other hand, extend farther towards the poles; they appear in great numbers in the tropical zones, but are also frequent in the sub-tropical, and even in the warmer part of the temperate zone. The Acaciæ predominate in the southern hemisphere of the Old, as well as of the New World; New Holland is especially their country, where they appear with leaves of a singular form, and thus give the vegetation a peculiar, or as we say, an Australian character. Some Acaciæ, as striking in form as those of New Holland, pass by New Guinea to the South Sea Islands; and even on the Sandwich Islands there is found a species, Acacia heterophylla, which is the proper representative of the New Holland Acaciæ, in the southern hemisphere. In South America, particularly on the western coast, Acacia caven, as well as several species of Prosopis, extend south beyond Conception; on the table lands in the centre of Chili, even above the altitude of 3000 feet, are sometimes whole forests of these plants of the Mimosa form, which, alas, must be gradually extirpated, for the inhabitants of Chili prefer these thorny shrubs for the making of dry fences round their fields and gardens. The Acacia caven grows very vigorously on the Cordilleras of San Fernando, at elevations of between 2000 and 3000 feet, where it forms a considerable tree, while at lower altitudes it is rather shrubby. The Gle-

* Compare von Martius' Reise, iii. p. 38.

ditschiæ and Robiniæ are the representatives of the Mimosa-form, which extend so high in the northern hemisphere. The North American Gleditschiæ and the most of the Robiniæ, which have been introduced into Europe, here even reach the sub-arctic zone, and are now of great use in ornamenting our gardens. The Gleditschiæ, which grow to an enormous size, give us an idea of the Cæsalpiniæ of the torrid zone.

The Papilionaceæ, or Leguminosæ with papilinaceous flowers, form a group of plants which are closely allied to the Mimosæ in the structure of their flowers and fruits, but are totally different in their general appearance. The Papilionaceæ are very numerous, and are distributed over the whole globe, even beyond the arctic circle, while the Mimosa form reaches its polar limit in the colder part of the temperate zone. Many of the Papilionaceæ have instead of the feathery leaves of the preceding form, only three-lobed leaves, but they are little inferior to the Mimosæ in the splendour of their flowers, which display almost every colour; yet the plants of this group are so seldom of considerable size, or of remarkable form, that they can influence the character of vegetation only when they grow in masses. A meadow in our zone, covered with clover or species of Medicago and Melilotus, at the time when they display their gay flowers, presents a lovely sight, which can scarcely be equalled in tropical regions. In the vast steppes of the temperate zone of the northern hemisphere, the Astragali grow in masses, and there, though in an opposite manner, take a share in determining the character of the vegetation. They give these regions a desert and repulsive character, while the blooming clover field inspires the greatest cheerfulness. The Papilionaceæ also appear sometimes in great numbers in the hotter, and even in the equatorial regions, and have there great influence on the character of the vegetation, chiefly by the brilliancy of their flowers.

The genera Hedysarum, Indigofera, Crotalaria, and

Dolichos, within the tropics, replace the Trifolium, Medicago, Astragalus, Aspalathus, and Lupinus of the colder zones, and some other large genera exclusively belong to this or that region. The Psoraliæ predominate in the sub-tropical zone of the southern hemisphere, while the Robiniæ are natives of North America and Asia.

### 9. *The Abietinæ.*

We placed the Mimosa form next the Ferns on account of their finely divided leaves, and next to them come the Coniferæ, on account of the slenderness of their stems. The dark green, the solemnity, and even the melancholy of the pine and fir woods of our cold zones, must certainly have struck every one who has compared them with the bright green, cheerful forests of our beech and oak. Among all the forms of vegetation, there are scarcely any which present a greater contrast than the Abietinæ and dicotyledonous trees when growing near each other. The numerous family of the Coniferæ is distributed over the whole globe, but the peculiar form of Abietinæ, viz., the genera Pinus, Abies, Larix, and Taxus, exclusively belong to the northern hemisphere, where they form a broad girdle round the earth throughout the temperate and sub-arctic zones, and even in the arctic zone ; and the farther north they extend, the greater is their influence on the character of the vegetation, and even on the whole physiognomy of nature. Our northern Abietinæ, under favourable circumstances, attain an enormous height, so that they must be considered among the loftiest trees which shade the earth. The persistent evergreen foliage of these trees renders them of great importance in the sub-arctic and arctic zones, where the winter is of long duration, and where the ground is long covered with a thick mantle of snow ; for they serve to animate the frightful dreariness which the severe cold spreads over all these regions.

After the birch, the Abietinæ extend farthest north, for in Europe they are still to be found in lat. 70°, and in Siberia in lat. 68°, although they are not so slender as when in lower latitudes. These trees also ascend high up the mountains, and generally form an upper tree limit, as is the case on almost all the mountains of the northern hemisphere. Species of Abietinæ of extremely similar habit, are spread over a very extensive range, for Pinus chinensis *Lamb.*, which is so extraordinarily like our Pinus sylvestris, forms at Macao, near the coast, as extensive forests as those of Pinus sylvestris with us; but man has begun to destroy them, and soon there will come a time when not a trace of these Chinese pine forests will remain. On the islands which lie at the mouth of the Tiger river between Macao and Canton, the sides of the lower mountains, quite down to the coast, are more or less clothed with forests of the Chinese pine, according as they have been more or less destroyed by man; formerly, perhaps, they would have been scarcely distinguishable from our pine-woods. As the Pyrolæ and Vacciniæ grow in our forests of Abietinæ, we find there small Crotalariæ resembling our Genista, and coarse Osbeckiæ, and Orchideæ, which supply the place of the Orchideæ of our forests. In general, however, the Abietinæ in the warmer regions do not appear in such vast close masses as with us, when, except the birch and a few other plants, there is little underwood amongst them, but stand farther apart with Ferns and Rhododendrons growing between them. On the whole, woods composed of a single kind of tree, such as our oak, beech, birch, fir, chesnut, and other forests, are not met with in tropical countries, where variety as well as luxuriance prevails.

It is most remarkable that all the true Abietinæ, viz., the genera Pinus, Abies, Larix, and Taxus, are wanting in the southern hemisphere, for as yet not a species of these genera has been found on that side of the equator. Pinus dammara, which, along with the lofty and majestic Podocarpus, appears at a height of 3000 feet on the mountains of Java, is still so imperfectly known, that it may be taken for granted that it belongs to a different genus. The true Abietinæ of the northern hemisphere are replaced by the genera Araucaria, Podocarpus, Cupressus, and by the Casuarinæ, unless we consider them together with the genus Ephedra as a separate form. The South American Araucariæ are trees of most remarkable form, particularly Araucaria imbricata with its large horizontally spreading branches. The Araucaria of Chili grows on the Cordilleras of Southern Chili; the most northern forests of this tree are in the latitude of Conception; it probably extends very far to the south, but its limit is still unknown. It is wanting on the coasts of the Straits of Magellan, but there is found there, according to Captain King, a tree resembling a Cupressus.

Molina has given us an excellent description of the appearance of this noble tree, and Poeppig has supplied all that was still wanting. Its trunk, which is said to attain a circumference of about eight feet, is from 70 to 100 feet in height, and bare, because, like our Abietinæ, it throws off the old branches and leaves. The crown, which includes about a fourth part of the whole height, is a perfect quadrangular pyramid. The lower branches are placed round the trunk in eights and twelves, the upper ones in fours and sixes, and the whorls are from four to six feet above each other. The branches spread horizontally, and are quite covered with leaves, which are three inches long and one inch broad, heart shaped, and as hard as wood. Every main branch, at certain intervals, forms smaller branches, which go off at right angles, and thus make the pyramidal crown thicker. The globular fruit of the Araucaria attains the size of a man's head, and is placed on the extremities of the twigs; each fruit contains from 200 to 300 seeds, which are twice the size of almonds, and are a pleasant and favourite food of the inhabitants of that country. There are frequently on a single tree from twenty to thirty of these fruits, which, at the end of March, when they are ripe, split up and scatter the seeds about.

According to the information collected by Poeppig, the forests of Araucariæ appear on the western side only of the Cordilleras of the south of Chili ;* and, indeed, as far south as Valdivia, they are found only at very considerable altitudes, not more than 2000 feet from the snow line, to which it is said they sometimes ascend.

In New Holland, Cupressus callitris forms extensive forests, which alternate with species of Eucalyptus, Acacia, and Casuarina ; particularly in those parts where the flat margins of the rivers are covered with the social Polygonum junceum. The noble Araucaria excelsa, a lofty tree, was discovered on Norfolk Island, and it occurs also in Van Diemen's Land below 40° south latitude.

The Casuarinæ, which reach their maximum in New Holland, appear there in far greater numbers ; in the structure of their fruit they come very near the Coniferæ, but are totally different from them in their form, and in the effect they produce on the character of vegetation. In the interior of New Holland the Casuarinæ grow scattered in the Acacia and Eucalyptus forests, and, therefore, do not form woods like our pines.

The singular form of Casuarina, which is found in the Eastern Peninsula, on several of the Indian islands, in New Guinea, and in great part of the South Sea Islands, where it points out the burial places, is represented by the genus Ephedra in the northern part of the Old World, as well as on the mountains of America. The form of the Casuarinæ is not very clearly seen in the small and bush-like Ephedræ of the northern hemisphere, nor in

* Reise in Chile, u. s. w. i. p. 403.

the stunted Ephedra americana, which appears on the heights of the South American Cordilleras; but on the Cordilleras of Chili, at altitudes of between 2000 and 4000 feet, grows the Ephedra chilensis, a lofty and beautifully slender tree, the foliage of which hangs down like that of Casuarina equisitifolia. Not a little striking is the contrast of the Ephedra with the singular pillars of the Cactus and the shining foliage of the Chilian dicotyledonous trees. The Mutisiæ climb up the trunks of the Ephedra, and a profusion of their scarlet flowers decorates the foliage as if they were the blossoms of the tree itself.

### 10. *The Forms of Protea, Epacris, and Erica.*

We have here grouped together three large families, which have a great resemblance to each other in the form of their individuals, and, when taken together, supply the place of the Coniferæ of the northern hemisphere in the southern. As the Coniferæ reach their maximum in the northern hemisphere, so do the genera Protea, Epacris, and Erica in the southern hemisphere; but though only single representatives are found with us, some of them extend even to the arctic zone, as the Ericæ in Kamschatka in Lapland, and the old Erica cœrulea in Greenland and North America show. The areas of the plants, which belong to the forms we have named, are in the southern hemisphere very strictly limited. R. Brown,* to whom we owe our accurate acquaintance with the Proteaceæ, says that none of the Australian species of the family have been found in any other part of the globe, and that not a single species appears both on the eastern and western side of New Holland.

The Epacridæ also, with the exception of a few species,

* General remarks, Flinder's Voyage to Terra Australis, London, 1814, ii. p. 568.

belong to the southern hemisphere only, and in it New Holland is peculiarly their native country. The Cape of Good Hope is in the same way the country of the Ericæ, which have as it were representatives only in the northern hemisphere.

The arborescent Ericæ, in the extreme south of Europe and in the sub-tropical zone, where they grow luxuriantly in the island of Teneriffe, have in their general effect a great resemblance to certain forms of the Coniferæ; but their small, needle-like leaves are beautifully adorned by a mass of pretty flowers, which are often of the most brilliant colours. Humboldt grouped with the heaths or Ericæ, the genera of similar form, Passerina, Phylica, Diosma, Gnidia, &c.; and through them the form of Ericæ is still further extended.

In the genus Protea the leaves are generally as fine as those of some Coniferæ; but in the numerous other genera of the large family of Proteaceæ the leaves are of greater or less breadth; they are then rigid, their upper surface shining, and as in Banksia, Dryandra, &c., the lower surface covered with woolly down.

The original form of Ericæ thus entirely disappears in those genera, and these trees in their general appearance come nearer the dicotyledonous trees with rigid and shining leaves. The forests of Banksia and Dryandra, in New Holland, unlike those of the Ericeæ of the Cape of Good Hope, are shady; the leaves are sometimes even broad and placed close together, while from the crown of foliage rise gay clusters of flowers, which resemble those of the myrtle form.

## 11. *The Myrtle Form.*

The plants of the Myrtle form directly border on the Protea form; the Banksiæ agree with the Melaleucæ and Metrosideræ, not only in the external figure of the clusters of flowers, but also, at least in some genera of these families, in the whole habit of the foliage, as well as in their social growth. The Melaleucæ, which, with their Conifera-like crown of foliage, are one of the principal forms of New Holland, contribute much towards forming the character of Australian vegetation; they are also almost exclusively peculiar to this country, for Melaleuca leucadendron and cajaputi only enter the territory of the Erica form, and appear in the southern extremity of Africa. Who is unacquainted with the splendour which the genera Melaleuca, Metrosidera, Beaufortia and Calothamnus have added to our collections of flowers? But to comprehend the beauty of such a vegetation, we must fancy these lofty trees laden with clusters of scarlet flowers growing near glossy Banksiæ, Hackeæ, Mimosaceæ, and Casuarinæ; a vegetation, each tree of which belongs to a peculiar characteristic form, which, though not so charming, is much more splendid than that of our bright green forest trees.

The beautiful forms of the genera Leptospermum and Bæckia also belong, almost without exception, to New Holland. The first strikingly resembles our Myrtle; but the Bæckiæ, for example Bæckia frutescens, which appears in such quantities on the southern coasts of China, show quite the Erica form. The numerous species of Myrtus, with their small-leaved glossy foliage, their round tops, and pretty white flowers, are most frequent in the New, but representatives of the genus are found in all parts of the Old World.

Our Myrtus communis is a native of the south of Europe, and it grows in the colder part of the temperate zone only when planted in favourable situations. The Myrtles in South America probably extend to corresponding southern latitudes, for they grow in the greatest luxuriance in the middle of Chili. In the province of San Fernando, in latitude 35° and at the height of between 1800 and 2000 feet, I have seen Myrtle trees,*

* Myrtus Luma *M.* and other species.

which were five, six, or even nine feet in circumference, and formed handsome wide-spreading bushes covered with innumerable white flowers.

Another remarkably beautiful and striking group of the Myrtaceæ is represented by the extensive genus Eucalyptus, which predominates to such a degree in a great part of New Holland, that it principally determines the character of the vegetation of that country. The Eucalypti, which belong exclusively to New Holland and Van Diemen's Land, sometimes reach an enormous circumference; for Eucalyptus globulus *Labill.* and another species peculiar to the south of Van Diemen's Land, not unfrequently attain a height of 150 feet, while the trunk near the ground is from 25 to 40 feet in circumference.* The foliage of the Eucalypti is very remarkable, and the sabre-like form of the leaves, the edges of which are turned towards the trunk, together with their bluish-green colour, gives them a sombre appearance. We may imagine how extraordinarily numerous the species of this beautiful genus are, since R. Brown tells us† that around Port Jackson alone more than fifty species have been discovered, the most of which are distinguished by different names by the rude inhabitants, who from the colour and texture of the bark, and the way in which it peels off, as well as from the form of the branches and the general appearance of the tree, distinguish the species far more surely than botanists have as yet been able to do.

The Eucalypti are indeed so numerous in New Holland, that they compose almost four-fifths of the forests of this continent, and yet they are met with in no other country except the adjacent Van Diemen's Land.

There are still a number of other Myrtaceæ, which, at least in the structure of their flowers and fruit, are

* See R. Brown, in Flinder's Voyage, ii. p. 547.

† L. C. p. 547.

naturally allied to those we have mentioned, but which in habit are totally different from the true myrtle form, and which, by the size of their leaves and their mode of branching, come nearer to the usual forms of the forest trees of our zone. The Guavas (species of Psidrum) which in both continents adorn every thicket; the Barringtoniæ, which grow near the springs and on the margins of the fresh-water lakes of India, and which so beautifully decorate with their grape-like clusters of flowers the edge of the dense and leafy thickets, together with the Jambosæ, and Eugeniæ, which have a similar habit, belong to this class.

## 12. *The Form of Dicotyledonous Trees.*

The ordinary form of dicotyledonous trees, which extend over almost the whole surface of the globe, through all zones and regions so far as arborescent vegetation appears, joins directly to the myrtle form. It is true that there are extraordinarily great contrasts between the several trees, which we comprehend in this form, but on a general consideration of nature, these contrasts in a great measure disappear. A willow, whose light foliage casts so little shade, when compared with a beech with its close branches, or this again with a laurel, shows no little dissimilarity; but when these forms appear in masses as forests, we pay less regard to the outline of the foliage or the form of the trunk, than to the general effect of the united masses with their shady foliage. Our present landscape-gardening, which seeks to imitate, in a small and limited space, the decided contrasts which nature presents on a large scale, can produce the most surprising effects by a judicious disposition of the dicotyledonous trees. What an interesting object is the weeping-willow, when planted on a declivity close to the brink of a small sheet of water! How beautifully the shining white stems of the hanging birch contrast with the dark massy foliage of the underwood, and the trembling aspen with the venerable oak with its bright green leaves! It is the same in nature on a larger scale; we step from a dense forest of oak or beech trees, and see on its borders a few birches, whose long drooping boughs are waved by the gentlest wind, or poplars, whose leaves are in perpetual motion on their slender leaf-stalks, and we can then estimate the powerful effect which these contrasting forms produce.

Of the many and various forms of Dicotyledonous trees, we may at least distinguish the following, which must be attended to not only by the painter, but by the landscape gardener, who wishes to imitate in his garden the most striking beauties of nature.

The Dicotyledonous trees with broad and tender leaves, chiefly belong to our colder zones, and indeed predominate in the colder half of the temperate zone, as the Abietinæ do in the arctic zone. Amongst them we distinguish those whose foliage is light and affords but little shade, such as the Birch, the Alder, the Poplar, &c.; those with irregular crowns, whose boughs spread widely in all directions, as the Oak, the Lime, the Elm, &c.; and lastly those, whose crowns are almost quite round, and whose thick close boughs and dense foliage cast a deep shade. The Beech, the Horse-chesnut, &c., are of this latter class.

The Dicotyledonous trees with thick leathery and shining leaves (often called evergreens), on the contrary, belong to the warmer zones, and also to the colder zone of the southern hemisphere; while only a few representatives of these beautiful trees, for instance, Ilex aquifolium, appear in the colder zone of the northern hemisphere.

In the south of Europe there are Chesnut forests, and Olive, and Laurel groves, which show this form, so that the physiognomy of the vegetation of that portion of

Europe is characteristically different from that of the north.*

The Willow, with its slender boughs and light, small, pointed leaves, is one of the most striking forms amongst Dicotyledonous trees; it is distributed over a great part of the northern hemisphere, but reaches its maximum in the colder part of the temperate and in the sub-arctic zone; it has also its representatives in the southern hemisphere. The influence of the willows on the character of the vegetation of our northern zones cannot be overlooked; they are fond of wet places, particularly the margins of rivers and ponds, but some species delight in the poorest soils. They grow on the side of our rivers, as the Mangrove forests do on the shores of the tropical seas. Many willows, in the form of low bushes, and in great numbers, grow socially in wet places of our zone; their foliage is often beautifully silvery, when the under surface of the leaves is covered with fine hairs. The willows are also remarkable for developing their flowers very early, long before the rest of vegetation, and afterwards putting forth their leaves. Many of these willows have much larger flowers than we usually see in our forest trees, and therefore when they appear in great masses, they have a singular, though very agreeable effect; particularly as at the time at which they flower, there are but few flowers in our fields and meadows. In summer also, when the female catkins are ripe and clothed with white wool, the willows have a peculiar appearance.

The most beautiful of all is the Weeping-willow (Salix babylonica), which, like the drooping Casuarinæ of the South Sea Islands, serves with us to point out the resting-place of our ancestors, and speaks more impressively than any epitaph.

* See some more special notices of the dissimilarity of vegetation in the northern and southern halves of Europe, in a paper by Willdenow, in the Magazin der Gesellschaft naturforschender Freunde zu Berlin. Berlin, 1811, p. 99.

The Dicotyledonous trees with large and remarkably beautiful leaves, all belong to the torrid zone, just as those with tender leaves are proper to our temperate zone. In the most striking of this form the leaves are more or less hairy and often of enormous size, as in Cecropia peltata in the Brazilian forests; or they are beautifully slashed, as in the Broussonetiæ and Artocarpus incisa; and from their covering of hairs, which is at least generally present and often of silvery brightness, they form striking contrasts with the dark green and generally feathery foliage of the neighbouring vegetation. The families Urticeæ, Euphorbiaceæ, and Malvaceæ, for the most part show this form. The beautiful leaf of the Bread-fruit tree (Artocarpus incisa L.), and the silver grey foliage of the Broussonetiæ, Boemehriæ, and Crotonæ, are of remarkably handsome form; and when growing in great masses near each other, they present a peculiarly striking spectacle. Frequently large and gorgeous flowers appear along with this form of leaf, and increase the beauty of these plants. This is especially the case in the Malvaceæ, amongst which the genera Sterculia, Lavatera, Hibiscus, and Ochroma, which Humboldt* has grouped under the Malva form, represent these Dicotyledonous trees. An Hibiscus chinensis, entirely covered with its splendid scarlet flowers, or the beautiful shady tree Hibiscus tiliaceus of the South Sea Islands, richly adorned with large and gorgeous flowers, can alone give a perfect idea of the beauty which this form of plants shows.

## 13. *The Cactus Form.*

No other group of plants shows forms so striking as that of the Cactaceæ, which, various as are their forms, impose rather by their singularity, than by the elegance and beauty which is proper to most of the

* Ansichten der Natur, ii. p. 31.

other chief forms of vegetation. But nature has tried to make up for the imperfections in the form of these plants by the profuseness and splendour of their flowers, for often it seems as if their whole effort was to produce the greatest possible quantity of the gorgeous flowers, with which they are so often completely covered. The extraordinary effect on the physiognomy of vegetation, which is produced by the contrast of the Cactus forms with the other groups of plants, is seen not only in nature, but also everywhere in our gardens. These would be without their gayest ornament if the Cactaceæ were wanting in them; and, of all the American families, this is the one which has been most generally diffused since the discovery of that new continent.

The Cactaceæ, with the exception of two species which have been found in the east of Asia, belong to the New World, where they extend from 40° north latitude to 40° south latitude; and from the level of the sea to the vicinity of the limit of perpetual snow. The maximum of these plants belongs to the torrid zone, yet certain forms of them rather prevail in the temperate zones, and in tropical regions grow on the mountains at elevations, where they find a cooler climate.

This striking family is certainly very rich in species, though as yet scarcely more than 190 of them have been described. We may confidently reckon that this number will be doubled when the mountains of America are more carefully examined; but it will be long before all the species are known, for it is very difficult and often impossible to transport them, and, besides, the traveller seldom finds them in flower. Modern botanists have divided the family into several genera, the characters of which are taken less from the structure of the flowers, than from the striking differences of their forms, so that the genus may be determined without seeing the flowers. The principal forms are,—

1. The Cerei; they have long pillar-like stems, which are sometimes 3, 4, 5, 6, 7, and many cornered, or even more or less round. They are either upright or creeping, and also either branched or unbranched. The upright Cerei grow like columns, which are collected in larger or smaller groups. The Cerei of the eastern and western coasts of South America, (forms which so closely resemble Cereus peruvianus, that it is difficult to distinguish them,) frequently grow to the height of 15, 20, or even 25 feet, as may be seen in Chili, on the Cordilleras of San Fernando, below the height of 3500 feet. Many plants of this lofty heptagonal Cactus grow near each other, and it shoots up from 10 to 20 columns from a single root. Some of these columns are dead; their fleshy covering has disappeared, and now there stands a symmetrical woody cylinder of a white colour in the midst of the green, sharp-cornered columns, which frequently display such a profusion of flowers, seven or eight inches long, that a great part of the column is concealed by them. Such groups of living columns are generally five or six paces distant from each other, and very few plants seem to approach these prickly strangers. In Chili and Peru there are wide plains, which, for vast distances, are covered solely by such groups of Cerei, and thus present a singular and striking, though cheerless prospect. In the Llanura de Rancagua, lying to the south of the Rio Cachapual in Chili, only a few Caven-bushes (Acacia caven *Mol.*) grow amongst these Cerei, which are often covered with the scarlet flowers of Loranthus aphyllus, from amongst which hang the long, white flowers of the Cereus. When travelling in the interior of Chili, we often bivouacked near these Cerei, and their dry woody cylinders furnished the best fuel for our fires. The wood of these plants attains a thickness of an inch or an inch and a half, and the whole cylinder is as much as 12 or 15 inches in circumference. The wood of the Cactus is applied to the most various purposes in the

treeless regions of the western coast of South America. As it is very light, it is carried up the Cordilleras, and on the plateaux, which lie far above all arborescent vegetation, doors and beams are made of it without any previous working.

The columnar Cacti ascend high up the Cordilleras. I have found them on the Cordilleras of Southern Peru, close to the equator, above the heights of 7000 and 8000 feet, and these are also the most beautiful forms which I have ever seen.* In the Cordilleras of Tacna and Arequipa, I found a true candelabra-like Cactus, which clothed the sides of the mountains in a singular manner, for scarcely a trace of vegetation was to be found there, only here and there rose one of these strange Cerei, the stem of which, at the height of eight feet and upwards, divided into a number of branches (from eight to twelve in number), which were round and twisted in various directions, sometimes upwards, sometimes downwards, or from side to side. The extent of the vertical range of this Cereus seems to be very limited; for on both the stations where I met with it, it very soon disappeared, and above the height of 7000 feet came in its place the Cactus senilis, which, with its long silvery hairs, could be placed by nature nowhere more appropriately than in these vast deserts. This singular Cactus is the more striking, as it always, wherever I have seen it on the western coasts, grows quite isolated in single stems, and never in groups, like so many of the other branchless Cerei. The Cactaceæ in general prefer desert, sandy, or stony soils, where the most excessive dryness prevails, to moist and fertile soils. Martius† has already drawn attention to this, and expressly remarks, that in the shady primitive forests of Brazil, the Cactus-plants are wanting, except a few species of Rhipsalis and Epiphyllum, which grow

* See Meyen's Reise um die Erde, i. p. 447.

† Reise nach Brasilien, iii. p. 26.

parasitically on the branches of the trees; on the contrary, they affect the bare rock of the province of Pernambuco. "In these dry regions," says this accomplished traveller, "over which is arched a pure and deep-blue sky, grow these unshapely plants, often exceeding the height of a man; the leafless, rigid masses stand irregularly, and their bluish green contrasts as strongly with the warm colouring of the landscape, as their harsh outline does with the soft, yielding forms of the other tropical vegetation."

The extensive country of Brazil, the greatest part of which possesses a very damp climate, produces an immense number of Cactus-plants, on the distribution of which Von Martius has published a very interesting treatise.* Stately Cerei, 30 or 40 feet high, are not unfrequent in Brazil; they appear there either branched to the very top, or in the form of a candelabrum with many branches, or arranged in espalier-like rows, and are sometimes a foot and a half in diameter at the base.

In perfect contrast to the long, slender form of the candelabra-like Cacti, are the spherical genera Melocactus, Echinocactus, and Mammillaria. These plants are spherical, having warts and spines arranged in a stellate form regularly distributed over the whole surface; they are often of enormous size, and are seated directly on the barren ground, or in the clefts of bare rocks. These singular plants seem to thrive best in the most desert regions, where all other vegetation ceases; and in hot countries, where almost all vegetation disappears during the dry season, they are as fresh and green as at the time when there was the greatest abundance of water. As their succulent tissue contains a great quantity of watery sap, they are eagerly sought after and sucked by the thirsty animals, which roam over the dry deserts of South America. In kicking off the prickly coat, these animals often in-

* Nova Acta Acad. Cæs. Leop. tom. xvi. p. 344, &c.

jure themselves so much that they can no longer move about, and at last die. At times when there is a want of water, travellers usually open these juicy plants, which have been called the springs of the desert, with their knives, and thus enable the animals to enjoy the sap without danger.

These globular Cactaceæ are somewhat less widely distributed than the columnar ; their maximum seems to be in the tropical zone of the northern hemisphere, but yet they are not unfrequent in the southern hemisphere ; and even in favoured Chili, within the sub-tropical zone, are found Melocacti of extraordinary size. We are perhaps still ignorant of the greatest altitudes at which the Melocacti appear, yet I think that they do not reach any great elevation, but are there represented by the Opuntiæ ; the Pereskiæ especially ascend to an extraordinary height, almost to the limit of perpetual snow. On the shores of the Lake of Titicaca we see tall .Pereskiæ with handsome reddish brown flowers, and at yet more considerable heights there are lower forms which are armed with proportionably longer spines. On the plateau of Southern Peru, near the limit of vegetation, there are seen mounds a foot or a foot and a half high of a reddish yellow colour, which are often mistaken from a distance for couching deer. But when more closely examined, the resemblance vanishes ; those little mounds are formed of low Cactaceæ, the leaves of which are pressed close to each other, and furnished with reddish yellow spines from two to three inches in length, which cover the whole surface of the mound, and give it the reddish yellow colour. The flowers appear amongst these spines, but do not rise above them. In these desert regions, where only similar tufts of Azorella, Bolax, Fragosa, dwarf Verbenas, and Lycopodia cover the ground and the rocks,* those singular plants contribute much towards forming

* See Meyen's Reise, i. p. 453.

the character of the country. In Chili also, even on the barest rocks of the Andes, Opuntiæ grow in turf-like patches.

Just as characteristic are the great unshapely and many-branched Tunas; the Cactus ficus-indica L. and Opuntia tunas, with its long spines, are the best known forms of this group of the Cactaceæ. These plants have found their way to Europe, in the south of which they are now naturalized. In Europe, as well as in America, they are used for making fences, and there is perhaps scarcely anything which answers this purpose better, for they have even been successfully applied to the purpose of military defences, viz., as chevaux-de-frise. Their cultivation is of the greatest importance in desert and barren countries, as they delight in the very dryest soils, and not only produce an abundance of edible fruit, but also a considerable quantity of fire-wood. These Tunas hedges are also used for breeding the American cochineal insect, and are planted in some provinces of Mexico to a very great extent for this purpose, as formerly the exportation of cochineal from that country was very large.

As we have hitherto, while describing the physiognomy of vegetation, regarded only the form of the plants, without paying any further attention to their natural affinity, we must here also consider, along with the Cactaceæ, those plants which show their form without standing in any nearer relation to them. I mean the family of the Euphorbiaceæ, which is so extraordinarily numerous both in the Old and New Worlds. In the genus Euphorbia there is a group of species, some of which are quite leafless, and some put forth a few leaves at the top; and thus so exactly imitate the Cactus form, that, without a knowledge of the flowers, they might often be confounded with them. It is a very striking fact that these Cactus-like Euphorbiæ are peculiar to the Old World, while the Cacti, whose form they imitate, belong solely to the New World; we must, therefore, clearly consider this as a representation, although the representing form does not belong to the same natural family as the form it represents. This appearance of the same singular form in two very different families is the more striking, as we find very distinctly amongst these leafless Euphorbiæ all the forms which we have already considered, as those of the various genera of the Cactaceæ. Euphorbia meloformis in the Old World represents the Melocactus of the New World; Euphorbia mamillaris, the Echinocactus; Euphorbia biglandulosa represents the genus Rhipsalis; Euphorbia trigona, the three-cornered Cerei; Euphorbia clava is likest the cylindrical Cerei, and the trees of the Euphorbia lactea and Euphorbia neriifolia of India have perhaps the most resemblance to the tall Pereskiæ. We might trace out much further this extraordinary parallel between the resemblances of these two families, but this is sufficient for our purpose. Euphorbia meloformis is placed on a cylindrical stem, and is like a Melocactus growing on a Cereus.

It is also very remarkable that some forms of the Cactaceæ are found in a third family, viz., in the Asclepiadeæ; and even among the Syngenesia. The numerous Stapeliæ of the south of Africa are not without influence on the physiognomy of nature; their angular stems in miniature resemble the candelabra-like columns of the Cerei; and large handsome flowers decorate these little leafless stems, in the same way as the large flowers which cause the splendour of the Cactaceæ. The genera Sarcostemma and Ceropegia, which also belong to the Asclepiadeæ, have a few species which strikingly exhibit the form of Rhipsalis; and among the Syngenesia there are some Baccharidæ, which resemble the Phyllanthus and other Cactaceæ.

### 14. *The Succulent Plants.*

The form of plants, which comes next the Cactus

form, is that of the succulent plants, which the families Ficoideæ and Sempervivеæ present to us; it contains only bushes and herbs, and influences the character of vegetation only where it occurs in great quantities, as in the extreme south of Africa. The best known genera are Mesembryanthemum, Sesuvium, Crassula, Sempervivum, Sedum, Cotyledon, Bryophyllum, &c., the great number of whose species so frequently excites astonishment. These plants, like those of the Cactus form, make less impression by their beauty than by their singular and various forms. The beautiful flowers of the Mesembryanthema, whose numerous ray-like petals open widest during the brightest sunshine, come near the Cactus flowers, and as they generally appear in great profusion, they may very beautifully adorn the dry regions of Southern Africa. It is evident that the greatest number of the plants of this form belong to the Old World, to replace, as it were in connection with the Aloes, the Cacti of the New World. At the Cape of Good Hope, and indeed over all the sub-tropical zone of Africa, this form predominates, particularly by the genera Mesembryanthemum, Cotyledon, Crassula, &c., which have only single representatives in the northern hemisphere; on the other hand, we are less poor in species of Sempervivum and Sedum. According to Von Buch, the Canary Isles possess fourteen species of Sempervivum; and Sicily, where the Semperviva seem to be entirely wanting, has according to Philippi eleven species of Sedum. A few Mesembryanthema also appear in the south of Europe, where, along with the Erica arborea, they represent the flora of the Cape.

## 15. *The Lily Form.*

We have already, when considering the Agave form, mentioned the tree-like Lilies, which the splendid genus Yucca shows; we then placed the Yuccæ, on account of their general appearance, next the Furcrœæ, though from the structure of their flowers they belong to the Lilies, of whose form we are now to speak. With the Liliaceæ, on the other hand, are to be joined the Amaryllidæ and Irideæ, the flowers of which have not only a perfectly similar structure in parts of sixes, but which also have the greatest resemblance to them in habit, and in the shape and position of the leaves. As we have separated the arborescent form from our liliaceous plants, and as these are herbaceous plants only, their influence on the character of vegetation is not perceived, except where they grow socially and display their splendid flowers in profusion, or where they produce single gigantic flowers, conspicuous either by their contrast with the vegetation around, or by their extraordinary magnificence and beauty. The large species of Pancratium and Crinum are imposing from their extraordinary size; they grow on the coasts of India with leaves three feet in length, and their beautiful flowers, which are of the most delicate colours and the most charming forms, spread a delightful perfume over the whole country.

The chief genera, which show the Lily form in remarkable beauty, are Lilium, Tulipa, Fritillaria, Hemerocallis, Crinum, Pancratium, Alstrœmeria, Amaryllis, Narcissus, Iris, Tigrina, Ixia, Gladiolus, &c.; they are diffused over the whole globe, from the extreme north to the extreme south, and from the level of the sea to the vicinity of the limit of perpetual snow; and thus through every zone and region. But certain parts of the northern and southern hemispheres, which form low plateaux and where there is a clay soil, are particularly rich in these liliaceous plants. Such places possess abundance of moisture at certain seasons, and are there for vast distances covered with splendid Liliaceæ, which often give the ground the appearance of a rich carpet; but in a few weeks all this splendour disappears; the leaves also decay in a short time, and with the commencement of the dry season every trace of the once gay covering vanishes. But in the soil are buried the germs of the next epoch of vegetation, which lie dormant during the summer, to be again awakened to active life by the first rain. It is extraordinary how hard the soil may be baked by the influence of the rays of the sun, and yet the bulbs remain in it uninjured; Lichtenstein* saw the soil of the tarrao grounds baked almost as hard as brick, but the bulbs lying in it were defended by a number of coats from its destructive influence; and I have seen the same in the lower plateaux and declivities of the Chilian Cordilleras. In the south of Africa the Ixiæ and Amaryllidæ predominate; in South America, on the contrary, Alstrœmeria is the principal genus, the numerous species of which show the most brilliant variety of colours; and in Asia whole plains are covered with tulips.

## 16. *The Lianas, or Climbing Plants.*

Climbing plants by themselves cannot be considered as one of the principal forms which determine the character of vegetation, as they appear only in the company of other plants, most frequently of high trees, and are dependent on them; yet their influence on the physiognomy of the principal vegetation is so considerable, that when adorned by Lianas, it assumes a new and extremely lively character. The Lianas are the plants which specially aid tropical vegetation in exhibiting the extraordinary luxuriance and profuse variety of the highly extolled primeval forests of the equatorial zone. They are almost unknown in our northern regions; the Hop, our Lonicera xylosteum, and the Bryoniæ can give us only a faint idea of the luxuriance of the Lianas of tropical countries; but our Convolvulus sepium, which often grows profusely over the highest bushes, by its beautiful leaf and the size of its flowers, may give us an

* Reise im südlichen Africa, &c., Berlin, 1811–1813, i. p. 197.

idea of the way in which the tropical Ipomœæ adorn the tops of the loftiest trees. The Vine is a native of the warmer parts of the temperate zone of the northern hemisphere; it is in them the queen of the woods, for it climbs with stems from three to six inches thick to the top of the highest trees, twines round them, and binds them to each other. Yet how different are the pliant, twining Lianas of the tropics and of the warmer regions generally, which are represented by the genera Passiflora, Bignonia, Banisteria, Paullinia, Aristolochia, Cissus, Aralia, Vitex, &c. On the Orinoco, says Humboldt,* the leafless shoots of the Bauhiniæ are often 40 feet long; they sometimes fall perpendicularly from the top of lofty Swieteniæ, sometimes are stretched obliquely like the ropes of a ship, and the tiger-cats show a wonderful agility in clambering up and down by them.

As the Bauhiniæ are peculiar to the New World, so the Old World has the singular palm-form of the genus Calamus, which supplies the place of the chief lianas of the New World. The number of species of these Cane-palms, which grow in such masses in the primeval forests of the Eastern Peninsula, and on all the islands of the Indian Archipelago, must be extraordinarily great. They mount to the top of the loftiest trees, with stems many hundred feet in length, sometimes extremely slender and smooth; often of great thickness and beset with shining prickles. It is vain to search for the ends of these twining stems, for they climb from one tree to another, or return without support to the earth, thence to ascend again. Their long tendrils often twine quite regularly round each other, and resemble cables, which bind the adjacent trees together, and the force of the most violent hurricane is spent in vain on such firmly bound masses of vegetation. Even when single trees decay, they are kept for a long time upright by

* Ansichten der Natur, ii. p. 38.

the net-work of climbing plants, and when they at last fall to pieces, the mass of climbers keeps its place without the original support. Often shoots like cords, 30, 40, and 50 feet long, hang down from the branches of the highest trees, and on account of their tenacity are used for packing. Before these cords reach the ground, they wave to and fro at the slightest breath of air. Other hanging shoots, of greater thickness, strike root in the ground, and are as tight as if they had been drawn by pulleys. Von Martius, who lived for many years in the primeval forests of Brazil, and who always conceives so characteristic an idea of the physiognomy of vegetation, gives an extremely interesting account of the climbing plants of Brazil, of which he has given excellent figures in the Atlas to his Travels.* "At first," says Von Martius, "they grow upright like weak shrubs, but as soon as they reach the support of another tree, they give up their original mode of nourishment and become parasites, which spreading out on the surface of the other stem, and moulding themselves to its shape, draw their support from it at first in preference to their own root, and at last from no other source." We shall become better acquainted with this singular group of Lianas, which occur in every natural forest of the torrid zone, when we enter on the consideration of tropical vegetation.

## 17. *The Pothos Plants.*

The Pothos plants, or Aroideæ, with their large and bright green leaves, rolled together like a horn, and their beautiful, large, shining, white flowers, which glance out so mysteriously from the surrounding verdure, for the most part grow parasitically on the bark of trees in tropical forests, and influence the character of the vegetation by themselves only when they grow socially in great

* See his Reise nach Brasilien, iii. p. 32.

masses; but in general their influence is visible merely in the greater luxuriance and variety of form which they give to the trees on which they fix themselves. The Pothos plants are genuine tropical forms, but they are frequently represented in the warmer parts of the temperate zone by the genus Arum; and by means of the beautiful Calla palustris, they even reach the sub-arctic zone. This marsh plant shows on a smaller scale the exact form of the Lily-of-the-Nile (Calla æthiopica), universally known from its frequent appearance in our gardens and windows; but the gigantic Pothos and Dracontium plants of the tropical forests of America are very poorly represented by it. The latter have always very large leaves, which are sometimes arrow-shaped, sometimes divided into lobes, and sometimes feathery. The leaves of some Pothos-plants, as Humboldt has observed, extend to so enormous a size that the diachyma of the leaves is perforated by larger or smaller holes. Dracontium perfusum is an example of this.

Martius* divides the Aroideæ, with respect to their physiognomy, into three groups, which we here mention. One group grow in the earth, and send out tuberous roots, which in some countries form the chief food of the people; these attain but a small height. Another group climb up the trunks of trees, twining more or less round them, and send out great masses of air-roots in all directions, thus to imbibe a greater portion of the moisture of the atmosphere, for which purpose these roots are furnished with peculiarly hygroscopical organs. The third group is represented in Brazil by the Calladium arborescens *Vent.*; its stems are white and shining, ringed across, and crowned with large arrow-shaped leaves, and they stand in close rows like palisades on the margins of sheets of water.†

* Reise, iii. p. 19.

† Ibid, iii. p. 19, and Tab. i. viii. 2*.

## 18. *The Orchideæ Form.*

The family of Orchideæ, which is distinguished above other families, as well by the greatest variety of forms as by the richest colours, reaches its maximum in those hot regions of the globe where moisture prevails in equal proportion. There the most of the Orchideæ grow on the bark of trees, often climbing up them and clinging to them with great white air-roots, as is the case with so many Epidendrons; or they fix themselves in the smallest chinks and angles of the boughs, where ever so small a quantity of soil has collected. The genera Oncidium, Stelis, Cymbidium, Vanilla, Dendrobium, Ærides, Epidendrum, &c., are the principal which, in union with the Pothos-plants and Lianas, adorn the primeval forests of the tropics with so astonishingly luxuriant a vegetation; for they enliven the smoothest trunks and the bark of the gigantic trees of these forests, which is as it were carbonized by the heat of the sun and by age. The Orchideæ of the colder zones grow in the earth and exhibit the great variety, which is a characteristic of this family, only in the forms of their flowers; Cypripedium calceolus is the only orchideous plant which, with us, shows anything of the tropical luxuriance of this family. "These flowers," says Humboldt, "sometimes resemble winged insects, sometimes birds, which the perfume of the honey has allured. The life of a painter would not be long enough to delineate all the magnificent Orchideæ which adorn the deeply excavated mountain valleys of the Peruvian Andes." And it is certain that there is as great a profusion of these plants in the damp woods of India, while they are almost entirely wanting in the South Sea Islands.

The insect-like form of our beautiful Ophrys has been long remarked, but the tropical Orchideæ show much more singular forms in their flowers; they have sometimes even feathery leaves and papilionaceous flowers.

## 19. *The Moss Form, and* 20. *The Lichen Form.*

Insignificant as are the little plants which are known under the name of Mosses, they are in certain regions not less important to the character of vegetation than the Orchideæ and Aroideæ in tropical countries. In the forests of the tropics the trees and rocks are covered with splendid Orchideæ and large-leaved Aroideæ, and in the same situations in the north appear the Mosses and Lichens, which present us, as it were, with a miniature picture of the rich profusion of tropical vegetation. When we visit the damp, shady woods of our regions, we find the trees often completely covered with these Cryptogamia; the Mosses form a thick matting on which other plants again frequently take root. The beautifully coloured Lichens, which cover the bark of the trees, as well as the surface of the rocks of our fields and mountains, and particularly the hanging Usneæ, have sometimes a very agreeable effect; but it is in the highest degree monotonous when Lichens or Mosses grow socially in vast quantities over wide tracts of country. Cenomyce rangiferina, Ceteraria islandica, Ceteraria spadicea, and several other Lichens, grow in this way in the north, often allowing no other plants to appear amongst them. In the arctic zone of North America there are some Gyrophora, which serve as the miserable food of the animals, and of the people also in times of dearth. The Mosses likewise sometimes grow on the ground in large patches, as, for instance, the well known Turf-moss, the genus Sphagnum, Dicranum glaucum, &c.; and when they grow on the thatched roof of a cottage, they give it a venerable aspect. The dampest places within the tropics are also rich in Mosses and Lichens; but it is rather the lower Jungermanniæ which grow there in such great numbers; they fasten even on the leaves and stems of other parasites, and by their exceedingly pretty forms give them a peculiar character of beauty.

### *B. General Phytogeographical Division of the Surface of the Globe, according to the Physiognomy of Vegetation.*

Now that we have learned to recognise the chief forms of plants which, principally by their general effect, constitute more or less distinct groups, we may proceed to the geographical division of vegetation. And as the presence of plants is closely connected with the distribution of heat over the globe, and as the distribution of heat from the equator to the poles runs in a certain parallelism with that from the level of the sea to the limit of perpetual snow, the vegetation must also be divided, first, according to the various zones, and secondly, according to the various regions; and then a similar parallelism will be clearly seen in the vegetation of the corresponding zones and regions.

There have already been made by other writers several geographical divisions of the vegetation of the earth, which, however, are founded on entirely different principles. Willdenow,* R. Treviranus,† De Candolle,‡ and Schouw,§ have laid down such divisions. Willdenow set out from the hypothesis that every primitive mountain has its peculiar plants, and that there are, therefore, as many chief floras or phytogeographical kingdoms, as there are primitive mountains. These plants are said to have descended from the mountains to the plains, and thus the earth has been peopled by them. The untenableness of such views is now universally proved by our more accurate knowledge of the stations of plants, as

* Allgemeine Bemerkungen über den Unterschied der Vegetation auf der nördlichen und südlichen Hemisphäre in den, ausser den Tropen gelegenen Ländern. Magazin der naturforschenden Freunde. Berlin 1811, St. 2, p. 96, and in several earlier works, as in Usteri's Neuen Annalen, St. 16, 1797, &c.

† Biologie, &c. ii. p. 85.

‡ Géographie botanique. Dictionnaire des sciences naturelles, t. 18, p. 411.

§ Grundzüge einer allgemeinen Pflanzengeographie, 1823, page 504.

well as from the more correct geographical views of the present day.

M. De Candolle and Schouw, on the contrary, in their division of vegetation into various geographical kingdoms, took the predominance of this or that characteristic form or family of plants as the principle of division, and then designated these different floras by the names of particular portions of the earth, or by the prevailing forms of plants, which characterize the region. To avoid arbitrariness as much as possible, Schouw fixed the conditions to be observed in laying down such a phytogeographical kingdom, viz.: there must belong to the portion of the globe which is to form a phytogeographical kingdom, at least a half of the known species; there must also be at least one fourth of the genera, either quite peculiar to, or so prevalent in this region, that they can only be considered as represented in other regions; and, lastly, individual families must be peculiar to this portion of the earth, or at least reach their maximum in it.

These phytogeographical kingdoms Schouw again divided into provinces* according to the smaller differences of the vegetation; one-fourth of peculiar species and some peculiar genera are sufficient to form such a province.

The whole geographical division of vegetation, according to Schouw, is the following:—

1. Kingdom of Saxifragæ and Musci (Alpine arctic flora.)
   *a.* Province of Carex (Arctic flora.)
   *b.* Province of Primulaceæ and Phyteumæ (Alpine flora of the south of Europe.)
2. Kingdom of Umbelliferæ and Cruciferæ.
   *a.* Province of Cichoraceæ (flora of Northern Europe.)
   *b.* Province of Astragali, Halophytæ and Cynarocephalæ (flora of Northern Asia.)

* L. c. p. 507.

3. Kingdom of Labiatæ and Caryophylleæ (Mediterranean flora.)
   - *a*. Province of Cistus (Spain and Portugal.)
   - *b*. Province of Scabiosæ and Salviæ (south of France, Italy, and Sicily.)
   - *c*. Province of the shrubby Labiatæ (flora of the Levant, Greece, &c.)
   - *d*. Province of the north of Africa.
   - *e*. Province of Semperviveæ.
4. The eastern temperate part of the Old World (probably the kingdom of Rhamneæ and Caprifoliaceæ.)
5. Kingdom of Asteriæ and Solidagines.
6. Kingdom of Magnoliæ.
7. Kingdom of Cacti, Piperaceæ and Melastomaceæ.
   - *a*. Province of the Ferns and Orchideæ.
   - *b*. Province of Palmæ.
8. Kingdom of Cinchonaceæ.
9. Kingdom of Escalloniæ, Vacciniæ, and Winteriæ
10. Chilian Kingdom.
11. Kingdom of Arborescent Compositæ.
12. Antarctic Kingdom.
13. New Zealand Kingdom.
14. Kingdom of Epacrides and Eucalypti.
15. Kingdom of Mesembryanthema and Stapeliæ.
16. Kingdom of Western Africa.
17. Kingdom of Eastern Africa.
18. Kingdom of Scitamineæ.*

* Schouw (Momente zu einer Vorlesung über die pflanzengeographischen Reiche. Linnæa viii. p. 625,) has in a later work of 1833 added to these phytogeographical kingdoms the seven following:—1. Himalayan kingdom (Wallich's kingdom) embracing the table land of India from 4000 to 10,000 feet in elevation. Now that we have so beautiful a work on the table land of India, we know that its vegetation by no means justifies the formation of a separate kingdom; and the same may be said with regard to the second, viz., the Alpine Japanese kingdom. 3. Polynesian kingdom (Reinwardt's kingdom). 4. Oceanic kingdom (Chamisso's kingdom). 5. Kingdom of the Balsam Trees (Forskal's kingdom). 6. Kingdom of the Deserts (Delile's Kingdom). 7. Kingdom of tropical Africa (Adanson's kingdom). 8. Kingdom of the table land of Mexico. 9. West Indian kingdom (Swartz's kingdom). I may here be permitted to remark

On reviewing this geographical division of vegetation, we find that half of the kingdoms are founded on those genera which appear to be characteristic of a certain country, and by their peculiar form determine the character of the vegetation, and also in a great measure the physiognomy of nature. Accordingly, these divisions coincide with the forms of plants in respect of their physiognomy, which I have treated of in the preceding section. A statistical classification of vegetation, if I may so express myself, is a very different thing from a physiognomical, in which the form and general effect of the plants determines everything. The former cannot lay claim to any degree of exactness until the greater number of the plants belonging to each country is known, while the physiognomics of plants has already attained its end, though there are still extensive tracts of country which have been little or not at all examined botanically; we may multiply the groups already formed without injury to the earlier ones, and enrich and justify them by new discoveries.

The general impression which the vegetation of a country makes upon us, by no means depends on the number of species and genera of plants, but on the mass, habit, and exact distribution of them. Accordingly, in order to be able to characterize more exactly particular portions of the earth, I have first specially considered the principal forms of plants, so far as they, by their masses, influence the character of vegetation; and now that the physiognomics of plants has prepared us for it, we proceed to the geographical division of vegetation, in which the respective appearance of different forms in different zones and regions shall serve as the basis of our considerations.

---

that Schouw has by no means laid down these kingdoms according to the rules which he had previously given, and that by proceeding in this way some twenty other kingdoms might be founded with equal reason.

### *a. Division of the Horizontal Range of Vegetation into Zones.*

The common astronomical division of the surface of the globe into three zones, the torrid, temperate, and arctic, is not sufficient for the purposes of botanical geography, for these zones are too extensive, and therefore include many various forms of vegetation, which might be more exactly marked out by smaller zones. I have therefore divided each hemisphere into eight smaller zones, though I have taken as the foundation of these the common division into three. We shall afterwards point out how these zones are characterised by their peculiar vegetation, and how they are repeated with the increasing height of mountains, in exact correspondence with the parallelism between the decrement of heat from the equator to the poles, and from the plain to the peaks of the mountains.

We begin with the description of the torrid zone, and only further remark, that all divisions of this kind are attended by great difficulties, as the ranges of the separate forms of plants are never so exactly defined as the limits of our zones must be, but at their frontiers run into each other.

### 1. *The Equatorial Zone.*

The equatorial zone embraces on both sides of the equator a zone of 15 degrees of latitude, and has a mean annual temperature of 26° or 28° Cels.; a heat which, in union with as high a degree of atmospherical moisture, calls forth an extraordinary profusion of vegetation, rendered more beautiful by the greatest variety of form and the most brilliant colours, and leaving a strong impression on the mind of every man of feeling.

The plants here are more juicy, their foliage seems fresher, and their stems are immensely strong. In this torrid zone, wherever local conditions do not oppose the invincible vegetation by the decrease of the powerful exciting causes, heat and moisture, there are developed those indescribably vast masses of plants, the delineation of which has been attempted by accomplished naturalists and distinguished artists.

The scale of the vegetation of the primeval forests of the equatorial zone is great in every respect ; trunks of enormous thickness rise more than 80 and 100 feet in height, their tops so closely interlaced, that not a sunbeam can reach the rich soil underneath, which is generally so thickly covered with lower plants, that one cannot take a step without first, axe in hand, hewing out a path. The atmosphere of these forests, where vapour is continually ascending, is oppressively hot and damp, and not unfrequently the air is filled with visible aqueous vapour. The shrill pipe of the large cicades high up in the tops of the trees, and the loud croaking of the horrid vampyre, the flying dog, and the bloodsucker, often for days accompany the wanderer in the forests of India.

The forms of Palmæ, Musaceæ, the arborescent grasses, Pandanus, Scitamineæ, Orchideæ, and Lianas, which are generally prevalent in the equatorial zone, determine the character of the vegetation ; but the Cotton-trees (Bombaceæ), with their gigantic trunks, often covered with warts and prickles of a peculiar nature, and the Fig-trees, whose trunks likewise swell to an enormous size, have also a considerable share in forming the character of the natural forests both in the Old and New Worlds, along with the Swieteniæ, Caesalpiniæ, Malpighiaceæ, Anonæ, Anacardiæ, Bertholletiæ, and Lecythideæ, for the New World ; and the Sapindæ, Caryotæ, Artocarpi, Sterculiæ, Ebenaceæ, Meliaceæ, Laurineæ, and many others, for the Old World. The immense diameter of some tropical trees excites universal astonishment. The Baobab, or Monkey-bread tree (Adansonia digitata L.), a native of Senegal, the Cape de Verd Islands, and even Egypt and Nubia, is undoubtedly one of the thickest trees. Trunks of it have been measured and found to be 77 feet and upwards in circumference, and the hollow in the interior is sometimes so considerable, that several negro families take up their abode in it. The Bombaceæ of the Old, as well as the New World, produce exactly similar unshapely trunks; from their excessive development of pith, they increase prodigiously in thickness, and lose the common cylindrical shape, instead of which they resemble huge casks 30 or 40 feet in height, and of proportionate circumference.* Not less astonishing is the immense height and bulk to which the plants of the Mimosa form, Swieteniæ, Hymeneæ, Caesalpiniæ, and several others, sometimes grow. But tropical vegetation is far from being exhausted by the production of such masses; the greatest variety and beauty of form, as well as the most brilliant colours and the most fragrant perfumes, are called forth by the glowing rays of the sun, and the excessive humidity of the atmosphere. As in the forests of the north, the bark of the trees is covered with humble mosses and lichens, so the trees of the tropical forests exhibit a profusion of the most luxuriant and beautiful forms; splendid Orchideæ grow in the clefts and crevices of the bark, on which climb Pothos plants, whose shining white flowers stand out from the midst of the beautiful bright green leaves. The most elegant forms of ferns, which sometimes belong to our well-known genus Polypodium, but chiefly to the tropical genera Hymenophyllum, Trichomanes, &c., climb up the trunks like our ivy; or they grow in tufts, which fasten on the branches, and contrast in a peculiar manner with the foliage. In the forests of the Philippines, there is a large, handsome Polypodiacea, which resembles Polypodium quercifolium, and sometimes entirely covers with its thick, bright brown, scaly roots, a large part of the branch on

* See v. Martius' Reise, iii, p. 29.

which it grows; this beautiful plant, the fronds of which are two and three feet long, the sooner attracts the eye, as its root-leaves, and indeed all the barren fronds, are of a bright yellow colour, contrasting very peculiarly with the reddish-brown roots, and the dark green around. If any little spot of the bark is still left vacant, lichens, mosses, and Jungermanniæ fasten on it; indeed, not contented with so limited a space, the prettiest forms of the latter, the beauty of which the microscope often first reveals, grow even on leaves of the other parasites. The leaves of the Orchideæ, in the natural forests of the islands of the Indian Archipelago, are rarely to be found without these Jungermanniæ; and even lichens and ferns are overgrown with them.

But not only do the trunks of the trees serve as the support of so luxuriant a vegetation, but high amongst the foliage are seen the scarlet flowers of Loranthus, shining Tillandsiæ, Pitcarniæ, and a whole host of climbing plants, which, taking root in the ground, at first twine up the trunks and branches, but afterwards forsake their parent soil, and continue to grow as parasites. Von Martius, during his long abode in the primeval forests of Brazil, has traced with extraordinary acuteness the manner in which these singular plants grow, and his description will give the best idea of it.* The stems of these parasites have a singular inclination to throw off their bark wherever they are irritated by contact, and to spread themselves like a fluid gradually over the foreign body. In the same way the branches of the parasites by degrees coalesce. In this process the strength of the original root is weakened, and as a counterpoise the stem sends down air-roots, and thus this tenacious and vigorous race continually gain fresh strength and space, by the destruction of their neighbours. We find this mode of growth in plants of the

* Reise, &c., iii. p. 32.

most different families, but most developed in many species of Guttifera. In tho forests of Brazil there aro Clusiæ, Havettiæ, Arrudaeæ, and the allied genera Ruyschia, Norantea, and Marcgravia, which by the confluxion of their branches and stem, bind themselves in the closest manner to the wood of the tree which supports them. On the banks of the Rio Guama, Von Martius saw whole rows of the Macauba palm (Acrocomia sclerocarpa *M.*) so overgrown by Clusia alba, that the parasite had formed round the trunk, which was 30 feet high, a cylindrical tube, which bore leaves and flowers on short branches, and from the top of which rose the noble crown of the palm. In the natural forests of the island of Luçon, I have seen similar envelopes of a sort of net-work of flattened Fig-trees, which grew like a trellis over the thickest trunks of other trees, and the origin of which seemed to me at first incomprehensible. I have already, in several parts of this work (page 151), mentioned the manner in which these climbing plants, like interlacing cords, bind the trunks and foliage of trees growing near each other so closely that the force of the most violent storm cannot separate the united masses. In the forests of the New World it is chiefly the Bauhiniæ, Paulliniæ, and Banisteriæ which form these living ropes, which often produce neither leaves nor flowers for a length of 30 or 40 feet ; and in the Old World, the Passifloræ, Aristolochiæ, and especially the Rattan-canes (species of Calamus). Large flowers of remarkably brilliant colours are indeed characteristic of the Lianas, or climbing plants ; but their pliant stems mount to the very top of the trees, and often a fallen flower or a peculiar odour first betrays the presence of these rare beauties, to which the botanist often looks up in vain.* Trees

* It is yet to be remarked as a peculiarity of the trees, and in general of the plants of the tropical forests, that, according to several travellers, for instance Wydler, Auguste de Saint Hilaire, and Pohl, they very seldom flower, and that they propagate themselves by suckers. Trees which

must be felled in order to reach the flowers of their parasites, for the trunks are either too thick, or armed with rough warts and thorns, or overgrown with prickly climbers, and refuse to allow the adventurer to ascend; while the Lianas, whose tight-drawn ropes would be such an assistance in climbing, are dreaded by the natives even, for their acrid juices and their unwholesome exhalations. The flowers of the Aristolochiæ are famous for their extraordinary size; on the shady banks of the Magdalena, in South America, Humboldt discovered Aristolochia cordifolia, the flowers of which are four feet in circumference, and are often worn in play as caps by the Indian boys; and the Aristolochia gigantea of Von Martius has flowers almost a foot long.

But the vigour of tropical vegetation is still not exhausted by these aërial gardens of such luxuriance and brilliancy, for from the roots spring up a variety of plants, often of gigantic size and of singular form, announcing as it were their obscure origin. The Rafflesiæ and Brugmansiæ of the Indian Archipelago are in appearance like large flowering mushrooms; the Rafflesia or Giant-flower attains a diameter of three feet. The tropical forests of America, the South Sea Islands, and even Africa according to some accounts, are rich in Balanophoræ of the most varied forms and colours.

In proportion to the majestic beauty of a primeval forest, is its fearful grandeur when in combat with the wild elements. To be in such a forest during a violent hurricane is described as more fearful than to struggle with the raging waves in the open sea; and even far less violent storms produce sublime spectacles. When the boisterous wind catches hold of the tops of the gigantic trees of these natural forests, and shakes the branches and trunks against each other, the air is filled

---

are incessantly growing and putting forth leaves and twigs, very seldom flower. A Qualea gestasiana remains barren for five years after once flowering.

with a fearful rushing, thundering, rattling, and crashing; even the strong Lianas are torn asunder, and the broken branches and stems fall to the ground. Great quantities of the parasites are thrown down from their lofty situations, and the trees are stripped of their fruit, which, generally cased in a hard shell, falls to the ground with a loud crash. The rain, at first warded off by the thick canopy of foliage, now falls in so much the greater masses, and adds to the horrors of the moment; almost all the inhabitants of the forest betray their fear by mournful howling and crying; the apes, the large bats, and the whole host of birds call loudly all together, and the croaking of the tree-frogs and others of this family, sometimes like the sound of a drum, discloses the great misery of the moment. The insects only, which long before announced the coming uproar, are now silent, and keep close on the under surface of the leaves until all is over, and the sun again shines brightly out.

Such are the wonders of the primeval forests of the equatorial zone; near their edges, and on the banks of lakes and rivers, the vegetation is not so rank, but therefore the more beautiful. Groves of the lower trees surround these sheets of water and open spaces, above the foliage of which rise majestic palms, sometimes, as the Piriguao at the mouth of the Guaiviare and Atabapo, adorned with the most beautiful fruit. This palm grows to the height of 60 feet, with sedge-like tender leaves, crisped at the edge, and bears a fruit resembling a peach, 70 or 80 of which hang down in enormous clusters, and offer a nutritious food to man. In the island of Luçon, on the banks of the rivers, where such masses of vegetation abruptly terminate, they are adorned by splendid climbing plants; elegant ferns, viz., a large leaved Lygodium, hang down from the tops of the trees in festoons 40 and 50 feet long, and panicles two and three feet in length hang from the flower-stalks of Bignonia grandiflora.

On the edge of such a grove the Indian usually erects his fragile hut; some bright green plantains, and the slender palm shooting up from the dark foliage of the neighbouring fruit trees, even at a distance indicate the dwelling of man, the erection of which, in India at least, is so much facilitated by the proximity of the arborescent grasses.

The noble form of plants which we have described under the name of the arborescent grasses, and which is most commonly represented by the genus Bambusa, occurs socially in the equatorial zone, and there forms woods as extensive as those of the Coniferæ in the northern zones. The Nipa palm also in social masses covers for miles the coasts of the islands of the Indian Archipelago; frequently bordering on the extensive Mangrove forests, in which the Mangrove, Avicenniæ, Bruguieræ, Dodoneæ, Tournefortiæ, &c., grow socially together in great numbers. The highest trees of these Mangrove forests are usually not more than 40 or 50 feet high; they are green throughout the whole year, which indeed is universally the case with the trees of damp tropical regions.

Such as I have here attempted to describe it, would be the vegetation of every part of the equatorial zone, if there, as well as in our zones, disturbing causes did not oppose the regular course of nature. In vain do we seek in the Savannahs of the Orinoco, in the Pampas of Southern Peru, or in the deserts of Africa, for that luxuriant vegetation which I have described as belonging to the equatorial zone. The degree of heat which these parts of the globe receive in consequence of their relative position to the sun, is the same in every longitude; but the variation in the supply of water is so great, that it causes the most striking differences.

I have already given an explanation of the causes, on which rests the difference between coast and continental climate, and therefore I may now refer to it; the same theories explain the great heat in such dry tropical regions during the day, and the great cold produced by radiation during the night. In those regions where the atmosphere and soil are destitute of the proper degree of humidity, there is a striking change in the relations of the different seasons. In summer, when with us vegetation is in its greatest splendour, it is dead in the parched countries of the tropics; the trees lose their leaves, and the herbs vanish without leaving a trace of their existence, from mere want of moisture; just as vegetation ceases with us in winter from want of heat, until the first warmth of spring calls it back to life. The light forests of Brazil (Catingas), which exhibit that singular phenomenon for the tropics, the fall of the leaf, and stand during the hottest season stripped of all their foliage, have been described at length by celebrated travellers; but this phenomenon is general, indeed universal, in all parts of the torrid zone, under similar circumstances. In the parched regions of the western coast of Peru, I have seen not only the native trees leafless during the hot season, but also our European fruit trees, which had been introduced there. We saw our Fig-tree standing next the Schinus, both leafless and apparently dried up; the fruit on the Schinus, and the thick buds on the boughs of the Fig-tree, alone gave token of the slumbering activity of these trees, which strikingly harmonized with the dead and perfectly bare country around. I cannot better describe these leafless woods of the tropics than by quoting a passage from Von Martius' Travels: "Every thing around us," says this celebrated traveller, "was stamped with a peculiar, and to us a strange character, and filled the mind with sadness. The thick forest seemed to us a vast tomb, for the dry season had stripped from it all the leaves and flowers; only here and there thorny species of Smilax or long shoots of Cissus, bearing a few leaves, climbed on the trees, or the stately flower spikes of the Bromeliæ rose up amongst the branches; but only the more visible seemed the trees in their whole immense extent, stretching out their branches, like gigantic arms, in the dark blue æther. Thorny Acaciæ, many-branched Andiræ and Copaiferæ, and milk-white Fig-trees were very frequent; but what struck us most, were the gigantic trunks of Chorisia ventricosa, which, contracted at the top and base, swelled out in the centre like enormous casks, and whose corky bark was set with strong, shining brown thorns. Here hung great bunches of parasitical mistletoe, generally distributed in such a way by careful mother Nature, that the female bushes were lower than the male. There myriads of ants had hung on the trees their dwellings full of labyrinths, which were several feet in circumference, and by their black colour singularly contrasted with the bright grey of the leafless branches. The autumn-like wood resounded with the cries of various birds, chiefly croaking Araras and Paraquitos. Shy armadillos and ant-eaters met us amongst the high mounds thrown up by the busy ants, and sloths hung stupidly on the white branches of the Ambauba (Cecropia peltata), which grew here and there amongst the other trees. Troops of apes were seen at a distance. The high dry grass was covered with crawling balls of little Carabatos, which, if we accidentally touched them, spread over us as quick as lightning, and caused a disagreeable itching."*

The deserts of the torrid zone, the physiognomy of which Humboldt has characterised so comprehensively,† exhibit still more striking phenomena. These more or less uniform plains are to be considered as local phenomena, the origin of which seems to be connected with the great geological revolutions, which have finally determined the configuration of the present surface of the globe.

Some of these tropical deserts, which consist of shift-

* Reise in Brasilien, ii. p. 499.

† Ueber die Steppen und Wüsten; Ansichten der Natur.

ing sand, are quite destitute of water, and neither rain nor vegetation is ever observed in them; such are the vast sandy deserts of Africa. Others of these plains are covered with a thin coating of soil, and exposed to the influence of periodical rains; they have a very different aspect at different seasons of the year; in the equatorial zone of America, for instance, they are during the dry season as desert as the Lybian wastes, but during the wet season they are covered with luxuriant grasses and the smaller Mimosæ.

## 2. *The Tropical Zone.*

This zone stretches on each side of the equator from the 15th degree of latitude to the tropics, and shows a mean heat of from 23° to 27° Cels. Some of the many instances of variation from this mean heat, to be found in this zone, I have mentioned in the first section of the book (see p. 18). In countries where the monsoons prevail, a summer heat of 27° and 28°, or even 30° is common, while in winter the temperature sometimes falls below the freezing-point. The mean annual heat of Canton is 17.5° R.,* or 21.87° Cels., while the mean summer heat is 22.2° R. (27.7° Cels.), and the mean winter heat 12.1° R. (15.1° Cels.). I have already spoken somewhat at length, in the first Part (pp. 9 and 18) of the variations in the progress of temperature at some of the chief points which lie near the tropic of Cancer.

As the course of the isothermal lines, as we have already seen, continually undulates, and here and there suddenly sinks and rises, we shall be able to point out some places in the tropical zone, where are found all the conditions which we formerly enumerated as belonging to the equatorial zone. As instances of this I might adduce the districts round Rio Janeiro and

* See Meyen, Bemerkungen über das Clima des südlichen China. Nova Acta Ac. C. L., vol. xii. p. ii. 903.

Canton, where we certainly find very little difference in climate and vegetation from the equatorial zone.

Besides the Palms, Musaceæ, Scitamineæ, Meliaceæ, Anonaceæ, Sapindaceæ, Orchideæ, and the plants of the Pothos form, Lianas, and others, which especially belong to the equatorial zone, and determine the character of its vegetation, the Ferns, Convolvulaceæ, Melastomæ, and Piperaceæ, are still more predominant towards the limits of the torrid zone. The arborescent ferns characterize this zone, just as the Palms, along with the Scitamineæ form, were specially proper to the region of the equator. A comparison of the vegetation of the Sandwich Islands with that of the Philippines, would alone be sufficient to enable us to comprehend the striking contrast between the imposing grandeur of vegetation in the equatorial, and its luxuriance in the tropical zone. In the forests of the Sandwich Islands, there is no want of profuse vegetation; an Acacia (A. heterophylla), and the splendid Aleurites triloba there form trees of immense circumference, and one cannot take a single step without first cutting down a path. An endless mass of arborescent ferns, Pandani, and Scitamineæ, is so closely interwoven by numerous plants of various species of Ipomœa, that all the climbing plants must be torn down in order to force a path. In these forests, there is also a great quantity of underwood, while in the forests of the equatorial zone the parasitical flora rather predominates, the Orchideæ and the Pothos plants, and especially the Lianas, which grow high amidst the foliage of the trees. In the thicker woods of the Sandwich Islands, the Pandani and plants of the Ananas form appear in greater numbers; they climb up the trunks of the trees, and surround them with hundreds of branches, so that their foliage is impenetrable, and the traveller must pursue his journey upon this matting of vegetation: at last he walks without perceiving it at a height of eight and ten feet above the surface of the earth, and it is only when he reaches a fissure in this mountain of plants that he can survey the enormous mass. Thick trunks of trees, decorated by variegated lichens,* together with the magnificent ferns which are grouped upon them, present the most beautiful sight which a botanist can desire. Immense Aspleniæ, the largest variety of Asplenium nidus, the leaves of which are from two to three feet long, and proportionably broad, little species of Pteris with linear lanceolate leaves, Piperaceæ in the greatest profusion, exquisite Jungermanniæ, foliaceous mosses, &c., and all growing on the same tree; what a magnificent sight! The singularly shaped Charpentiera obovata *Gaud.* hangs its bunch of flowers negligently over the arborescent Lobeliaceæ, and the great profusion of Urticeæ, the large leaves of which are more or less conspicuously covered with white hairs, here characterize the lofty vegetation, while the ground is entirely covered with ferns four, five, and six feet high. The beautiful and large tree Metrosideros polymorpha and Jambosa malaccensis, close to which pretty Dracænæ and tall, wild plantain-trees so frequently shoot up, contribute not a little to the beauty of the forests of the Sandwich Islands. The masses of the splendid scarlet flowers of these trees, so often visited by little humming birds, beautifully contrast with the white hairy foliage of the surrounding Urticeæ.

It is remarkable that Orchideæ are totally wanting in the luxuriant vegetation of the Sandwich Islands, and the Umbelliferæ also are extremely rare. I may here take the opportunity of noticing a singular peculiarity in the fauna of these islands. It is well known that in the damp forests of the equatorial zone in the Old World, as well as the New, insects are extraordinarily numerous, so that one can seldom turn up the leaves of a branch without finding some beetles, &c., on them;

* Parmelia perforata var. melanoleuca and var. ulophylla, Usnea australis Fr., Sticta lurida n. sp., &c.

but in the Sandwich Islands insects are almost entirely wanting, and, singularly enough, they are, as it were, replaced by land-snails, which are so numerous that there is scarcely a little plant, or a branch of a tree which has not several of them upon it.*

Unfortunately, great part of the countries within this zone lies under such conditions that, from want of sufficient moisture, and from having too poor a soil, they are destitute of almost all the beauties of tropical vegetation. We found this to be the case on the western coast of South America, where in the latitudes of this zone there is the most miserable vegetation that can be imagined. A very few palms, some Acacias, and tropical fruit-trees, are the only signs which betray the position of the country. The extreme south of China, the northern part of the Philippines, Cochin China, &c., also lie in the northern tropical zone, and show some deviations from the vegetation of the equatorial zone, but here what is characteristic is often suppressed by the influence of the monsoons ; indeed, in China and Cochin China, culture for ages, and the dense population, have so modified the vegetation that little that is characteristic remains. Among the peculiarities of the vegetation of these countries is the social growth of the Chinese pine, which forms woods as extensive as those of our common pine. These pine forests strikingly contrast with the airy groves of the arborescent grasses which the Bambusa arundinacea forms over such extensive tracts, and which are such an ornament to the landscape. The woods of the arborescent grasses in the Old World continue almost without interruption to the equator, but in the equatorial zone, in place of the pines, appear the Casuarinæ, which extend to the southern tropic. Cupressi also grow in the tropical zone of India along with these pines and Casuarinæ, and even in New Caledonia they occur along with the latter

* See Meyen's Reise, ii. p. 142, &c.

plants. This large island, which lies within the southern tropical zone, can boast of but little of the luxuriant tropical vegetation which every traveller has found in the south of Brazil and in India, but, on the contrary, it is bare and even may be called destitute of wood, though in a few places, where there is probably a more abundant supply of water, appear many of the beautiful tropical plants which we have mentioned in the preceding section.*

Mangrove forests in this island also grow on the shores, and near them occur remarkable Fig-trees, whose beautiful foliage is so thick that even the burning rays of the sun cannot penetrate it, and it thus offers the inhabitants an agreeable shade, which is rendered still more delightful by the melodious song of a number of birds. These Fig-trees, says Forster, have an extremely singular form, for their trunks, at a height of fifteen or twenty feet from the ground, rest on a number of long roots, which are like cords, descend perpendicularly to the earth, and are also as round as if they had been turned, and as elastic as the string of a bent bow. The cocoa-nut, yam, arum, plantain, and sugar-cane, are the chief plants which are cultivated for food; but the soil is so unfruitful that the natives are sometimes obliged to content themselves with the roasted bark of trees, for instance, of Hibiscus tiliaceus. The splendid Melaleuca leucodendrum, one of the group of the Protea form, grows here in great quantities, and its bark is used to cover the inside of the walls of the native huts. A beautiful example of the Myrtle form, viz., an Eugenia, also grows in New Caledonia, and rows of it, as well as of the plantain, are planted between the fields of the yam, arum, and sugar cane.

The West Indian Islands appear to be very rich in Ferns and Orchideæ, and have therefore been considered as distinct provinces of the American flora.†

* See Cook's Second Voyage, ii. p. 309, &c.

† See Schouw, l. c. p. 516.

We must regret that as yet only a few points of this zone in the Old World have been described in such a manner as to enable us to judge of the physiognomy of its vegetation.

### 3. *The Sub-tropical Zone.*

The sub-tropical zone stretches, in both hemispheres, from the limits of the torrid zone, that is, from the tropics to 34° of latitude. It embraces a tract of land, the inhabitants of which rejoice in the happiest clime; the mean temperature of this zone is from 17° to 21° Cels., but it has a summer temperature of from 23° to 28° Cels., and therefore a number of the tropical fruits, and many annuals, which properly belong to the equatorial zone, succeed in it. The winters also are so mild, that man scarcely needs a substantial dwelling to shelter him from the rudeness of the climate.

We shall immediately see that the sub-tropical zone of the northern hemisphere is much less known than that of the southern hemisphere, and that the peculiar configuration of the land in the southern hemisphere is particularly advantageous for making comparative inquiries into the flora of the sub-tropical zone.

For the sub-tropical zone of the northern hemisphere, we have first an accurate knowledge of the vegetation of the Canary Isles, obtained from Leop. v. Buch.* In this excellent work there is not only a very complete flora of the Canary Isles, but the native and introduced plants are accurately distinguished from each other; their stations in different regions, and those which are common to the flora of the neighbouring continent exactly given, as well as excellent general phytogeographical descriptions.

In the sub-tropical zone the vegetation is green throughout all the year, like the forests of the damp

* Physicalische Beschreibung der Canarischen Inseln, Berlin, 1825, 4to.

regions of the torrid zone. From the great heat of the sun, palms as well as bananas grow here in the plains; indeed in Egypt the banana is cultivated in gardens to the 34th degree of latitude, while Cucifera thebaica, that remarkable dome-palm with a branched trunk, reaches only the 30th degree.* The date-palm belongs to the whole western part of the sub-tropical zone of the Old World beginning on the Canary Isles; in India, however, for example between Delhi and Serampore,† Phœnix sylvestris, and Phœnix humilis appear as its representatives. But in North America, for instance at New Orleans, Chamærops palmetto grows socially in the marshy plains over extensive tracts, and sometimes reaches a height of six fathoms.

The appearance of a number of succulent plants which belong to the genera Sempervivum, Aizoon, Cotyledon, Crassula, Mesembryanthemum, Portulaca, &c., is one of the chief peculiarities of the sub-tropical zone, in which the Canary Isles lie; the genus Sempervivum has here even arborescent species, such as Sempervivum arboreum in the island of Madeira, which have an exceedingly strange character. But the most peculiar plants of all are the arborescent Euphorbiæ, which, with their prismatic succulent stems, here imitate the Cacti of the New World. Euphorbia balsamifera, the milk of which is so innocuous and sweet that it is thickened to jelly and eaten by the inhabitants, is a very remarkable tree, which Von Buch has fully described.‡ "The trunk, though very crooked, at first rises without branches, but afterwards divides into a great number, which again divide into innumerable smaller branches. Leaves are

* See N. Bove, Relation abrégée d'un Voyage Botanique en Egypte, dans les trois Arabies, en Palestine, et en Syrie. Ann. des Scienc. Nat. 1834, tom. i.

† Royle, Illustrations of the Botany and other Branches of the Natural History of the Himalayan Mountains and of the Flora of Cashmere. London, 1833, Fasc. i.

‡ L. c. p. 115.

nowhere to be seen but at the extremity of the branches round which they are placed. They are short, lanceolate and narrow, grey, and furnished with a little thorn at the point. Those which directly bear the flowers are somewhat broader, egg-shaped, paler, rather fleshy, and fall off after the flowers, &c. "But," says Von Buch, "the Cordon (Euphorbia canariensis, the juice of which is as acrid as that of the other Euphorbiæ) is a still more wonderful production of nature. Its dark green branches, completely leafless, rise all at once from a common root, bend in a semicircle down to the ground, and then rise again perpendicularly at various distances from the first root, so that the tree resembles an enormous lustre with a number of branches bearing lights. Each branch is as much as half a foot in circumference, and is a prism of four, or more generally of five sides. Their edges along the whole length are set with pairs of short prickles. At the extremity of these thick, angular, fleshy branches break forth the scarlet flowers, which at a distance are like burning coals. Higher up the older branches divide, and form separate smaller lustres on the larger one. Or the tree stands at the edge of a rock, down the face of which the branches fall in the strangest curves, and rise again perpendicularly. Or it grows on level ground, and the branches, pressed to the earth by age and weight, do not rise again till at a great distance from the central point, and hence arises the singular spectacle of a little wood of living five-sided prisms. There is nothing in this which can recall to us the ordinary form of a bush or tree; even the flowers at the extremities do not, for they might be mistaken for knobs with which these wonderful branches are set." Notwithstanding all these peculiarities of the flora of the Canary Isles, it cannot be overlooked that it bears some resemblance to the vegetation of Southern Africa, which lies in the same zone of the southern hemisphere. The numerous succulent plants which appear there, are something more than mere representatives of the flora of the corresponding zone. But it is perhaps striking that the flora of the Canary Isles shows so extremely few tropical forms of vegetation, and it would, therefore, be the more important to become acquainted with the flora of the adjacent continent of Africa, in order to learn whether the same proportion is found in it. Besides the Palm and Banana forms, we have yet to name the Dracænæ, the genera Pancratium, Saccharum, Rottbœllia, and a few others which reach their maximum in the torrid zone, and seldom pass beyond it. Of the large genus Ficus, Ficus carica alone appears, and it has been introduced. On the contrary, in the sub-tropical zone of Egypt grows the Ficus sycamorus, an extremely vigorous tree, whose trunk attains a diameter of from nine to twelve feet, and is fifty or sixty feet high. From its vigorous branches and beautiful persistent foliage, this tree affords a delightful shade.

As the flora of the Canary Isles and Madeira shows but few tropical forms, so is it with the vegetation of the western part of the Himalaya mountains, for instance of the country round Delhi, lying in the 28th degree of latitude, and at an elevation of 800 feet. In summer there is here a tropical heat which ripens almost all the fruits of the equatorial zone, while in winter the temperature is so low that often old trees of the well-known noble tropical fruits perish. During the summer, which is here at the rainy season, there are cultivated in the province of Delhi, rice, indigo, the cotton-tree, maize, Holchus sorghum, some species of Panicum, Paspalum, and Eleusine ; of leguminous plants, species of Phaseolus and Dolichos ; gourds, Sesamum, some species of Solanum with edible fruits, ginger, turmeric, Crotalaria juncea, and Hibiscus cannabinus, for the materials of clothing. But the appearance of the inhabited districts of this country is totally different in winter when the cereals of the north are cultivated, such as wheat, barley, oats, millet, and also beans, vetches, mustard, coriander, carrots, tobacco, flax, saffron, &c.*

But, as was to be expected, the vegetation of the uncultivated places, as well as of the cultivated soil, exhibits these different characters at the different seasons; that is, in summer it resembles the vegetation of the warmer zones. In winter, on the contrary, only old, well-known genera belonging to our colder part of the temperate zone appear. We then find the genera Potentilla, Campanula, Arenaria, Spergula, Lithospermum, Tradescantia, and Poa; and the following species, Malva rotundifolia, Veronica hederifolia, Fumaria vaillantii, Anagallis cœrulea, Sonchus oleraceus, Antirrhinum orontium, Silene conoidea, Saponaria vaccaria, Avena fatua, Lolium temulentum, Verbena officinalis, &c., are identical with those which appear with us; however, several of them have certainly been introduced along with our cereals.

The aquatic plants of this country, and those which grow in the vicinity of water, have also for the most part a northern character, for we find there our genera Herpestes (H. monniera), Gratiola (G. juncea), Marsilea (M. quadrifolia), Sagittaria, Butomus, Polygonum, Rumex, Trapa (T. bispinosa), Nymphæa, Utricularia, Potamogeton, Lemna and Vallisneria, and even Ranunculus sceleratus and Ranunculus aquatilis. The tropical genera among these aquatic plants, on the other hand, are Hydrolea zeylanica, Sphenoclea zeylanica, Limnophila gratioloides, Coix, Leersia, Pontederia, Nelumbium speciosum, Euryale ferox, and Damasonium indicum.

The chief plants, which in summer adorn the district round Delhi with a more southern character, are Dalbergia sisso, Acacia serissa, A. arabica, and A. farnesiana, Cedrela toona, and various species of Melia, Ficus, Morus, Trophis, Bauhinia, Cordia, Gmelina, the two species of Phœnix previously named, &c.†

* See Royle, l. c. p. 10.

† Among the shrubs and herbs I may name the following species and

In all the plants we have named we perceive a receding from the equator; there is no longer a trace of the excess of tropical forms; but so beautiful a country, the climate of which unites the advantages of the torrid and temperate zones, will, in the possession of an active nation, soon become the rendezvous of all the cultivated plants of the various zones, and even now there is grown there a variety of the beautiful fruits of the torrid and temperate zones, such as hardly another country can boast of.

Some important forms of plants, through which vegetation assumes another character, first appear on the eastern side of the Himalaya mountains, in the sub-tropical part of China, and in the extreme south of Japan. The arborescent grasses here, in the vicinity of the sea, extend far north; indeed the Scitamineæ, Musaceæ, Cycadeæ, and the Palms also extend northwards in far greater numbers than is the case in the western part of the old continent; but it is chiefly the genera Camellia, Thea, and Aucuba, which appear with their beautiful large, dark green, glossy leaves in so great a number, that they are amongst the most characteristic forms of the vegetation of China and Japan, which is still more the case, as the cultivation of some of these shrubs is one of the principal branches of the agriculture of these countries. Camellia Sasanqua *Thunb.* is the olive of the Chinese; I saw the high banks of the Tiger river planted with it, just as the vine is on the banks of our Rhine. We have already named the Tea; the Olea fragrans is grown on account of the fragrance of its flowers. In habit these shrubs belong to the myrtle form, which, it is well known, is so prevalent in the sub-tropical zone of the southern hemisphere under the same meridian.

---

genera, Zizyphus, Capparis, Carissa, Vitex negundo, Buddleia neemda, Guilandina bonduc, Cassia, Hedysarum, Justicia, Barleria, Cucurbitaceæ, Euphorbiaceæ, Sida, Cissampelos, Vallaris pergulana, Plumbago zeylanica, Cardiospermum halicacabum, Boerhavia, Aneclema, Aloe, Gloriosa superba, Costus nepalensis, &c. See Royle, l. c. p. 8.

We know but little of the character of the vegetation of the sub-tropical zone of North America, yet it seems to be very different from that of the Old World. The magnificent evergreen trees and shrubs, which have large, shining, and deep green leaves, and sometimes astonishingly large and fragrant flowers, are well known. The splendid Magnolias (M. grandiflora, M. glauca) which are now famed throughout the world, Calycanthus floridus, Kalmia hirsuta, K. cuneata, Halesia tetraptera, H. diptera, Laurus catesbeyana, L. carolinensis, Diospyrus virginica, Olea americana, Ilex vomitoria, and species of Pinus and Quercus form the characteristic vegetation between 30° and 36° of latitude. On the Lower Mississippi there are vast forests of Cypress (Cupressus disticha), the trees of which are covered with the tropical parasite, Tillandsia usneoides, which appears in Mexico under similar circumstances, but at greater altitudes. Dicotyledonous trees are less frequent in these regions, and the Fan-palms appear in greater or smaller numbers, often in extensive social masses where a marshy soil favours their propagation. Salix nigra, Populus deltoides, Diospyrus virginica, grow on the banks of the Mississippi above New Orleans, and also evergreen shrubs, such as Laurus sassafras *L.*, and Myrica caroliniensis, as well as impenetrable forests of tall arborescent grasses, composed of Miegia macrosperma *P.* and Ludolphia mississippensis *W.*, which are related to the Bambusæ, and attain a height of from 36 to 42 feet; but they are not so tall in the 34th degree of latitude. In the swamps on the borders of the Mississippi, species of Rubus appear in great numbers, and Vitis riparia, and Ampelopsis bipinnata occur as climbing plants. We owe these copious details almost entirely to the learned account of his travels which Duke Paul Wilhelm von Würtemberg has given to the public.* At the junction of the Ohio and Mississippi, the banks are covered with magnificent

* Erste Reise nach dem nördlichen Amerika in den Jahren 1822 bis 1824. Stuttgart und Tübingen, 1835, p. 82–117.

pyramidal Poplars (Populus deltoides) and Willows (Salix nigra); and near the banks of the Lower Mississippi, along with the magnificent forests of Cypress we have already mentioned, and the beautiful Magnolias, occur Juglans pacan, J. rubra, Laurus borbonia, Acer negundo, and the impenetrable thickets of Miegia macrosperma, which from 30° 40′ to 32° 2′ north latitude grows as high as 36 and 40 feet.*

We have already seen that the flora of the northern sub-tropical zone has, as it were, a double aspect, according as the summer or winter vegetation is most fully displayed. We have now become acquainted with the great number of northern plants which even predominate in the sub-tropical zone in winter; we also find something similar in the sub-tropical zone of the southern hemisphere, for I can account for the great number of European plants which R. Brown† found to be common to Australia and Europe, only by the winter climate of the former country, which is similar to that of the summer of the north of Germany. From the peculiar configuration of the land in the southern hemisphere, the greater part of Australia, the southern extremity of Africa, and a small zone of South America, belong to the sub-tropical zone, and it is worthy of notice that the floras of these tracts of land not only show little similarity to the corresponding floras of the northern hemisphere, but are totally different from each other, and present mutual contrasts, which are quite unknown in the northern hemisphere. We have already remarked, when describing the Myrtle, Protea, Epacris, and Erica forms, that these forms play the chief parts in the flora of New Holland, and we, therefore, now refer to that place (p. 134). The forests of New Holland consist of leafless Mimosæ, species of Casuarina, Eucalyptus, Banksia, Callitris, Melaleuca, Olax, Xanthorrhoea, and

* See the account in Alexander von Humboldt's Naturgemälde, p. 87.

† Allgemeine geographische und systematische Bemerkungen über die Flora Australiens—in R. Brown's Vermischten Schriften, i. p. 131, &c.

Exocarpus, all trees of such various forms, and of such remarkable beauty, that the landscape there is certainly very different from ours. On the shores of the bays of that country grow Eucalyptus resinifera, and E. amygdalina, species of Angophora, Leptospermum, and Metrosideros, and some other trees of gigantic size and spread. Zamia spiralis, Mimosa sophora, M. saligna, M. nigricans, Hæmodorum teretifolium, Drosera pedata, Marsdenia suaveolens, Stackhousia monogyna, Samolus littoralis, Hibbertia volubilis, H. diffusa, Juncus vaginatus, Lycopodium uliginosum, and many other plants grow near and under the shade of these trees.* Further in the interior of the country the Eucalypti, Casuarinæ, Mimosæ, and Bánksiæ grow to a larger size, the Melaleucæ appear in greater numbers, and Loranthus and Viscum show themselves on the trees, up the branches and trunks of which climb beautiful species of Billardiera, Chorozema, and Kennedia. In low damp places grow Dianella, Cæsia, Anthropodium minus, fimbriatum, and paniculatum, Stylidium graminifolium, lineare, and many others ; on places which are often inundated appear Lobelia fluviatilis, L. inundata, and L. purpurascens, Dichondra repens, Epilobium, Lepidosperma gladiata, and L. lateralis. On the margins of rivers and other sheets of water, appear scarcely any but European genera, as, for instance, Alisma, Triglochin, Actinocarpus, Najas, Lemna, Cyperus, Scirpus, Schœnus, Carex, Myriophyllum, Mentha, &c. The Azollæ also grow along with Lemna. On the wide plains of Bathurst and Macquarie, Gaudichaud† found a number of plants which agreed with those of France, and gave the country the aspect of the cold temperate zone of Europe. At Sydney, our beautiful garden fruits, apricots, apples, pears, watermelons, &c., succeed well. The vine, which is grown in great numbers, also thrives well, and a few years ago

* See Gaudichaud, Freycinet Voyage autour du Monde, Part. botanique. Paris, 1826, p. 115, &c.

† L. c. p. 119.

wine made there was imported into London, though its excellence is said not to be remarkable. In the interior of New Holland, Polygonum junceum grows socially, and spreads over wide tracts of country; and the Kanguru-grass (Anthistiria australis) is said to appear in great masses, and with Mesembryanthemum æquilaterale to be the most widely diffused plants in New Holland.*

The character of the vegetation of the southern extremity of Africa is totally different from the flora of New Holland. In the district of Cape Colony to the Karroo fields, the four genera Protea, Erica, Diosma, and Restio so decidedly predominate, that they produce the general character of the vegetation, and this is the more striking as these forms, which we have already (page 134) more fully considered, are so peculiar. Besides these, the Compositæ prevail by number of species of the genera Gnaphalium, Elichrysum, Eriocephalus, Calendula, Othonna, Arctotis, Corymbium, Senecio, &c., as also do the beautiful genera Virgilia, Aspalathus, Polygala, Lobelia, Indigofera, Agathosma, Phylica, and the splendid Gladiolus, Morea, and Ixia. In Lichtenstein's Travels in South Africa† there are accurate descriptions of the character of the vegetation of that country. Together with the forms of Erica and Protea, it is said there,‡ that the greatest part of the genera Gnaphalium and Elichrysum exclusively belong to South Africa, and also Galenia africana, Halleria lucida, and Halleria elliptica. The mountains of Zwellerdam, where the Ericæ cease, contain Blæriæ and the genera Struthiola, Passerina, Phylica, Podaliria (P. buxifolia, myrtillifolia, vulgata), Polygala (P. oppositifolia), Aspalathus, Liparia, Rafnia, and Cleoma.

The following are the plants, which, according to Lichtenstein, constitute the forests in the sub-tropical zone

* See R. Brown, Journal of the Royal Geographical Society of London, 1830–1831, viii. p. 19.

† Reisebeschreibung über das südliche Africa. Berlin, 1811, 2 Bande, 8.

‡ II. p. 201.

of Africa. There are there no extensive masses of individuals, but an extraordinary variety of species; neither are there found such lofty forest trees as those of New Holland. The trees are species of Diosma, Barrosma serrulifolia, Cluytia pulchella, C. tomentosa and C. gnidioides, Agathosma serpyllaceum, A. linifolium, Anthericum, Bulbine, Adenandra uniflora, A. villosa, Diosma pectinatum, D. obtusatum, Myrsine africana, Cliffortia juniperina, Laurophyllus capensis, Ekebergia capensis, Podocarpus elongatus. As underwood in these forests appear Royenæ, Bryoniæ, Cluytiæ and Cynanchum obtusifolium, which here twines round the branches of the trees; and Galium glabrum, which resembles our Galium aparine, Plectranthus fruticosus, Hebenstreitia dentata, Ornithogalum parviflorum, Crassula sylvatica, &c.

In the mountain streams of Southern Africa, Lichtenstein observed Acorus palmita growing in such quantities that it sometimes impeded the current of the water; thus it is here a very social plant. In general, we may say that in the south of Africa, notwithstanding the wonderful variety, single species and genera have a very limited range. According to Burchell's* account, the four characteristic families of the Cape flora, viz., Erica, Diosma, Protea, and Restio, disappear in the latitude of the Karroo-pass, and thus these extremely large genera have a very small circle of distribution.

It is remarkable that probably all the Palms are wanting in South Africa; Schouw† alone mentions Phœnix reclinata as belonging to Cape Colony. For New Holland also, R. Brown gives but one species of Palm, which appears beyond the tropic, and extends as low as 34°. In New Zealand a species of Areca has been found below 38°. The Cycadeæ, which were formerly included in the genus Zamia, but are now sepa-

* See his Travels, p. 146.

† Grundzüge einer Pflanzengeographie, p. 312.

rated from the Zamiæ of South America, and form the genus Encephalartos,* are to be regarded as the representatives of the Palms in the south of Africa. The thick, unshapely, pithy trunks, which these African Zamiæ form, have a very singular appearance, and as they grow in the desert and barren table lands of the south of Africa, where the ostrich and gazelle take up their abode, they exercise the greatest influence over the character of the vegetation in these places. I refer to the figures of these singular plants which Lehmann has given in the paper already mentioned.

The Zamiæ of New Holland seem to belong to the same genus as those of Southern Africa, and there will probably by and by be found individual species which are common to both these countries. In general, though the physiognomy of the vegetation of these two countries is very peculiar, and though it is different in each, yet there are not wanting forms which belong in common to both, and they possess a still greater number of genera which mutually represent each other. The Restiaceæ and Proteaceæ are common to them. Burchell also found Metrosideros angustifolia in Cape Colony at the Rodezard Pass.

The Flora of the sub-tropical zone of South America is entirely different from the floras of the south of Africa and New Holland, and the resemblance which it shows, according to some authors, merely consists in the appearance of a few species and genera which are common to these three tracts of land ; in other respects the physiognomy of the vegetation of this zone in South America far more resembles the flora of the south of Europe, after the exclusion of those genera and families which are known to belong only to the American continent. According to Schouw,† of 109 genera which belong to Buenos Ayres, 70 appear in Europe, and 85

* Lehmann, Ueber die Cycadeen des südlichen Afrika. Allg. Gartenzeitung. Berlin, 1834, N. 11.

† L. c. p. 430.

in the northern temperate zóne. Although the accounts of the vegetation of the extensive country of Chili were very imperfect ten years ago, yet Schouw then formed the territory of Chili, and the province of Buenos Ayres, together with the other countries bordering on the Rio de la Plata, into separate phytogeographical kingdoms. Believing that the country on the western side of the Cordilleras possessed a very different vegetation from that of the eastern side, he divided the sub-tropical zone of South America into these two kingdoms, and called the country lying east of the Cordilleras, that of the arborescent Compositæ. Now that we possess a much more accurate acquaintance with the floras of these two countries, the separation of these two kingdoms cannot be maintained. It is well known that the eastern coast of South America possesses a warmer climate than the western, and the consequence is that when the vegetation of the two sides in corresponding latitudes is compared, that of the eastern coast is more luxuriant and tropical, than that in Chili. The whole country east of the Cordilleras is low; even at Mendoza it does not reach the elevation of 2500 feet,* therefore we must compare the vegetation of this country with that of the lowest region of Chili only, and we shall find a close agreement between them. The tall, woody bushes of Compositæ, which are so numerous in the province of Buenos Ayres, are also numerous in Chili; the few species of Calceolaria which appear in the lowest region of Chili, are indeed characteristic forms which are wanting in the eastern kingdom, but the greater number of this beautiful genus belong to higher regions, which no longer occur to the east of the Cordilleras.

The Myrtles are the characteristic form of the arboreous vegetation of sub-tropical Chili, but it is remarkable enough, that in Chili the trees and shrubs have almost universally very strong, thick, leathery and shining

* See Meyen's Reise um die Erde, i. p. 330.

leaves. The numerous shrubby Compositæ which, often adorned with the gayest flowers, characterize the flora of Chili, universally show such stiff glossy leaves, and are thus strikingly different from the shrubby Compositæ of South Africa. They are also almost always rich in resinous and often fragrant sap, in which also they differ from those at the Cape. At the Cape of Good Hope the principal genera are, Gnaphalium, Xeranthemum, Arctotis, Othonna, Osteospermum, Calendula, &c., while in the sub-tropical zone of America the genera are Baccharis, Eupatorium, Proustia, and the remarkable Mutisiæ, which are for the most part climbing plants. In the vicinity of the coast Mutisia ilicifolia *Cav.* covers tall shrubs and trees, and in the brilliancy of its flowers rivals the neighbouring Compositæ, as, for instance, the splendid flower-heads of Proustia pungens and Proustia pyrifolia. The Myrtles and Fuchsias, which are covered with lovely flowers throughout the year, grow everywhere, and from the ground spring the beautiful Calceolariæ, Oxalideæ, and the magnificent pyramidal Lobelia tupa. The shrubby and arborescent species of Psoralea and Cestrum appear in great numbers, and are overgrown with great masses of Cuscuta, which here, like all the other vegetation, grows to a large size. Other bushes again, in particular the dry Acacia caven, are brilliantly adorned by climbing Loasæ and Eccremocarpus scaber, for the golden flowers of the latter plant, in themselves remarkably beautiful, contrast splendidly with the pale yellow Loasæ. The genera Salpiglossis and Malesherbia also specially belong to the Chilian kingdom. But before all must be named the Cactus form, which here appears in the sub-tropical zone from the coasts as far as the second region. It is principally the Cereus form which descends still farther south, while that of Melocactus probably does not extend beyond 32° south latitude. I have already remarked the peculiar effect which the Cerei, often covered with the scarlet Loranthus aphyllus, give to the vegetation of Chili. I have also already fully discussed the characteristics of the majestic, tall arborescent grasses of this zone, when describing this form of plants (p. 108), to which place I now refer. The Chilian flora of the sub-tropical zone would certainly be much more luxuriant, if the plain, or the lowest region of plants were not so destitute of water, and if so many other obstacles were not in the way of the further extension of the vegetation there. Besides, the lower region of this country is very limited, and for the most part covered with sand and other dry and barren soils; but, at least from the 31st degree, there are several extensive plateaux, which lie like terraces above each other, and in them there is a greater abundance of water and a more luxuriant vegetation; the second of these plateaux, the Llanura de Mapocho or the valley of Santiago lies at the height of 1600 feet above the level of the sea, and therefore the vegetation there is still that of the lowest region. Here the Acacia caven and the Prosopis siliquastrum appear in forests, and resinous shrubs, almost without exception belonging to the Compositæ and Labiatæ, thickly cover the wide plain, which at the rainy season is adorned with thousands of gay liliaceous plants. But when the moisture has disappeared, when the sun has beat for months during summer on this plain, all its splendour is gone; not a trace of these beautiful lilies is to be seen, and the bushes even seem dead; their leaves lie in heaps round the stem, and in the leaf buds only, we perceive the dormant life of these plants.

### 4. *The Warmer Temperate Zone.*

The Warmer Temperate Zone, according to the division which I am here attempting to make, includes the countries of the warmer part of the temperate zone after deducting the sub-tropical zone, which we must separate on account of its more luxuriant and tropical vegetation.

This zone embraces the space between 34° and 45° of latitude; in Europe, including the flora of the south of Europe, as far as the Pyrenees, the mountains in the south of France, and those in the north of Greece. Asia Minor, the tract between the Black Sea and the Caspian, the northern part of China, and Japan lie in this zone, the average mean temperature of which is between 12° and 17° Cels. It is true that there are in the southern parts several places which possess a sub-tropical climate; for the course of the isothermal lines begins to be particularly irregular in this zone. Palermo, where the mean temperature is 17.5° Cels.,* and Catania, where it is 20° Cels., enjoy a coast climate, and from the mild winter have so high a mean temperature that they possess all the advantages of the sub-tropical zone. As Catania is protected towards the north by Etna, and thus enjoys a warmer climate than Palermo, so the island of Majorca is sheltered in the north by a chain of mountains, and in consequence the orange and the cotton tree can be cultivated there. In the plains of Majorca the carob tree and the olive grow with the greatest luxuriance; the latter even ascends to the altitude of 1500 feet. From this height to that of 2100 feet, the forests are chiefly composed of Pinus halepensis, and the oak ascends as high as 2400 feet. Clematis cirrhosa and Hypericum balearicum form the brushwood above the height of 3000 feet. The dwarf-palm covers the coasts and the lower hills; under its broad leaves the genera Cyclamen, Polygala, Ononis, and Anthyllis shelter themselves.† Grain, pulse, the almond, and the fig are cultivated in the rich plains of Palma and Manacor; the date-tree shades the dwellings, while Cactus opuntia fences the gardens. The vine

* See Philippi's Communications on the Vegetation of Etna. Linnæa, 1832, p. 733.

† See J. Cambessedes Enumerat. plant. quas in insulis balearibus collegit, earumque circa mare Mediterraneum distributio geographica. Mém. du Muséum, vol. xiv. p. 173–339. 1827.

covers the sides of the mountains, and even the Cheremoya (Anona cherimolia) is cultivated there. But in Minorca, where the shelter from the north is wanting, the carob-tree and olive almost entirely disappear.

On the whole, says M. Cambessedes, there is on the coast of the Mediterranean a great uniformity in the vegetation, as well as in climate and soil. The same Jura limestone appears almost everywhere ; sometimes in bare ridges, sometimes clothed with wild olives, Aleppo pines, oaks, pistachios, myrtles, and numerous Cistineæ. The date appears on the southern coast only, consequently in the sub-tropical zone, while the dwarf-palm on the northern side is spread over Spain and Naples. Pinus halepensis grows on the sandy steppes and on the sea shore, alternating with oaks and olives, along with which, on the rocky coasts, grow myrtles, pistachios, and other evergreen trees. The whole flora of this warmer temperate zone has a different aspect from that of northern Europe. In the south of Europe there are a great number of trees and bushes, with rigid glossy leaves, which remain green throughout the year ; many of the herbs and shrubs are set with numerous thorns and prickles. The flora of southern Europe shows above 300 woody plants, great part of which retain their leaves during winter.*

It is true that the trees there, as with us, have small insignificant flowers, but those of the shrubs are large and handsome, and secrete odoriferous oils and gums. Several of the beautiful tropical plants grow in the south of Spain with the greatest luxuriance. Erythrina corallodendron, Schinus molle, Phytolacca dioica, and the Banana are frequent on the banks of the Guadalquiver.

* See Willdenow Allg. Bemerkungen über den Unterschied der Vegetation auf der nördlichen und südlichen Halbkugel, l. c. p. 201, and Mirbel's Untersuchungen über die irdische Verbreitung der phanerogamischen Gewächse in der alten Welt von Æquator bis zum Nordpol. Mém. du Muséum, t. 14, p. 350–477. Translated in the Literaturblättern der Botanik. Nürnb. 1828, p. 1, &c.

All the oranges are here indigenous.* The sugar-cane, coffee, indigo, and other well-known colonial products might be cultivated, but it seems that the inhabitants are too indolent to do so. The splendid Gum-cistus (Cistus ladaniferus) is found only in the south of Spain and Portugal, where it forms extensive groves, and it does not occur either in Italy or Greece.

Schouw, as we have seen,† has named the flora of Southern Europe, together with that of the sub-tropical part of Northern Africa, the kingdom of the Labiatæ and Caryophyllaceæ, as he says that these portions of land are principally characterized by the great number of species of these families. M. Mirbel, on the contrary, asserts that the Compositæ and Leguminosæ form the largest part of the Mediterranean flora, viz., a fourth of the species, and that then the Cruciferæ, Gramineæ, Labiatæ, Caryophyllæ, and Umbelliferæ follow.

These different accounts of two so distinguished botanists are sufficient to show that in this transition-flora there are no families which predominate so considerably as to give their names to it. I believe that in general the characteristic forms of the prevalent families are what define the physiognomy of nature ; but from the Caryophyllæ and the little Labiatæ, in however great quantities they may appear, we shall hardly be able to form an idea of the aspect of the landscape in the south of Europe. Meadows are less frequent here than in the north ; evergreen dicotyledonous trees with glossy leaves are very numerous, and shrubs with beautiful flowers, such as the Cisteæ, and a great number of liliaceous plants, are also seen. The beautiful representatives of the large families Ericeæ, Laurineæ, and Myrtaceæ, viz., Erica arborea, Laurus nobilis, and Myrtus communis, here first occur ; beautiful oaks (for example,

* The China orange, according to Link, does not extend beyond 40° north latitude ; the citron can bear more cold than the common orange, but the latter bears still less than the China orange.

† Grundzüge, p. 512.

Quercus cerris), Ilex, Suber, Castanea, Prunus laurocerasus, Punica granatum, Viburnum tinus, Arbutus unedo, Arbutus andrachne (which extends westward only as far as Greece), Ruscus aculeatus, Phyllireæ, Rosmarinus, Nerium, Ephedra distachya, and many other shrubs and trees, appear with glossy evergreen foliage.

This warmer temperate zone is the native country of the vine; Parrot* describes it as the queen of the forest trees in the woods of Mingreli and Imereti. The stem of the vine there attains a diameter of from three to six inches, and it mounts to the top of the loftiest trees, festooning them, and binding them to each other. I shall have an opportunity of considering this more fully when treating of the culture of the vine, and, therefore, I may now pass it over.

We have fixed the equatorial limit of the warmer temperate zone at 34° of latitude, but for the western part of the Old World I particularly remark, that the flora of the north of Africa has quite the same aspect as that of the extreme south of Europe, and that it is not until we reach Mount Atlas that we perceive such a change as alters the character of the vegetation.

We can say but little concerning the character of the vegetation of this zone in the eastern countries of the Old World; great part of these tracts of land lies far above the lowest region of plants, and a still larger portion must have been completely changed in the thousands of years during which it has been cultivated by man. Figs, oranges, pomegranates, and all our cereals are there exceedingly productive. The flora of Japan is, perhaps, particularly striking; in the southern part, which belongs to the sub-tropical zone, grow some of the most remarkable tropical plants, while the northern flora of this country contains a number of plants which belong to our northern temperate zone. Thunberg's well-known Flora Japonica furnishes proofs of this.

* Reise zum Ararat, p. 247.

Among the chief alimentary plants of Japan are, Triticum sativum and hybernum, Avena sativa, Eleusine coracana, Panicum verticillatum, Holcus sorghum, Trapa natans, Beta vulgaris, Daucus carota, Oryza sativa, some species of Convolvulus and Dioscorea, Polygonum fagopyrum, Castanea vesca, Punica granatum, our European fruit trees, Nymphæa nelumbo, Arum esculentum, Cycas revoluta, Sesamum orientale, oranges, melons, &c.*

From want of descriptions we cannot say much as to the character of vegetation in this zone of North America. From the consideration of the predominance of the genera Solidago and Aster in the lists which have appeared of the plants of this country, Schouw has here founded the kingdom of these genera. As further characteristics of this kingdom he gives numerous oaks and pines, a deficiency of Cruciferæ, Umbelliferæ, Cichoraceæ, and Cynarocephali, an absence of heaths, and an excess of Vacciniæ. In the most southern parts of the United States, Schouw has formed the kingdom of the Magnoliæ, in doing which he has evidently been led by the striking aspect of these beautiful broad-leafed trees, with large and splendid flowers, while in other instances he has founded these phytogeographical kingdoms on the families predominant by number. Some Magnolias, for instance the Tulip-tree, and also a number of noble plants of the Mimosa form, which are forms almost entirely foreign to the south of Europe, extend into the warmer temperate zone of this continent.

The recently published travels of Duke Paul Wilhelm of Würtemberg, have here also filled up a sensible gap. Thorny shrubs are plentiful beyond the sub-tropical zone of North America, just as in the corresponding zone of the Old World; Smilax china, S. hastata, and S. walteri *Pr.* here come in the place of Smilax mauritanica, and the gigantic reeds which we formerly mentioned (page 182) extend into the warmer temperate

* See Thunberg, Flora Japonica, p. 34, &c.

zone, corresponding to the Arundo in the south of Europe. The Gleditschiæ appear on the banks of the Ohio quite overgrown with climbing Bignonias, and here also are seen the evergreen forests which so decidedly fix the character of vegetation in the south of Europe. The chesnut appears, and vast woods of oak, hazel, beech, and ash, and the Platanus occidentalis, whose pale green foliage beautifully contrasts with the other dark green trees, grows to an immense circumference.

In the forests of Missouri above St. Louis, appear thorny roses, which ascend to the top of the highest trees, and adorn them with countless rose-red flowers.*

The warmer temperate zone of the southern hemisphere includes New Zealand, Van Diemen's Land, the south of Chili, and the country south of Buenos Ayres as far as Patagonia.

The vegetation of New Zealand was still in its virgin beauty when this island was visited by Cook; it exhibited a luxuriance such as cannot be found in the corresponding countries of the northern hemisphere, which have been under cultivation for thousands of years. The lofty and vigorous trees in the forests of New Zealand are covered with climbing plants, which mount high amidst the foliage, and when the tree is felled in order to reach the flowers, remain hanging from the tops of the other trees, even after being cut away from the root. In every part of New Zealand is found the most luxuriant vegetation; climbing plants and shrubby ferns, which prefer a damp climate, grow profusely; and the number of the most remarkable forms of true tropical plants would alone be sufficient proof, that New Zealand enjoys a climate which agrees with that of the South of Europe. The two Forsters planted in New Zealand a number of roots and seeds, which bore the winter there, and they say that these plants would not have stood the winter of our colder

* Paul Wilhelm, Herzog von Würtemberg: Erste Reise nach dem nördl. Amerika, p. 120–204.

temperate zone.* They even found a cabbage-palm (Areca oleracea) in New Zealand below 41° south latitude.†

The splendid tree, the broad-leaved Dracæna australis, which is so like a palm, here represents the Pandanus form, and Phormium tenax, the New Zealand flax, the Ananas form. There is also no want of representatives of the Mimosa form (Sophora microphylla), of Myrtaceæ, Proteaceæ, and other forms, which produce some resemblance between the flora of New Zealand and those of New Holland and South Africa ; the genera Protea, Restio, Epacris, Melaleuca, Oxalis, Passerina, Gnaphalium, Mesembryanthemum, Tetragonia, Wintera, Weinmannia, &c., prove this. The flora of New Zealand seems to be rich in trees with dark evergreen foliage ; but dicotyledonous trees with tender green leaves also occur, just as in our forests of beech and oak, for both the Forsters frequently describe the pleasing contrast, which these two forms of dicotyledonous trees present when growing together.

Since the travels of the Forsters, it is universally known that several of the arborescent ferns are used for food by the natives of New Zealand, and it is repeated in book after book, that it is the inner part of the root which is eaten. But this is not the case ; it is the juicy farinaceous pith, which resembles that of the Cycadeæ, and yields a sago-like substance, which is eaten as bread when roasted. Forster‡ himself is very accurate, for he says that the edible part of these plants is a white pulpy substance which is inside the wood, and forms the pith of the stem. In the Sandwich Islands I have frequently seen the Kanacas eat the pith of the shrubby ferns, which is very nutritious, and generally of a sweetish taste.

* See Cook's Second Voyage round the World, vol. i. p. 372.

† See Forster, l. c. iv. p. 354.

‡ L. c. i. p. 384.

It is particularly worthy of remark that there is in the flora of New Zealand a lofty tree of the noble Pandanus form, with broad glossy leaves, viz., Dracæna australis.

The flora of Van Diemen's Land, so far as it is known, is very similar to that of the sub-tropical zone of New Holland, yet it seems as if the Myrtle form, and especially the Eucalypti, were here more prevalent, and the Acaciæ begin to disappear.

The flora of the southern part of Chili is totally different from that of New Holland and Van Diemen's Land, which in part extends into this warmer temperate zone ; but, as regards its physiognomy, we possess only some small fragments from which to form an idea of the vegetation of the southern part of Chili. In New Zealand Areca oleracea passes beyond 41° south latitude, yet in 36° the Chilian palm is no longer wild ; but arborescent grasses, allied to Bambusa, still appear growing socially in great masses, which is said to be the case in New Zealand also. In the district of Talcahuano evergreen woods prevail, just as in the corresponding zone of the south of Europe ; two or three trees only lose their foliage in winter.* One of the splendid climbing plants there is the Lapageria with its large evergreen and glossy, dark green leaves, and bright red liliaceous flowers. Fuchsia, Arbutus, Weinmannia, Coriaria, and Myrtus compose the brushwood and lower thickets, close to which grow tall timber trees, chiefly of the genera Fagus, Persea, Laurelia, &c., which even in the latitude of Conception are not unfrequently covered with mosses. In this comparatively well watered zone of Chili there is so extraordinarily luxuriant a vegetation, that the forests of this country could supply the whole western coast of South America with wood and charcoal.

* See Pöppig, Reise in Chili, Peru, &c. Leipzig, 1835, i. p. 317, &c.

## 5. *The Colder Temperate Zone.*

The colder portion of the temperate zone includes a belt which begins at 45° lat., and ends at 58°. In Europe this zone begins at the northern boundary of the warmer temperate zone, viz., on the northern side of the mountain chains in the south of Europe ; in Asia it embraces Caucasus, a great part of the Ural mountains, the Altai and Daurian chains, and continues its course to the shores of the Pacific, where the temperature is much lower than in the corresponding latitudes on the western side of this continent ; and consequently, the vegetation of the eastern coast exhibits a far more northern character than the western, so that the flora of Kamschatka, the greater part of which lies in this zone, already shows the character of the sub-arctic zone.

The mean temperature of this zone stands between 6° and 12° Cels. England, the north of France, and Germany, as we are best acquainted with them, shall furnish us with the characteristics of its vegetation in the northern hemisphere.

From their comparative prevalence Schouw has named this belt of land in the Old World, the kingdom of Umbelliferæ and Cruciferæ,* in which he includes all the tract as far as the arctic circle. What I remarked concerning the preceding zone (p. 193) holds good here also ; my division of vegetation rests on the general impression, which the physiognomy of the different zones makes on us ; but the little herbaceous plants contribute little to this, in comparison with the forms of trees and shrubs, and the manner in which these groups of plants are distributed. The dicotyledonous trees, along with the Abietinæ, form the characteristic forests in this zone, whose polar limit may, in the western part of Europe, be taken as also the polar limit of the beech (Fagus

* See Grundzüge, &c. p. 50.

sylvatica); the cultivation of wheat, too, is seldom of any consequence above 58° north latitude.

In order to comprehend the characteristics of the vegetation of the colder temperate zone, we must think of the condition of the countries in it several centuries ago, viz., at the period when man had not yet destroyed their natural features; for the south of Germany, for instance, when the Romans invaded it, must have had a very different aspect from the present.

Whoever has compared the north with the south of Europe must have been struck with the dissimilarity between the vegetation beyond the Alps and Pyrenees, and that of the north of France and Germany; this dissimilarity is caused by the difference of the trees, and by the proportion of the masses of trees to the fields and meadows. Although, as Schouw has shown, the Umbelliferæ and Cruciferæ are comparatively prevalent in Northern Europe; and the Labiatæ, Caryophyllæ, or, according to M. Mirbel, the Syngenesia show the comparative maximum in species in the south of Europe, yet these plants can by no means exhibit the characteristics, by which we are to distinguish the Italian landscape from that of our country. Schouw himself admits this, and explains that the kingdom of Umbelliferæ and Cruciferæ is not so decidedly separated from that of the Labiatæ and Caryophyllæ, as is the case with his other phytogeographical kingdoms. More than half the species of this zone of Europe appear in the south of Europe also, to which extremely few of the genera of these predominant families are peculiar. Some families are more numerous in the north, others in the south of Europe, but all the distinctions, which rest on this, are trifling.

The frequent occurrence of our noble meadows, the wide heaths covered with Erica vulgaris, amongst which grow the juniper (Juniperus communis *L.*), Ledum palustre, Andromeda polifolia, and here and there a few dwarf willows, and the vast forests of dicotyledonous trees, with tender, pale green leaves, together with masses of the social pines; these are the chief great features in the distribution of the vegetation of our zone. The forests of our dicotyledonous trees lose their leafy covering in winter; only the mistletoe (Viscum album *L.*) is here and there green amidst the naked branches; the ground and the trees are covered with snow, and the dark green foliage of the Abietinæ alone shows that vegetation is not quite dead. But in spring, when our northern vegetation awakes again, Nature displays a loveliness which is wanting even in the torrid zone; the bright green, fresh foliage of our beautiful trees, such as it is in the month of May, can be found no where else. Our forests are poor in comparison with the luxuriant vegetation of the tropics; and in place of the shining Tillandsiæ, the bark of our forest trees is overgrown by Usneæ, Ramalinæ, and other foliaceous lichens and mosses. Instead of the Lianas of the tropics, the honeysuckle (Lonicera periclymenum) climbs to the top of the lower trees of our woods; and the ivy (Hedera helix) clothes the trunks, on which, under a tropical sky, grow fragrant Orchideæ, glossy Aroideæ, and numerous ferns in the greatest profusion. The hop (Humulus lupulus) is the most considerable twining plant of our colder temperate zone.

This zone is particularly rich in shrubs, and the most of them have large and beautiful flowers. Our numerous roses, our species of Rubus and the snowball tree (Viburnum opulus) are some of the conspicuous plants of northern countries.

If we proceed further east to Asia, and compare the vegetation there with ours, we find extraordinarily little difference between them; that is, allowing for the influence which the ruder climate of the east has on the vegetation, on account of which a more northern vegetation has naturally pressed further south. Our German vegetation continues almost unchanged to the Wolga; until we reach this river we find Trapa natans, Chara vulgaris, Salvia pratensis, Thesium linophyllum, and Limosella aquatica (on the muddy banks of the river); but beyond it these plants disappear, and on the other hand, Cucubalus tataricus occurs in the forests.* The Steppes, which also lie beyond the Wolga in 50° east longitude, are local phenomena, and exhibit a vegetation quite peculiar to them; the genera Anabasis, Halocnemon and Brachylepis are, according to Lessing,† peculiar to the salt Steppes. Pallas for such soils gives the following plants; Salsola prostrata, Statice tartarica, Glycirrhiza hirsuta, G. lævis and G. echinata, Lathyrus tuberosus, Medicago sativa, Vicia sylvatica, Lotus corniculatus, Serratula arvensis, and Inula britannica. On dry places grows Anabasis aphylla, and Artemisia absinthium, Tamarix gallica, Cynanchum acutum, Senecio linifolius, &c. also appear. The character of the vegetation of the sandy Steppes of this region is formed, according to Lessing, by grasses with rigid, rolled up leaves, by Atriplicinæ and Chenopodiæ.

Lessing gives 75° east longitude as the eastern limit of our oak in latitude 55°; Pallas found it, along with Corylus avellana, in gardens, as far east as 80°, in latitude 59°. Gentiana pneumonanthe, G. amarella, and G. cruciata also grew there.‡ An inestimable treasure of observations, by means of which we can compare the vegetation of these countries with ours, is to be found in Pallas's Travels, to which I must expressly refer. From its position the whole of Kamschatka belongs to the colder temperate zone, but, from the severe climate which we noticed before, when speaking of the isothermal lines, the flora of this country is that of the sub-arctic zone, which we shall afterwards describe. In Petro Paul's Haven, in 52° lat., the birch alone grows to the height of a tree, but from Langs-

* See Pallas' Reise, Band i. p. 15 and 168.

† Linnæa, 1834, Number ii.

‡ See L. c. ii. p. 273.

dorf's account* the vegetation in other places is not so poor.

But if we proceed still further eastwards, we arrive at Sitka on the western coast of America, in 57° north latitude, and we here again find a vegetation which corresponds with that of Western Europe under the same parallels. We owe to Mertens,† who has died too early for our science, an extremely interesting account of the vegetation of this country. On the coasts of this part of the North Pacific Ocean grow Arenaria peploides, Glaux maritima, some species of Carex and Juncus. Veronica serpyllifolia and Veronica anagallis grow inland. Potentilla anserina, P. ruthenica, and a beautiful Sisyrinchium are native here. The genera Plantago, Triglochin, Dodecatheon, Pedicularis, Elymus, Bartsia, Campanula, Angelica, Heracleum, Fritillaria, &c., show species very similar to those in Europe; but Pisum maritimum, Cochlearia danica, Ranunculus acris, Galium boreale, Geum intermedium, Turritis hirsuta, and T. glabra, are indigenous here as well as in our native country. The forests of Sitka are composed of colossal fir trees (Abies), and species of Alnus, Sorbus and Cratægus also appear, while Rubus odoratus, with white flowers, forms the underwood. The chief plants which we need mention here, are Cornus suecica, Rubus spectabilis, Ribes, a tall Azalea, a Calla, Linnæa borealis, Lathræa stelleri, Cymbidium, Trientalis, a Salix and the characteristic Panax horridum. The last plant is a very remarkable climber, which renders the forest so dense that it is difficult to penetrate it.

But as the vegetation of Kamschatka on the eastern coast of the continent is far inferior to that of the western coasts under the same parallels, so the vegetation of Labrador on the eastern coast of America is far behind that of Sitka in the very same latitude.

* See his Bemerkungen auf einer Reise um die Welt, Bd. ii. p. 224.

† See his Botanisch-wissenschaftliche Berichte vom October, 1827. Linnæa, 1829, p. 43–73.

It is well known that in the southern hemisphere there is very little land, which extends into the colder temperate zone; but it is fortunate for us that some of these points have been pretty accurately examined. We shall first consider the Falkland Islands, which, indeed, as treeless islands in the open sea, present an appearance very different from that of the vegetation proper to this zone. But all the differences may be explained, partly by the peculiarities of insular climate, partly by the sterility of the soil, which was probably still greater a thousand years ago. The Falkland Islands lie in 52° and 53° south latitude, and at a very small distance from the east coast of South America. Their climate is on the whole mild; on the eastern island, East Falkland,* the temperature during the severest winter is never below — 2.67° R., and in the hottest summer does not rise above 19.11° R. On an average it varies between — 0.89° and 8° R. in winter, and between 8° and 19° R. in summer. These islands, therefore, have a much milder winter than we have in the northern hemisphere; they, however, know nothing of the agreeable heat of our summer months, and, therefore, their vegetation, as well as their agriculture, must be totally different from that of our northern continent. The weather in East Falkland is indeed changeable, but falls of rain, snow, or hail are always of short duration; the snow disappears in a few hours, except on the tops of the mountains, and ice is seldom found above an inch thick. In summer north-west winds blow, and in winter south-west winds.

At the present day the soil of the Falkland Islands is very suitable to agriculture, as it consists of a layer of black mould from six to eight inches deep. Wheat and flax are grown of equal, if not better quality, than the seed which was brought from Buenos Ayres; and pota-

* Beschreibung der Ost-Falklands-Insel von Vernet. Berghaus, Cabinets-Bibliothek, Berlin, 1834, i. p. 158.

toes, cabbages, and turnips produce excellent crops. Trees do not grow naturally on the Falkland Islands, and planting, which would certainly be successful, has scarcely been attempted, for fruit trees, which were sent from Buenos Ayres, perished on the way.

In the Falkland Islands there are extensive meadows and moors, which give the landscape a character similar to that of our country; if the forests of the continent were not wanting here, the resemblance would be still greater. We possess an extremely valuable description of the vegetation of these islands by M. Gaudichaud,* which we owe to the French voyages of discovery. With the greatest judgment this distinguished botanist has arranged the plants for the use of botanical geography, so that we receive a correct idea of the physiognomy of these remote islands. The meadows there are formed by Agrostis magellanica *L.*, Agrostis cæspitosa, and Aira flexuosa, and Avena redolens, A. phleoides, Festuca magellanica, F. erecta, Arundo, Carex, Scirpus, and Juncus make up the rest of the gramineous vegetation. The damp and more moory places produce a number of plants, which are also quite similar to ours under similar conditions; as, Marchantia polymorpha, Sphagnum acutifolium, Lysimachia repens, Caltha appendiculata, Sagina procumbens, and S. crassifolia, Calitriche verna, and Misandra magellanica; and in the numerous pools appear Limosella tenuifolia, Azolla magellanica, Caltha sagittata, Montia linearifolia, Myriophyllum elatinoides, and M. ternatum, &c. On the sides of damp mountains the beautiful Lomaria setigera grows in the greatest profusion, and also the singular Bolax glebaria, which forms thick, close, green bushes, often three feet high and from eight to ten feet thick; thus this plant appears under a form very similar to that of the little social Umbelliferæ on the heights of the Cor-

* Freycinet, Voyage autour du Monde, Part. Botan. p. 123–143.

dilleras of Chili and Peru, of which we have already spoken in several parts of this book (see, for instance, p. 85). The rocks on the mountains of these islands are, just as with us, covered with a great quantity of lichens, the most of which are identical with ours. As has been already said, there are no forests in the Falkland Islands; but there are thickets of bushes from four to five feet high, and these also belong to genera, which principally belong to our northern zone; viz., to Rubus, Arbutus, Andromeda, and Empetrum, the berries of which, here as well as with us, are for the most part agreeable fruits. M. Vernet* speaks of a tea plant on East Falkland, which grows close to the ground and produces berries as large as a strawberry, which are white, streaked with rose colour, and extremely well tasted. The wood needed for fuel is easily brought from the Straits of Magellan, but there is abundance of turf in many parts of the Falkland Islands; and some of the bushes have stems as thick as a man's arm, and are used for fuel. M. Vernet names three of these bushes, but adds that they bear no fruit there!

From what has been said, it is easy to perceive the great similarity between the vegetation of the Falkland Islands and that of our northern temperate zone, and the absence of trees must be regarded as a local phenomenon only. A great number of the plants of these remote islands perfectly agree with those of our native country. The other plants partly belong to the southern continent of America, and partly are closely allied to the Alpine plants of Chili. The Nassauviæ, Perdiciæ, and the singular Mulineæ are quite those of the Alpine regions of the Chilian Cordilleras.

M. Dumont d'Urville† has given in his flora of the Falkland Islands 139 genera, containing 214 species,

* L. c. p. 159.

† Flora des Malouines: Mém. de la Société Linnéenne de Paris. Paris, 1826.

94 of which are cryptogamic, and 120 phænogamous plants.

It would be extremely interesting to compare with this description of the vegetation of the Falkland Islands, that of the opposite continent, but unfortunately a few isolated portions only of it are known. The Mesier Canal is the most northern point (48° and 49° south latitude), of whose vegetation Capt. King has given us any description.* On both sides of this Strait, the coast is hilly, but not very high, and in many places the land is very low and always thickly wooded. The trees, it is said in this account, are here the same species as are found over the tract between Cape Tres Montes and the Straits of Magellan ; the most common are Fagus antarctica, Fagus betuloides, Winterana aromatica, and a tree which has quite the appearance of a Cypress. Although the trees are of very considerable thickness at the base of the trunk, they never grow very high ; yet the woods are said to be so thick that the sun's rays cannot penetrate them. On the west coasts the ground in these forests is always covered with damp moss.

The Straits of Magellan in their course from east to west show very great differences in the configuration of the land, as well as in the vegetation. The west end and the centre, says Capt. King,† are of primitive character, rugged and very mountainous, while the eastern part is of a newer formation and low. About the middle of the Strait the rock is clay-slate ; the mountains are higher, steeper, and more abrupt in their outline. Their average height is about 3000 feet, but some rise above 4000 and even 6000 feet. The limit of perpetual snow appears to be at the altitude of 3500 or 4000 feet. The character of the vegetation is as varied as that of

* Some Remarks on the Geography of the South of America, viz., of Terra del Fuego and the Straits of Magellan. Translated in Berghaus, Cabinets-Bibliothek, i. Bl. 1834, p. 134.

† L. c. p. 146.

the country; and that not so much from variety of plants, but rather from their habit. In the western part of the Straits of Magellan vegetation is very stunted; in the middle it is in its greatest luxuriance, and on the eastern side there is a total absence of trees. The trees on the granite soil of the western part of the strait are low, and at most nine or ten inches thick; from the want of mould the vegetation is very poor. But it is otherwise with the luxuriant vegetation in the middle of the Strait; Fagus betuloides here grows in great numbers; trees three feet in diameter are frequent, and even a diameter of four feet is not uncommon; and there is a tree here which is seven feet in thickness, 17 feet above the root; it there divides into three branches, each of which is three feet thick. I consider this evergreen beech as the representative of ours in the southern temperate zone, but we probably never meet with such large trees of our beech. There are in the Strait few other trees besides this beech which could be used as timber. Only two other species of beech and the Winterana aromatica are fit for this purpose. The latter tree,* which is also an evergreen, is found mixed with the first in all parts of the Strait, so that the land and mountains from the height of 2000 feet above the sea to high-water mark are clothed with a perpetual verdure, which presents an extremely remarkable spectacle, particularly where the glaciers descend to the sea. In this country Capt. King observed species of Fuchsia and Veronica (?), the stems of which were six or seven inches in diameter, and yet the country there is covered during the winter, that is from April to August, with a thick mantle of snow.†

* King, l. c. p. 149.

† At Port Famine, in the Straits of Magellan, in 53° 58′ south latitude, the temperature during the winter months was 34.5° Fahr. (44° Cels.); the maximum was 49.5° Fahr., and the minimum 12.6° Fahr. The mean temperature of the autumn (February, March, and April), was 47.2° Fahr. (8.4° Cels.); the maximum 68° and the minimum 28°.

Our knowledge of the vegetation of Terra del Fuego and Staaten Land is still not much greater than at the time of Cook's voyages round the world. Since the well-known botanical excursion which Banks and Solander, on the first of these memorable voyages, made to the coast of Terra del Fuego, when, in the middle of summer, several men perished from the severity of the climate,* this southern region of America has been much decried, though there is no want of wood, and the character of vegetation in general resembles that of the vegetation of our northern countries. The base of the mountains is clothed with wood; wet moors are covered with low birch trees, and the fertile plains are adorned with beautiful turf. Sparrman and the two Forsters gathered here Pinguicula alpina, Ranunculus lapponicus, Galium aparine, Statice armeria, Dactylis cæspitosa, D. glomerata, a Sanguisorba, Fagus antarctica, Winterana aromatica, &c. In places exposed to the wind, these latter trees never attain any great height. In the interior of this island the climate is by no means so dreadful as Banks and Solander have described it, for in their account they have forgotten, that they were on a high mountain, the Bell in the Bay of Good Success, and far above the region of trees.

The other large islands of the southern hemisphere which belong to this zone, South Georgia, Kerguelen's Land, &c., are solitary, extremely barren islands, and their vegetation could not give us any correct idea of that which is proper to this zone.

### 6. *The Sub-Arctic Zone.*

The sub-arctic zone is of less extent than the preceding one, and perhaps in the interior of Asia it may not be so clearly distinguishable from it, as it is in Europe. This zone stretches from 58° latitude to the Arctic Cir-

* See Cook's Reise um die Welt, Berlin, 1774, i. p. 45, and so on.

cle, that is to latitude 66° 32′. The average mean temperature of this zone is between 4° and 6° Cels., but, as the course of the isothermal lines in this region is extremely irregular, this degree of heat will also vary according to the different localities.

In the northern hemisphere, the sub-arctic zone is the zone of the firs and willows, and properly begins at the polar limit of the beech. In the southern hemisphere only a few barren islands extend into this zone, for instance, New Scotland, which, from some hitherto scarcely explained cause, has as rude a climate as Spitzbergen, which lies within the arctic circle.

In the preceding zone, the pines (Pinus sylvestris) and the noble dicotyledonous trees, such as oaks and beeches, form the most extensive forests, and predominate by their masses; in the sub-arctic zone, on the contrary, these trees are found only on the southern border, scarcely above 60° latitude, and even here they show little of their grandeur and luxuriance in the thick forests of Germany and England. Even the pine (Pinus sylvestris) entirely disappears, and though it still grows on the western coast of Norway, which possesses the coast climate of Scotland, in the interior of the country the noble fir appears in place of it. The aspen, birch, service tree, and juniper there compose the forests, along with the lofty and dark green firs. The ash (Fraxinus excelsior) the lime and the elm (Ulmus campestris) still grow at Christiania,* with a mean temperature of 4.96° R. In the gardens of Christiania are grown apples, cherries, pears and apricots, and even the vine† ripened fruit in the open air for several successive years. This, however, is by no means to be regarded as the normal vegetation above 60° north

* See Lessing, Reise durch Norwegen nach den Loffoden durch Lappland und Schweden. Berlin, 1831.

† The limits of the principal fruit trees in the Scandinavian peninsula are, according to Schouw, for the apple and plum 63½° latitude; for the cherry 63°, and for the pear 62°.

latitude, but is a most striking exception, which can only be explained by the peculiarities of the climate on the western coast of the continent. Further east, neither in the Old nor the New World can there be found a second point which shows so distinguished a vegetation. We possess an excellent work on the difference of vegetation on the eastern and western sides of Norway, which makes the influence of the coast climate very apparent; this work is by the prematurely deceased botanist, Christian Smith,* and it is unfortunately known only from the extract which L. von Buch has given from it.† It is said in it, that, on the eastern side, after the severe winter follow a few weeks of summer with constantly bright and clear days. The sun of an almost perpetual day calls forth a profusion of leaves and flowers, which could scarcely be expected in such northern latitudes. Beyond the mountains, on the other hand, the always open sea diminishes the severity of the winter, and the constant winds from the west and south, blowing over the sea, warm the coast. But they, at the same time, envelope it in thick mists and clouds, which obstruct the beneficent influence of the sun, and therefore the heat of the summer is of shorter duration and has less effect.

Although the birch bears even the rigorous Siberian winter, it requires during its time of growth a warm summer, and therefore its limit comes always nearer the plain and farther south, as the heat of the summer decreases. Along with the pine-forests on the western coast of Norway appear, according to Smith, the beautiful Digitalis purpurea, Hieracium aurantiacum, Bunium bulbocastanum, Sedum anglicum, Chrysosplenium oppositifolium, Hypericum pulchrum, Erica cinerea, Rosa spinosissima, &c., plants which, as Smith remarks, we

* Topographisk-statistiske Samlinger, udgivne af Selskabet for Norges. 2den Deels, 2det Bind. Christiania, 1817.

† Physikalische Beschreibung der Canarischen Inseln, p. 38.

should look for in vain where the birch ascends to the height of 3000 feet.

It will be most accordant with our plan if we now proceed to consider the vegetation of the Faroe Islands, for which the recent work of Mr. Trevelyan* furnishes us with excellent materials. On the Faroe Islands only barley thrives, and even it does not always ripen; but the turnip and the potato succeed very well.† There are no trees in the Faroe Islands, but willows and Amentaceæ are not wanting. On the mountains of these islands the region of alpine plants begins at the elevation of 1500 feet; many of the mountains are covered with long, cæspitose mosses.

The vegetation of Iceland is quite the same as that of the Faroe Islands; there is in it also a total want of trees, for the birch and the alder attain a very small height, although the mean temperature is by no means so low as in more eastern parts of the old continent in the same latitude, where there is even the greatest superabundance of forests of fir, birch, and poplar.

Schouw remarks,‡ that the vegetation of Iceland very closely agrees with that of Norway, where also, towards the sea, trees disappear, not on account of the lower temperature, but rather of the very damp sea air. In the middle of the last century the trees in the birch woods of Iceland were generally from three to four ells high, and three or four inches thick; the tallest were from six to ten ells high.§ There are now no forests in Iceland; a few clumps of stunted birches are all the

* On the Vegetation and Temperature of the Faroe Islands. The Edinb. New Phil. Jour., October, 1834–January, 1835, p. 154–164.

† The mean temperature of the Faroe Islands, according to four years' observations, is as high as 54.6° Fahr. In the year 1821 only 51.6° Fahr. The highest temperature observed was 72.5° Fahr., and the lowest was 18.5° Fahr.

‡ Europa, p. 18.

§ See Olafsen's and Pavelsen's Reise durch Island. Kopenhagen and Leipzig, 1774, 4to, p. 89–126, &c., which contains besides very valuable contributions to our knowledge of the vegetation of Iceland.

wood of that miserable country, but there are many facts which prove the earlier existence of high birch forests there. These are fully enumerated in the work by Olafsen which we have just named, and it may still be observed that the ground on which these birch forests stood is now changed into moors and bogs. The young birch trees, which at present grow in Iceland, are rather bushes, and never attain the height which they must have reached in former times, so that we may believe that the climate of this island must be considerably changed since the extirpation of the forest. It is also a fact highly worthy of remark, and confirmatory of the opinion that the distribution of vegetation is regulated by climate, that Iceland, though from its position it belongs to the sub-arctic zone, possesses none of the trees of this zone except the birch and alder. The whole island, therefore, has the coast flora of Norway; the juniper is the only coniferous tree which is found in it, but a great number of the plants agree with the flora of the northern temperate zone. The green meadows of Iceland are almost the same as ours; Agrostis arundinacea, Aira cæspitosa, A. flexuosa, Poa pratensis, P. trivialis, P. compressa, P. annua, &c., grow there as well as with us, and Trifolium arvense, T. pratense, and T. repens flower there as gaily. The heaths in Iceland are also covered with Erica vulgaris, and Juniperus communis grows on them. The pools show Chara vulgaris, Chara hispida, and our species of Callitriche, and on their margins grow Hippuris vulgaris, Veronica anagallis, Arundo phragmites, Comarum palustre, Limosella aquatica, &c. In the fields of Iceland, as in ours, grow Serratula arvensis, Thlaspi bursa-pastoris, and T. campestre, Draba verna, Prunella officinalis, Thymus serpyllum, Lychnis flos-cuculi, Spergula arvensis, &c., and the delicious berries of Vaccinium myrtillus, V. uliginosum, V. oxycoccus, and Arbutus uva-ursi are in Iceland as with us an agreeable food.

Corn is cultivated to a very trifling extent in Iceland ; often not at all for a long series of years, for towards the end of summer the weather is so changeable that the grain seldom ripens ; yet in more early times, about a hundred years ago, rye was grown in several places. On the other hand the principal vegetables, such as the varieties of the cabbage, even the cauliflower, the potato, various kinds of turnip, cummin, &c., are very successfully cultivated. Fucus saccharinus *L.*, F. esculentus *L.*, F. palmatus *Gm.*, and a few other plants of the family, were in earlier times, particularly before the introduction of the potato, almost the sole vegetable food of the inhabitants of Iceland. These marine plants were eaten either fresh or dried, and were even an article of inland trade ; and certainly they are a very nutritious and palatable food. We know that on many of the coasts of the main sea various other marine plants are used either as an ordinary article of food or as a luxury ; for instance, Fucus antarcticus *Cham.*, at the southern extremity of America, several of the large Laminariæ and Fucus pyriferus on the coasts of Chili, Fucus cartilagineus in India, China, Japan, and the whole Archipelago of that sea, &c.

Besides these Algæ there are yet to be named the following plants, which are considered articles of food in Iceland, viz., the Iceland moss (Lichen islandicus *L.*), Pisum maritimum, and Arundo arenaria. The Iceland moss grows in many parts of the coast of this island in great quantities, and it is collected from the same ground every three years. The bitterness of the plant is extracted by soaking it in water, and it is then eaten, generally boiled with milk ; a kind of bread also is said to be baked of it. The time of gathering the Iceland moss is like a merry harvest time. The pease of Pisum maritimum are said to be very well flavoured ; and the little seeds of Arundo arenaria are in some places ground to a fine meal.

We now return to the continent and proceed eastward to the remote regions of Siberia, which belong to the sub-arctic zone, where we find almost everywhere the same plants which appear in the western parts of the sub-arctic zone, viz., in Norway and Sweden, on Iceland and the other adjacent islands. The trees only are different; immense forests of Coniferæ appear, and extend as far as the east coast of Kamschatka, where they again give place to the birch.

Langsdorf* names all the plants which he met with in his travels in Siberia, between Ochotsk and Jakutck, which I shall also do here, in order to compare them with the floras of Norway, Sweden, and Iceland.

The forests of those parts of Siberia consist of Pinus cembra, P. larix, P. abies, Platanus orientalis, Populus alba, P. balsamica, Betula alnus, B. nana, and B. fruticosa; and Rhododendron tauricum, R. chrysanthum, Stachis palustris, Stachis sylvatica, Scutellaria galericulata, Schwertia perennis, Sanguisorba officinalis, Tanacetum vulgare, Trientalis europæa, Valeriana officinalis, Vaccinium vitis-idæa, V. uliginosum, Anemone narcissiflora, A. sylvestris, Atragene alpina, Andromeda polifolia, Antirrhinum linaria, Arbutus uva-ursi, Euphrasia officinalis, Potentilla anserina, the beautiful Pyrolæ of our woods, Galium boreale, Sedum palustre, Lysimachia thrysiflora, &c., speak very plainly for the similarity of the vegetation of these eastern regions to that of the western part of Europe.

We shall in conclusion compare the flora of Kamschatka with what we have already said concerning the vegetation of the sub-arctic zone. The flora of the southern part of Kamschatka is still that of the preceding zone, but on the whole it is of sub-arctic character. That corn is not cultivated in Kamschatka, is a departure from the law observed in the west of the continent, which may be explained in the same way as the ill success of agriculture in Iceland. I name the following plants of Kamschatka from the list which Langsdorf has

* L. c. ii. p. 316.

given in his travels ;* Rubus chamæmorus, Vaccinium vitis-idæa, V. uliginosum, Berberis vulgaris, Ribes rubrum, Empetrum nigrum, Lonicera cœrulea, Prunus padus, Sorbus aucuparia, Rubus arcticus, Arbutus uva ursi, and Betula alba, B. nana, Pinus larix, P. abies, P. cembra, Populus alba, Platanus orientalis, Betula alnus, Salix arenaria, S. pentandra, Juniperus communis, Cratægus oxyacantha, Rosa canina, R. spinosissima, Lonicera cœrulea, &c., formed the woods and brushwood.

### 7. *The Arctic Zone.*

The arctic zone embraces a still smaller belt than the sub-arctic zone, of which we have just been treating ; it extends from the northern arctic circle (60° 32′ lat.) to the most northern points of the Scandinavian countries bordering on the Northern Ocean, and, therefore, to 72° latitude, which seems to be the limit of trees, and at the same time of all culture of the soil. We have already seen that the course of the isothermal and isotheral lines becomes always more irregular as we recede from the equator, and consequently vegetation also, following the irregular progress of the distribution of heat, presents always more and more anomalies as we approach the pole. In the most western countries of the old continent, the culture of the cereals extends above 70° to a region, which in the most eastern parts of Siberia, if the descriptions do not deceive us, must be bound in perpetual ice. This inequality in the distribution of heat and, consequently, of vegetation, must not, however, deter us from attempting a general description of the whole zone ; even though here and there the limit of trees appears to be in lower latitudes.

The mean temperature for the arctic zone may be, at the maximum, about 2° Cels., but in the colder part of this zone it is certainly much lower, and often even be-

* L. c. ii. p. 224.

low the freezing point. In the southern hemisphere there are known only a few islands lying in 68° latitude, which extend to this region, and we as yet know nothing of their vegetation.

We possess the most masterly works on the vegetation of separate portions of the arctic zone; particularly that of Wahlenberg,* in which the botanical geography of a particular country has been first worked out with extraordinary success.

The most predominant of all the plants of this zone is the birch, and after it some of the Abietinæ, as the pine (Pinus sylvestris), and the spruce fir (Abies excelsa), which here still form extensive forests. According to Schouw,† the birch almost reaches North Cape, the spruce fir extends to Alten (60°–70° latitude), and the beautiful firs to 69°, or even above 70°, on the eastern side of Norway and Sweden. The poplar and the service-tree are the only other trees which extend along with the three already named beyond the arctic circle. The Juniper, Rubus chamæmorus, Cornus suecica, and numerous species of willow form the brushwood here; and besides them the genera Diapenzia, Azalea, and Andromeda are peculiarly characteristic of Lapland. The dry and barren fields are covered with incredible quantities of lichens, amongst which the rein-deer moss (Cenomyce rangiferina) covers the most extensive surfaces with a matting over which it is very fatiguing to travel in summer, when the plants are dried up by the perpetual sunshine. We have already, in the preceding zone, observed the appearance of lichens in as great quantities, for example the Ceteraria islandica on Iceland; but we have seen at the same time that the vegetation of this island is so much kept down by the coast climate, that it assumes quite the character of the arctic zone. In those parts of North America which belong to this zone,

* Flora Lapponica. Berlin, 1812.
† Europa, p. 8.

the appearance of vast masses of lichens is also frequent; they are here chiefly the Gyrophoræ, which extend even to the shores of the Arctic Ocean, and are used as food in times of scarcity.

The green turf of our zone is still not quite wanting in the arctic zone, where Aira cæspitosa and Aira flexuosa help to form it;* even Milium effusum clothes with the greatest luxuriance the mountain sides of the Loffoden islands, but the beautiful carpet of moss which so frequently adorns the forests of our country, is no longer to be found, though there is certainly here no want of Mosses and Jungermanniæ. The Polytricha especially prevail in the most luxuriant beauty in the extreme north of Norway and Sweden.

In the Scandinavian peninsula, agriculture extends over all the arctic zone, though it is confined to a very few objects. Of the cereals, only barley and rye are cultivated; the first of which, according to Schouw,† extends as high as 70° north latitude, while rye on the western side extends to 64°, and on the eastern side to 65° or 66°. At Enontekis, in latitude 60°, and 1350 feet above the level of the sea, some corn is still grown, though it ripens at most only every three years.

At Hammerfest, in latitude 71°, experiments on the cultivation of plants have been made, and it has been found that cabbage, turnips, carrots, potatoes, spinage, and salad succeed very well even at this high latitude.

It has been observed in several parts of this arctic zone, that a so-called Alpine vegetation appears quite at the level of the sea, and this may be very easily explained, without bringing down this Alpine flora from the mountains. As we shall by and bye see, the treeless vegetation of the polar zone first corresponds with that on the summits of the Alps; on the contrary, on the coasts of the arctic zone, particularly where the weather is very

* Wahlenberg, l. c. p. 59.

† Europa, p. 9.

inconstant on account of the prevalent north wind, the temperature of the summer is so considerably lowered that it corresponds with the summer temperature of the polar zone, and on this account a number of Alpine plants descend in this zone to the level of the sea, where they occur along with the shore-plants. Thus Lessing* observed on the west coast of Norway, in the neighbourhood of Kunnen, in the meadows, Silene acaulis, Saxifraga oppositifolia, Potentilla aurea, Thalictrum alpinum, Erigeron alpinus, Gentiana nivalis, Alchemilla alpina, Arbutus alpina, Empetrum nigrum, Astragalus alpinus, and close to them, on the sandy sea-shore, Arenaria peploides, Lotus siliquosus, Silene maritima, Cochlearia danica, &c.

The most eastern countries of the old continent which project into the arctic zone, are unfortunately quite unknown; however when they are examined, they will certainly exhibit very considerable differences from the vegetation of this zone which we have now described.

We are now acquainted with a very great number of plants from the arctic zone of America, and the recent flora of these regions which Mr. Hooker is issuing, raises the highest expectations; but there is still a great want of descriptions which would enable us to depict the physiognomy of its vegetation.

But when we compare the flora of this arctic zone which we have received from Dr. Richardson,† we not only perceive the most extraordinary accordance between the plants of these two countries, but we even find only two plants which occur in America and now also in the extreme north of Norway and Sweden; that is, if we leave out from that flora the varieties marked A and B. In proof of this I give the names of the principal plants which characterize this arctic zone.

* L. c. p. 44.

† See his Flora of the Polar Regions, which appeared as a supplement to Franklin's Narrative of a Journey to the shores of the Polar Sea. London, 1823, 4to.

Rhododendron lapponicum here grows in the plain, and its companions are the little bushy plants, Andromeda tetragona, A. polifolia, A. caliculata, Vaccinium vitis-idæa, Oxycoccus palustris, Azalea procumbens. But it is especially to be remarked that the birch (Betula glandulosa), and the alder (Alnus glutinosa) appear here just as in the arctic zone of Europe. The chief polar or alpine plants which, from the peculiarity of coast climate, descend to the sea-coast in the arctic zone, are Saxifraga aizoides, S. oppositifolia, cernua, groenlandica, Polygonum viviparum, Arnica montana, Dryas integrifolia, Holcus alpinus, Pedicularis lapponica, P. sudetica, and P. hirsuta. Besides these there have been observed Plantago lanceolata, Cerastium viscosum, Oxyria reniformis, Triglochin maritimum, Tofieldia borealis, Epilobium palustre, latifolium, angustifolium, &c.

The great number of lichens which often cover extensive tracts in this region is also worthy of remark; the Gyrophoræ especially, as G. proboscidea, G. hyperborea, G. pensylvanica, and G. muhlenbergii, which in case of need may be used for food, prevail in every rocky place.

## 8. *The Polar Zone.*

To the polar zone belong all the tracts of land which lie above 72° of latitude. No trees or bushes grow in these cold regions, where cultivation of any plant for food is impossible, for the mean temperature is far below the freezing point, and the summer at most lasts only from four to six weeks. At present the mean temperature of a single point only of this zone is known, viz., of Melville's Island, where observations have been made during ten months. We have given the curve of temperature for this place in the Table accompanying the first part; the mean temperature is equal to — 16.9° Cels.; that of the summer to 3.1°, and that of winter falls so extremely low as — 3.33° Cels. In the month of July only, the temperature rises to 5.8° Cels., and in

August it again falls to 1.2° Cels.; with so low a temperature the vegetation cannot be of any great importance, and from all observations hitherto, it consists of mere Alpine plants; that is of those little plants which adorn the highest peaks of the mountains to the very limits of perpetual snow.

Even low bushes are wanting in the polar zone, and only a few species of this group appear as herbaceous plants. Though only a few points of this zone have been visited by travellers, we possess remarkably excellent works on the floras of these few points by Mr. R. Brown* and Dr. Hooker.†

On looking over these floras we see that the vegetation of the polar zone is extremely scanty in comparison with that of the warmer zones, but in number of species, as well as of genera, and probably in number of individuals also, this flora is perhaps not superior to the corresponding vegetation of the highest mountain tops. Want of water during the short summer, and a rocky, barren soil are also obstacles to the development of a richer summer vegetation in this dreary region.

But on comparing the catalogues of plants belonging to Spitzbergen, Greenland, Baffin's Bay and Melville's Island, we must be astonished at the exact conformity of the vegetation of these different places. It is true that several plants have hitherto been found on Melville's Island only, but it must be remembered that no other part of this zone has been so long and carefully examined, and that it is therefore to be expected that several of the plants, as yet peculiar to Melville's Island, will, at some future time, be found in other parts of this zone.

* See his List of Plants collected on the Coasts of Baffin's Bay, &c.; his Flora of Melville's Island; his List of Plants found in Spitzbergen; and his Supplement to Richardson's Polar Flora. All these works translated into German are contained in the first volume of Brown's Vermischten Schriften, published by N. von Esenbeck.

† List of Plants from the Eastern Coast of Greenland—as Appendix to Scoresby's Journal of a Voyage to the Northern Whale Fishery, &c.

Extreme poverty is the sole characteristic of the vegetation of the polar zone; whole countries within it are, on account of the barren soil, perfectly destitute of vegetation; and in others, the little and, for the most part, exceedingly pretty plants grow in turf-like patches, or at least socially. The principal genera which characterize the zone, or whose species, however few, are scarcely ever wanting, are Saxifraga, Dryas, Papaver, Andromeda, Juncus, Cochlearia, Cardamine, Pedicularis, Eriophorum, Ranunculus, Pyrola, Silene, Potentilla, Salix, &c. The flora of this zone possesses these genera, and even many of the species of these genera, in common with the region of Alpine plants, however great the distance between the mountains and the Pole may be. The following genera are considered peculiar to the polar zone; Parrya, Eutrema, Platypetalum, Phippsia, Colpodium, Dupontia, Pleuropogon, &c. Several species of these genera appear within the arctic zone, but they are not found on the mountains of more southern countries.

The relative proportion of the species, genera, and families of this polar flora to each other, is in all probability quite different from that in the arctic zone, but our materials are yet too scanty to enable us to determine this. The arboreous and shrubby vegetation, which alone can materially affect the physiognomy of nature, is totally wanting in the polar zone, and the culture of any plant for food is there impossible; but to give as complete a view of the vegetation of these countries as our means will allow, we have only to mention the principal plants of the several countries of this zone. According to the lists of plants in the works of Phipps* and W. Scoresby,† the following phænogamous plants have been found in Spitzbergen; Phippsia algida, Juncus campestris, Tillæa aquatica, Cochlearia danica, and C. groenlandica, Cardamine bellidifolia, Draba alpina,

* A Voyage towards the North Pole. Lond. 1774, p. 200–204.
† An Account of the Arctic Regions, &c. Edinb. 1820, viii. p. 75–76.

Dryas octopetala *L.*, Salix polaris, and Salix herbacea *L.*, Pedicularis hirsuta, Papaver nudicaule *L.*, Cerastium alpinum, Andromeda tetragona, Saxifraga oppositifolia, S. cernua, S. nivalis, S. rivularis, and S. cæspitosa. The number of Cryptogamia is proportionate to this; there are 19 species of lichens, which predominate in the flora of Spitzbergen by number of species and probably by their masses also. On Melville's Island, besides the above-named plants, grow a number of Ranunculaceæ, Compositæ, and Gramineæ, the most of the species of which appear also in the arctic zone as alpine forms. Eriophorum capitatum, E. angustifolium, Alopecurus alpinus, Phippsia algida are found on Melville's Island, and in the arctic and sub-arctic zones.

It is to be desired, that the vegetation of that part of Siberia and Nova Zembla, which projects beyond the arctic into the polar zone, was known; a comparison of the vegetation of this region, which is mostly in connection with the mainland, or even a continuation of it, with the insular flora of Spitzbergen, Melville's Island, &c., would be extremely interesting.

### *b. Division of the Vertical Range of Vegetation into Regions.*

We shall now divide the vertical range of vegetation into regions, just as we have already, in the preceding section, divided the horizontal range into different zones.

When ascending from the plain to the top of a mountain, we easily find in the different regions forms of plants similar to those which belong to the different zones from the latitude of the mountains to the pole. A mountain of the sub-tropical zone, for instance, which rises above the limit of perpetual snow, naturally exhibits the vegetation of those zones only which succeed each other from the sub-tropical zone to the polar regions; and so the mountains of the north of Norway and Sweden, as well as the Loffoden Islands, which lie above the Arctic Circle, can only show two different regions.

Baron v. Humboldt, in his later writings on botanical geography, divided the superficies of tropical mountains into three regions, the torrid, the temperate, and the frigid, corresponding to the common division of the surface of the globe, and he, at the same time, pointed out the principal sub-regions of this or that region. In these sub-regions are indicated almost all the divisions, corresponding to the eight zones of the surface of the globe, which I shall go over one by one, and I am convinced that this division into regions may be applied to the mountains of every zone. I also think that, in phytogeographical descriptions of various mountains and countries, it is necessary to proceed on one and the same principle; therefore one mountain ought not to be divided into a greater number of regions than another in the same latitude. By following this method, and always beginning the consideration of mountain-vegetation at the top of the mountain, that is, at the limit of perpetual snow, we shall always be able to follow out with accuracy the comparison of its vegetation with that of other mountains of different heights.

If all mountains rose to the limit of perpetual snow, it would certainly be much better to give the elevation and area of the different regions always from above downwards, viz., according to the various distances from the limit of perpetual snow; thus to adopt just the opposite method to that which has hitherto been followed.

From a comparison of the different heights of the snow line in the eight zones, we arrive at the result that, beginning at the polar zone, the snow line rises 1800 or 1900 feet for each zone, so that in the equatorial zone we find it at the elevation of 15,000 or 16,000 feet. This increase in the height of the limit of vegetation of 1900 feet for each zone, therefore, exactly corresponds to one of the eight regions into which I intend to divide the mountain vegetation of the equatorial zone, and therefore one region after another must disappear from the mountains of each successive zone from the equator to the poles. I think that the following Table will most clearly explain this. (See pages 226, 227.)

This division of the mountain flora into regions which correspond with the zones of the horizontal range of vegetation, may perhaps not meet with general approval, and it will therefore be necessary for me to combat here the chief objections which probably may be raised against it.

It is true that there is no such exact and regular distribution of vegetation into different regions, as is assumed in this table, but that there is often, on different mountains in the same latitude, a difference of several hundred feet in the height of the same vegetation; nay, that even the height of the snow line on mountains in the same zone varies not only sometimes some hundred, but even some thousand feet. The snow line in America on the Cordilleras beneath the equator, is fixed by Humboldt at the height of 15,736 Paris feet, but on the Cordilleras of Southern Peru, in 15°–18° south latitude, the limit of perpetual snow frequently lies above 17,000 feet; indeed the Volcan of Arequipa rises to the height of 18,373 feet, and yet I have observed snow only on one side of its peak. Hall* observed the lowest snow line on Cotopaxi at the height of 15,646 feet; on Antisana, of 15,838 feet, and on Chimborazo of 16,000 feet, while close at hand he observed a large field of snow on Cayambe at the height of 14,217 feet. So great are the variations between points lying close to each other! Nay the pass leading from Arequipa to La Paz, los Altos de Toledo, lies at the height of 15,600 feet, and yet we find there a vegetation of alpine plants and dwarf bushes, corresponding with the vegetation of

* Excursions, &c. Hooker's Journal of Botany. London, 1834, i. p. 343.

*Comparative Exhibition of the Different Zones*

| Distance of the regions from the plain. | Name of the Zones. | Equatorial Zone. | Tropical Zone. | Sub-Tropical Zone. | Warmer Temperate Zone. |
|---|---|---|---|---|---|
| | Area of the Zones. | lat. 0-15° | lat. 15-23° | lat. 23-34° | lat. 34-45° |
| | Mean Heat. | 26-28-30° Cels. | 23-26° C. | 18-21° C. | 12-16° C. |
| 15200′ | | 15200′ | 13300′ | 11400′ | 9500′ |
| 13300′ | Region of the Alpine Plants. | | 11400′ | 9500′ | 7600′ |
| 11400′ | Region of the Rhododendrons. | | 9500′ | 7600′ | 5700′ |
| 9500′ | Region of the Abietinæ. | | 7600′ | 5700′ | 3800′ |
| 7600′ | Region of the European Dicotyledonous Trees. | | 5700′ | 3800′ | 1900′ |
| 5700′ | Region of the Evergreen Dicotyledonous Trees. | | 3800′ | 1900′ | |
| 3800′ | Region of Myrtles and Laurels. | | 1900′ | | |
| 1900′ | Region of Tree Ferns and of Figs. | | | | |
| 0′ | Region of Palms and Bananas. | | | | |

*with the Corresponding Regions.*

| Colder Temperate Zone. | Sub-Arctic Zone. | Arctic Zone. | Polar Zone. | Distance of the regions from the Snow Line. | Mean annual heat of the regions. |
|---|---|---|---|---|---|
| lat. 45-58° | lat. 58-66° | lat. 66-72° | lat. 72-82° | | |
| 6-12° C. | 4-6° C. | 0 to −2° C. | — 2° C. and under | | |
| 7600′ | 5700′ | 3800′ | 1900′ | | |
| 5700′ | 3800′ | 1900′ | | 1900′ | 3-4° C. |
| 3800′ | 1900′ | | | 3800′ | 7° C. |
| 1900′ | | | | 5700′ | 11° C. |
| | | | | 7600′ | 14° C. |
| | | | | 9500′ | 17° C. |
| | | | | 11400′ | 20-21° C. |
| | | | | 13300′ | 23-5° C. |
| | | | | 15200′ | 27-30° C. |

the arctic zone. Similar inequalities may be pointed out on the different mountains of the Himalaya chain, where the snow line, though the mountains lie in the sub-tropical zone, rises, on account of the peculiar form of the mountain-masses, to the same height as on many parts of the Cordilleras beneath the equator. These are obviously great and most important differences in the height of the upper limit of vegetation, and the exact boundaries of the different regions which we have assumed in the Table by no means suit such anomalies. But all these exceptions from the rule are sufficiently explained in the present state of meteorology, and therefore it will still be the best method of proceeding to divide the mountain flora into different regions according to a general comprehensive principle, and to mark all exceptions from the rule as we go on. Though there may still appear to be great difficulties in the way, they are certainly only apparent, and in every special investigation of a mountain they may be brought under the rule laid down.

It is, besides, very easy to show, that vegetation is just as little strictly divided into different zones, since the course of the isothermal lines is in itself extremely irregular and becomes always more so as we approach the poles. But as the course of the isotheral lines, which chiefly determine the distribution of annual plants, is also irregular, it becomes extremely difficult to fix the boundaries of the several zones of vegetation, and there are everywhere exceptions to and variations from the rule ; but yet no one will hold that such a division of vegetation into zones is superfluous, it being absolutely necessary in order to have some means of arranging our information. Since, however, as has already been often proved, the gradual change of climate on a mountain from the plain to the limit of perpetual snow exactly corresponds with the changes of climate from the equator to the polar regions, and since the whole state of vegetation is regulated by the conditions of climate, no division of the mountain flora into regions can be better founded, than that which attempts to mark out regions corresponding with the different zones. It will indeed not unfrequently be found, that on some particular mountain this or that region is so faintly marked that it can scarcely be recognized ; but we also find this when considering vegetation according to its horizontal range, of which examples enough have been given in the preceding section.

It is to be wished that the geography of plants were treated somewhat more generally, than it is in recent works. It is certainly highly interesting to know exactly the upper and under limit of the station of a tree or plant, particularly when specially investigating the flora of a mountain ; but the determining them within a few feet is of much less importance to the general science, as these limits are found to vary much, even on different sides of the same mountain. I am of opinion that the extreme exactness, which it is intended to reach by such means, is only apparent, and of local interest only. It has long been proved, that the vegetation of the different regions of mountains corresponds with that of the corresponding zones of horizontal range ; and the sole aim of the present investigations into this subject is to learn and explain the apparent anomalies and exceptions from the rule, as well as to draw attention to the differences, which characterize the physiognomy of the vegetation of the corresponding regions and zones of different mountains.

As we have attempted, in the preceding section, to exhibit the physiognomy of vegetation in the different zones of the globe, from the equator to the poles, so we shall now, when delineating the vegetation of the regions, begin with the plain of the equatorial zone, and ascend to the region of perpetual snow.

### 1. *The Region of Palms and Bananas.*

The region of Palms and Bananas begins at the level

of the sea coast, and extends to an altitude of 1900 feet; where the atmosphere in temperature and humidity differs but little from that of the plain; the vegetation of this region accordingly is that of the equatorial zone, and I may, therefore, refer to the description of it, page 161 to page 171. A short recapitulation of the outlines of that description may, however, be desirable, in order to place somewhat more clearly before us the transition of the vegetation of this region into that of the one next following.

We have seen that the vegetation on the sea shore and at the mouth of rivers, is characterized by mangrove forests throughout the torrid zone,* and that the barren coasts are covered with species of Sesuvium, Portulacastrum, Heliotropium, and Convolvulus, with Lythrum maritimum and Roccellæ, while the adjacent more fertile tracts are adorned by species of Pandanus, Tournefortia, Dodonea, Sonneratia, and Barringtonia, above the thick foliage of which shoot up at intervals the slender trunks of palm trees. Next to these grow the social marsh-palms or vast forests of the graceful arborescent grasses; and then appear the primeval forests, where the ground cannot hold all the masses of vegetation, but one plant grows upon another, and climbers bind the branches and tops of the trees together with so close a net-work, that often not a sunbeam can pierce through the foliage. Here grow the colossal masses of single species and individuals, which we have already noticed in various parts of this work; here we find single fig-trees, the innumerable off-shoots from which form quite a forest, which yet remains in the closest union with the parent-tree. Reinwardt saw on the island of Semao a great forest, all the trees of which had been produced from a single tree of Ficus

* These mangrove forests are said to extend often far inland, and to be adorned by noble and splendid trees, or, as on the coasts of Brasil, by Sagus tædigera Mart.

benjamina.* These fig-groves have the same general appearance as the mangrove forests; they send down roots from their branches to the earth, which take root there and shoot up new trunks, while the Rhizophoræ germinate even on the parent plant and send roots to the ground, from which proceed new trees.

In this region, beginning at the sea-coast, whenever there is a rich, good soil, to the height of 1000 and even 1600 feet, the forms of Palma, Musa, Heliconia, Urania, Alpinia, and of the Scitamineæ are especially characteristic of the vegetation; in the New World the Cereus form, and its representatives, the Cactus-like Euphorbiæ, in the Old, begin here. The graceful Mimosa form here shows the most elegant shrubs, and the most gigantic trees, and the dicotyledonous trees are adorned with large and often beautifully shaped leaves. As the height increases, the Palms become less frequent, the Bananas appear of smaller size, the Scitamineæ begin to disappear, but the Orchideæ and Pothos plants become more numerous, and Peperomiæ grow on the bark of the trees, and at last show that we have entered the following region.

## 2. *Region of the Tree-Ferns and Figs.*

The second region on the mountains of the equatorial zone is that of the Tree-ferns, which are as characteristic of it, as the Palms and Bananas are of the lower region. Commencing at the height of 1900 feet, this region stretches as high as 3600 and 3800 feet, possessing a mean temperature of from 22° to 23.5° Cels. The extremely interesting form of the Tree-ferns is found only in a very damp climate; in barren soil, or even where the atmosphere is very dry, they are entirely wanting. Humboldt† extols the extraordinarily fine climate of

* Ueber den Charakter der Vegetation auf den Inseln des indischen Archipels. Berlin, 1828, p. 9.

† De Distributione Geogr. Plant. p. 91.

this region, where there is abundance of water, and where a luxuriant vegetation clothes the sides of the mountains. The delightful climate and the beautiful vegetation of the islands, which lie in the ocean near the confines of the torrid zone, where the lowest region runs almost parallel to the second region of the equatorial zone, have often and justly been extolled. Near the coast the Sandwich, Cape de Verd, and Ladrone Islands, as well as New Caledonia, the Isle of France, the Isle of Bourbon, and the most southern of the Friendly Islands in the southern hemisphere, possess, particularly in summer, almost the same climate as we find at the equator, and consequently the same vegetation prevails here as in equatorial regions, only somewhat less luxuriant, since there is generally a deficiency of soil and water. But in these islands the palms and bananas quickly disappear when we ascend above the level of the sea; and at the height of 300 or 400 feet we enter the region, where the shrubby and arborescent ferns predominate.

I have already (page 125) attempted to depict the beauty of this interesting form of plants. In their shade I saw splendid Strelitziæ in flower, and in the island of Luçon the singular Marantha raises its glossy shaft amidst the slender stems of the Cyatheæ, the trees of which in Java are tall and slender, like those in our pine forests. In the Brazils Von Martius observed trees of the handsome Alsophila excelsa and the Didymochlæna 25 feet high and six or eight inches thick. These Tree-ferns, however, are but seldom the characteristic form of this region, and when they are so, they predominate only in small tracts by their masses. In the New World the Cinchonæ appear along with the Tree-ferns on the Cordilleras of the northern part of South America; however these trees, which furnish the medicinal Peruvian bark, have a pretty extensive area; some species of them ascend nearly as high as 9000 feet,* while the ferns keep very closely between 1200 and 3000 or 4000 feet. The same altitude has lately been given by Von Martius for the station of the ferns in Brazil, for this traveller found the Tree-ferns, with the exception of a few species, almost always between the altitudes of 1200 and 3000 feet. On the mountains near Rio Janeiro, the first slender Tree-ferns I saw grew at the height of 1000 or 1100 feet; on the mountains of the island of Manilla these slender trees appear at the height of 1200 feet, while the station of the shrubby species lies as low as 300 and 400 feet. In the South Sea Islands of the tropical zone appear the magnificent trees of the family of Urticeæ, as the genera Broussonetia, Artocarpus, Boehmeria, Neraudia, &c., which have large and partially hairy leaves, and which furnish the natives of these islands with the materials of their clothing. But in India and in the islands of the Indian Archipelago, these genera are, as it were, replaced by the numerous species of Ficus, a genus of the same family. Most of the species of this genus grow in less elevated forests, the general characteristics of which, says Reinwardt, are the imperviousness and gloom, the density, and height of the wood; the damp close atmosphere within it, the enormous thickness, irregular shape and wide spread branches of the trees, their uncommonly rapid growth, and the soft, often spongy substance of the trunk, the great variety of parasitical and climbing plants which are supported on the trees, the deep, porous, damp parent soil, the numbers of Quadrumana, which leap screaming amidst the boughs, and the numerous choir of various birds which enlivens the thicket. Only a few species of Ficus, viz., those lower species with variegated and yellow leaves, ascend to a greater altitude, and they at the same time gradually diminish in size.† Shrubs, bushes, and herbs

* See A. v. Humboldt, Naturgemälde der Tropenländer, p. 62.

† See Reinwardt, l. c. p. 10.

of the genera Grevia, Elæocarpus, Phyllanthus, Ruellia, Justitia, Dimocarpus, Solanum, &c., form the underwood, and the beauty of the forests is increased by some plants of the Dracæna form, Dracæna terminalis and others, as well as by a great number of Aroideæ and Orchideæ, species of Cissus, Pepper plants, and wild Bananas. In the New World the exceedingly numerous species of Melastoma belong to the region of Tree-ferns, and their handsome, glossy leaves and large, blue or violet flowers give these trees a remarkably beautiful appearance ; they predominate by their greatest number of species in America, but they also appear in India, southern China and the islands of the Indian Archipelago not only as shrubs, but as lofty trees. Humboldt mentions that the shrubby species of Bocconia, Alstrœmeriæ of various colours, and Passifloræ, which are as tall and thick as our oaks, peculiarly belong to this zone on the Cordilleras of South America. Several slender and reed-like palms also belong to the region of Tree-ferns, for instance, Kunthia montana, Oreodoxa montana, Chamædorea gracilis, Martinezia caryotæfolia, which however show forms less characteristic than those of the other palms, whose station we have already pointed out.

It is remarkable that several species of Calceolaria, which belong to higher regions in the more southern zones of South America, appear in this region on the Peruvian Cordilleras.

### 3. *Region of the Myrtaceæ and Laurineæ.*

The third region on the mountains of the torrid zone corresponds with the sub-tropical zone, in which Myrtles, Magnoliæ, Camelliæ, and dicotyledonous trees with glossy leaves generally predominate, and in which the Proteæ, Eucalypti, Acaciæ, and Ericæ attain their maximum. At the equator this region, which I may call that of the Myrtaceæ and Laurineæ, begins at the height of 3800 or 3900 feet, and stretches to 5700 feet, where the laurels especially predominate. In the northern parts of Chili, lying in the sub-tropical zone, the vegetation of which quite corresponds with that of the region of Myrtaceæ, the Myrtles appear in great numbers and take possession of the whole lowest region of vegetation, where they grow most luxuriantly to the altitude of 1900 or even 2000 feet.

In this region of the mountains within the tropics, in the Old as well as the New World, predominate the genera Melastoma, Liquidambar, Styrax, Eugenia, noble oaks with glossy leaves, Ingæ, and often a number of lofty ferns or even Coniferæ. Reinwardt* remarks the beauty of the Rasamala woods of Java, which clearly appear to belong to this region ; the trees of these woods are probably a species of Liquidambar, which yields the true storax. Its beautiful, strong, lofty, and straight white trunk, which is less overgrown than in the Figs, and its more regular, thick crown of bright foliage point out, says Reinwardt, the higher woodland, which receives its character from this beautiful tree. A dense, thorny brushwood of several species of Calamus, and a great variety of Rubiaceæ, whose powers of secreting peculiar juices are discovered by strong odours diffused far around, fill up the spaces between the aromatic trees.

The Coniferæ appear at the height of 3000 feet on the mountains of Java ; the beautiful Podocarpus, rising majestically above the surrounding forest trees, which are already diminishing in size, grows along with Pinus dammara, on the trunks of which climbs the Nepenthes with its singular pitchers, while at their base appear handsome flowering Rhododendrons, tall, shrubby Ferns, Eugeniæ, Myrtles, Gardeniæ, Magnoliæ, and oaks, together with great masses of Orchideæ.

Schiede has given us a description of this region of Myrtaceæ on the Cordilleras of Mexico in the account

* L. c. p. 11.

of his excursions round Jalapa.* This town is situated at the height of 4200 feet, consequently in the region of Myrtaceæ and Laurineæ, and yet, as we learn from the description, not only do a number of the most luxuriant ferns ascend into this district, but there prevails in it such a variety and profusion of noble plants, that one would imagine he was again surveying the vegetation of the equatorial zone, if some trees quite foreign to that zone did not appear. The woods round Jalapa consist of species of Liquidambar, oaks, Ingæ, Clethræ and feathery Mimosæ, in the shade of which rise the broad fronds of arborescent ferns, whose trunks, three and four fathoms high, are covered with delicate mossy Trichomanes, and between them are groups of elegant dwarf-palms, with feathery leaves and black clusters of fruit placed on coral-red stalks ; Melastomæ, Rhexiæ, Myrtaceæ, and Laurineæ, from which hang climbing Sapindaceæ and Banisteriæ, with purple and orange flowers, while the ground is covered with the fresh verdure of mosses, Lycopodiæ, and Anemiæ. The culture of the banana extends far within this region ; it is also carried on very successfully throughout almost all the subtropical zone, and is even to be found in the south of Spain.

I have ascended the Peruvian Cordilleras within the tropical zone in two different latitudes, but I have found on them, at the altitude of this region, so scanty a vegetation that the tropical character could not be perceived. Tall, candelabra-like Cacti, Schinus molle, a number of Mimosæ, Bignoniaceæ, Loranthi, and especially Solaneæ, a few beautiful grasses and Cyperaceæ appeared in well watered places, while close at hand there was not a trace of vegetation.†

* Linnaea of 1829, page 218.

† I may here take the opportunity of remarking, that so full a description of the mountain vegetation, according to the different regions, as may be given in a special examination of a particular mountain, is by no means to be looked for here. Here, the character of the vegetation of the vari-

### 4. *The Region of the Evergreen Dicotyledonous Trees.*

The fourth region on the mountains of the torrid zone is that of the evergreen dicotyledonous trees ; it commences about the height of 5700 feet, and extends above that of 7600 feet. This region of evergreen trees, which corresponds to the warmer temperate zone, consequently to the vegetation of Southern Europe, generally possesses the most delightful climate, a mean temperature of 16°–17° Cels., and an abundant supply of water. In the south of Europe and in the north of Africa, the evergreens, as we have already abundantly shown in page 192, form the character of the vegetation, and our laurel appears here, representing as it were this numerous family of the hot zone.

On the mountains of Java, the laurel forests ascend to the height of 7000 feet, and above this elevation we first remark that the trees are no longer of their usual size and beauty. On tropical mountains a great number of laurels appear in the region of tree-ferns, and some are even found in the plain. Those dicotyledonous trees with rigid and shining leaves, such as Melastomæ and oaks, which appeared in great numbers in the preceding region, are by no means wanting here also. In the Cordilleras of the northern parts of South America the oaks generally occur at an elevation of almost

---

ous regions can only be indicated by some of its principal features, in order to draw attention to its conformity to that of the corresponding zones, which we have always treated of more fully. However, the greatest difficulty is the want of materials available to this work ; it is to be hoped that future travellers will pay more attention to the physiognomy of vegetation, and then a more accurate description of the various regions over the whole globe will become possible. It is also not to be denied, that in this way of exhibiting mountain floras, in which it is always attempted to compare the various zones with the corresponding regions, we partly lose sight of the changes of the vegetation of a mountain with the increasing height, and that we can become acquainted with them only by a special description of a single mountain. In this latter respect I can only draw attention to the beautiful and excellent works which specially treat of the vegetation of single mountains, and which have been so frequently made use of in this work.

5000 feet;* but on the mountains of the Philippines I observed noble oaks, with handsome, glossy, indented leaves, at the height of 1400 or 1500 feet. On the mountains of Mexico, at the confines of the torrid zone, and consequently in the tropical zone, according to our division, they also appear within the second region, which there corresponds with the sub-tropical zone. These oaks alone, says Humboldt, sometimes exhibit to the inhabitant of the tropics a faint picture of the awaking of nature when spring returns, for, in consequence of the drought, they lose all their leaves at once, and at the commencement of the rainy season the fresh and tender green of the new shoots contrasts very agreeably with the variegated flowers of the Epidendrum, the roots of which closely encircle the black fissured branches of the oaks. The celebrated Cheiranthostemon in Mexico, the gigantic tree of Toluca, which, with the Baobab, the far-famed Dragon-tree, and the immense Cotton-trees, is amongst the giants of the vegetable world, also grows in the region of laurels, but more properly belongs to the following region, in which the oaks predominate in yet greater numbers.

In the Canary Isles in the sub-tropical zone, the second region from the altitude of 2000 to 4000 feet, is the region of evergreen trees which Leopold v. Buch† designates as the region of dense forests. The laurels, viz., Laurus nobilis, L. fœtens, L. indica, and L. barbusano, together with Ardisiæ, Visnea mocanera, Ilex perado, Arbutus callicarpa, Olea excelsa, and Myrica faya, form dense forests, and in their shade grow Ranunculus teneriffæ, Geranium anemonifolium, Convolvulus canariensis, and species of Digitalis, Dracocephalum, and Sideritis.

Humboldt, making use of the observations of V. Buch and Chr. Smith, has given us a beautiful representation

* See A. v. Humboldt, Naturgemälde, &c. p. 71.

† L. c. p. 129.

of the distribution of plants on the Peak of Teneriffe.* On this Table the lower limit of the beautiful Erica arborea, and E. scoparia is marked below 3000 feet, and this height corresponds with the station of these plants in the warmer temperate zone of Europe, where they grow most luxuriantly in the lower hilly districts. An attentive inspection of Humboldt's Table of the distribution of vegetation on the Peak of Teneriffe, is to be strongly recommended in every respect ; no other representation can give so clear an idea of the changes which the vegetation of a mountain undergoes with the increasing height.

In Northern Chili, which belongs to the sub-tropical zone of the southern hemisphere, the region of evergreen trees, in which the Laurineæ play a chief part, lies, as in the Canary Isles, in the second region, viz., above the height of 1900 feet. In some districts, for instance on the Cuesta de Zapata, the Laurineæ, for example Laurus peumo and Laurus aromatica, appear at the height of 1500 or 1600 feet, along with Drimys chilensis and Smegdadermos quillay ; but on the banks of the Rio Tinquiririca, in the province of San Fernando, latitude 34° 30′, and above the height of 2000 feet, I saw for the first time the noble vegetation of the evergreen forests of this zone. Myrtles thirty and forty feet high, and with trunks three and four feet thick, still grow in this region, where Laurelia serrata sends up ten, twelve, and thirteen stems, each a foot in thickness, from a single root, and the Espino (Acacia caven), which in the plain is a shrub only, is here a high and beautiful tree. Lofty Escalloniæ alternate with thick columns of Cerei ; and amidst prickly Colletiæ, completely covered with red and white flowers, grow tall trees of Ephedra americana *Humb.* ; Mutisiæ cover the foliage of these singular vegetable forms with scarlet flowers, and species of Cissus form liana-like wreaths between the trees of

* Voyage de Humboldt et Bonpland. Prem. Part. Atlas Géogr. et Phys. du Nouveau Continent. Tab. 2.

Smegmaria, Peumus (Peumus fragrans) &c., and a profusion of plants with large gay flowers belonging to the genera Schizanthus, Alstrœmeria, Loranthus, Lobelia, &c., grow on the borders of these beautiful forests, which bear some resemblance to our beech woods. The barks of these trees are, as with us, covered with pretty lichens, many of which are the same as ours.

### 5. *The Region of Dicotyledonous Trees.*

The region of dicotyledonous trees is peculiarly difficult to characterize. On many tropical mountains it is perhaps but faintly defined; at least in authors we rarely find any notices respecting it. In the equatorial zone this region stretches from 7600 to 9500 feet above the sea, where a mean annual temperature of 14° Cels. prevails. In those mountainous parts of the torrid zone where the declivity of the mountains is very steep, a cool climate prevails at this elevation, and luxuriant arboreous vegetation does not ascend above the height of 8574 feet.* But on the mountains, which at great heights spread out in table lands, lofty trees ascend far above this and the following regions, and even to the region of the Rhododendrons.

The region of dicotyledonous trees corresponds with the colder temperate zone, where northern oaks and majestic beeches are the ornaments of the forests. This region seems to be entirely wanting on the Cordilleras at the equator, but the oaks there ascend far above 9000 feet, and are still predominant in several places. Nothing of this kind is to be found in those parts of Peru, from the 16th to the 19th parallel, which I have visited; from the dryness and extreme sterility of the soil and the total want of rain in those districts, they are perfectly devoid of vegetation; trees are rare, and always grow singly, indeed in many cases it seems uncertain whether

* A. v. Humboldt's Naturgemälde, page 73.

they have been planted by nature or by the hand of man.

At the northern limit of the torrid zone, on the Cordilleras of Mexico, the dicotyledonous trees of the colder temperate zone appear in great masses at corresponding elevations ; 1000 feet above Jalapa, at the altitude of 5000 feet, not only do the Mexican oaks predominate, but alders grow in company with them, and Ternstrœmiæ, Melastomæ, and Crotonæ compose the underwood of these forests. Nay, on the Sierra Colorado, above the village San Andres, which lies at the height of 5000 feet, Schiede found a wood of beeches (Fagus sylvatica), alders, oaks, and Clethræ, in which grew Melastomæ and Rhexiæ even to the very peak of the mountain.* Hyperica, Vaccinia, a Fuchsia, an Ascyron, Eryngium, Botrychium, Carex castilleja, and many other plants grow in the shade of these forests. On the Cordilleras of Northern Chili, which belongs to the sub-tropical zone, I found above the region of laurels where the Chilian Cereus had disappeared, and the Espino was still only a low bush, a small region which was covered with dicotyledonous trees resembling our beeches, but on account of the rapidly increasing altitude of the Cordillera it soon ceased, and after it appeared a forest of bushes, which are characteristic of this region on the Peruvian Cordilleras beneath the equator, as well as in Chili. Humboldt† names this region in the Peruvian Cordilleras, the region of Barnadesia, or of Duranta ellisii, and Duranta mutisii, as these three plants and the Berberis are said to characterize the high and rugged plains of Pasto and Quito. These plants, however, ascend far above the limits prescribed to the region of dicotyledonous trees, and have a peculiarly alpine appearance. Along with the splendid Barnadesiæ grow Castillejæ (C. integrifolia and C. fissifolia), Columellæ,

* L. c. page 220.

† Naturgemälde, page 73.

the silver-leaved Embothryum emarginatum, Clusiæ, and Calceolariæ; the latter genus is said not to extend beyond 1° 40′ north latitude. On the Chilian Cordilleras another species (Barnedesia flavescens *nob.*), which has little parchment-like leaves, and large brilliant yellow flowers, and which covers large tracts on the declivity of Monte Sillo,* corresponds with the Peruvian Barnadesia; the well-known Macraea rosea, forming a pretty bush, rivals this Barnadesia in the profusion and splendour of its flowers, and Wendtia gracilis *nob.*, a bush about the same height, adds to the variety of the plants which adorn the country.

As in the colder temperate zone the Coniferæ appear in great masses along with the forests of oaks and beeches, so we observe on the sides of the Peak of Teneriffe a region covered with Coniferæ, corresponding with that zone, and ascending far within the succeeding region to the height of about 6000 feet. Pinus canariensis here forms dense forests; all the large-leaved trees keep below the limit of this pine; Erica arborea only, according to Von Buch, ascends to the highest elevations. On the Cordilleras of the torrid zone of America, as well as on the Himalaya mountains, there are found Coniferæ, which, in the region of the European dicotyledonous trees, represent the Coniferæ of the colder temperate zone. In Chili and Peru the Ephedra, and in the north of South America some Cypresses, appear in this region. But the want of Coniferæ on the Cordilleras of South America is to be connected with the universal deficiency of this family in the southern hemisphere. The mountains of the East Indies, as well as of Mexico and the East, are richly clothed with firs and pines, but the vast forests of Abietinæ which appear on the plateau of Mexico belong to more elevated regions.

* See Meyen's Reise, &c. i. p. 307.

## 6. *The Region of Abietinæ.*

In an earlier section, in which we delineated the vegetation of the various zones, we remarked that it was extremely difficult to divide the vegetation of the colder from that of the warmer temperate zone, particularly with respect to the characteristics of the herbaceous vegetation, but that the separation of the sub-arctic from the colder temperate zone was still more difficult; and we drew attention, in several places, to the manner in which the characteristic features of the vegetation of these two zones partly disappear and partly pass into each other. It is still more difficult to separate, on the same principle, the vegetation of the higher regions of mountains; but future observations will certainly perfect what we have here been able to indicate by a few of the main features only.

The region of Abietinæ, on the mountains of the equatorial zone, stretches from the height of 9500 to 11,500 feet, and it possesses a mean temperature of 11° Cels. But on extensive plateaux, such as that of Mexico, which lies on the borders of the tropical zone, and where, therefore, this region lies 1900 feet lower, consequently between 7600 and 9500 feet, the mean temperature sometimes rises much higher, and changes and peculiarities in the vegetation are thus produced which are not at all in conformity with the vegetation of the corresponding zone.

The Coniferæ are wanting on the Peruvian Cordilleras, but the Escalloniæ predominate in place of them. According to Humboldt, the region of Escalloniæ begins at the height of 8900 feet, and ascends as high as 10,400 feet, and is characterized by the appearance of Wintera grenadensis. "A few trees of the orange Peruvian bark (Cinchona lanceifolia), a few Rhesiæ and Melastomæ, with dark violet, nearly purple flowers, are scattered over these deserts. Alstonia, from the leaves of which is extracted a sweetish, but very wholesome and strengthening tea, Escallonia tubar, and some species of Andromeda here shade Lobeliæ, Basellæ, and the Swertia quadricornis, which is constantly in flower."

The vegetation of this region appears to be extremely characteristic on the plateau of Mexico, which is fully described in the paper by Schiede, already so often named.* On that plateau, which lies at the height of 7400 feet, Schiede found forests of Abietinæ, consisting of Pinus occidentalis and a species of Cypress, which covered the greater part of the plain and the mountain slopes. But still more striking were the woods of arborescent Liliaceæ, viz., the Yuccæ, which rose with simple stems only branched a little at the top, and sedge like rigid leaves to the height of 30 feet and upwards. I have already (page 118) remarked that these Yuccæ and the junipers of this region are covered with filamentous silvery Tillandsiæ, in a manner similar to that in which the Usneæ sometimes appear in our damp forests. In this region, wherever these lofty trees are wanting, the dry ground is covered with herbs and low bushes. The host of Astragali, Daleæ, Synantheræ with yellow flowers, a little Croton with silvery leaves, and Cisti, cover wide spaces, and above them rise blue Lupins and Buddleæ with white leaves, Solanum, Tunas and other forms of Cactus, such as Melocactus and Mamillaria, and Agaves also are not wanting. Somewhat higher appear forests of oak and species of Arbutus.

Schiede† gives a very excellent description of the vegetation near the upper limit of this region, viz., at the height of 8000 feet, at the base of the Peak of Orizaba, and at the same time compares the natural beauties of this country with the forests of the Tyrol, which appear in similar circumstances. The occidental pine here also occupies wide tracts, in which grow scattered oaks and alders ; but, says Schiede, the rushing

* Linnæa of 1829, p. 224, &c.

† L. c. p. 226.

streams which refresh the ground are wanting, and, consequently, the profusion of herbaceous plants, which are the peculiarity of our Alps. Purple Steviæ grow here amidst isolated tufts of tall grasses, and species of Eryngium, Arenaria, and Hypoxis, along with violets and Ranunculi, break up the monotony. Lastly, forests of oaks, alders, and Coniferæ appear, the underwood of which is composed of shrubby species of Cineraria, Ribes, and Rhododendron; Vaccinia and Castillejæ grow here, and still higher up species of Pedicularis. This region of Abietinæ, therefore, in the Cordilleras of Mexico, stretches to the height of about 10,000 feet, and melts into the region of Rhododendrons, which of itself is there of small extent.

In the sub-tropical zone of the northern hemisphere, there properly remains for our consideration only the Peak of Teneriffe; for, as yet, we have not sufficient notices of the vegetation of the Himalaya chain. As the Peak of Teneriffe increases in steepness with the increasing height, the temperature on it must sink with corresponding rapidity; and, accordingly, we find a sinking of the various regions of vegetation on the upper part of it. We have already seen that the region of Pinus canariensis appears on the Peak in the region of dicotyledonous trees, and these forests of Abietinæ, which ascend only as high as 5900 feet, therefore extend but a short way into the region, which ought to correspond with the sub-tropical zone. But another form of vegetation appears at this altitude, and ascends to the height of 8000 feet, which partly fills up a portion of the region of Abietinæ, and partly occupies the place of the Rhododendrons; this form is Spartium nubigenum (Retama blanca), along with which grow Spartium microphyllum, Juniperus oxycedrus, &c.

Although Sicily, especially Etna, properly belongs to the warmer temperate zone, yet from the peculiar situation of this beautiful country it possesses so warm a climate, that the vegetation of Etna runs almost parallel with that of the Peak of Teneriffe. According to Philippi* the forests of pine (Pinus laricio) and of Betula alba on Etna ascend to the height of 6200 feet; Juniperus hemisphærica, on the contrary, grows at the height of 7100 feet, thus in the region of Rhododendrons. Genista ætnensis, which is peculiarly characteristic of this region, Juniperus hemisphærica, Astragalus siculus, and Berberis ætnensis *Presl.*, are here very frequent, but become predominant in the succeeding region.†

In the warmer temperate zone, the region of Abietinæ, according to the division we have made, ought to lie from the height of 3800 to 5700 feet; and the observations of Ramond and De Candolle‡ exactly coincide with this. The region of Abietinæ there begins at the height of 4000, and ascends as high as 5544 feet (viz. Pinus uncinata); at that elevation begins the region of bushes, which corresponds with our region of Rhododendrons, and lastly, the region of alpine plants appears above the height of 7800 feet.

On the Apennines, the region of Abietinæ exactly corresponds with our assumptions of the height of this region; we here make use of Schouw's description of the vegetation of this chain,§ the 3d and 4th regions of which we may join together, and place the lower limit of the Abietinæ at the height of 3800 feet. The beech predominates on the Apennines in the region between 3000 and 5000 feet, and Pinus picea, P. sylvestris and Taxus baccata rarely appear there, but it is said that at the height of 5000 feet the beech is no longer upright. It grows as a creeping bush as high as 7000 feet, and this is also the case with Pinus sylvestris. This equality

* Linnæa, vii. p. 745, &c.

† The vegetation of Etna has no resemblance to that of the Alps or of the Canary Isles. All the alpine plants of Etna, except Genista ætnensis, have been found in Sicily and the adjacent countries.

‡ See A. v. Humboldt, De Distribut. Geograph. Plant. p. 122, &c.

§ L. c. p. 475.

in the station of the Abietinæ and the beech is quite peculiar to these mountains, and deserves to be carefully inquired into.

The observations on the altitude of these upper regions of vegetation on the Swiss Alps do not coincide with the heights we assume for the mountains of the colder temperate zone, but on them these regions lie at almost the same elevations as on the mountains of the warmer temperate zone, viz., the Pyrenees. This may be perfectly explained by the trifling difference of latitude, and especially by the greater masses of highly elevated land in Switzerland. The region of Coniferæ in Switzerland extends from 4000 to 5500 feet, a range which almost exactly corresponds with that of this region in the warmer temperate zone. On the Silesian Riesengebirge, and on the Harz mountains, which lie within the colder temperate zone, the limit of the Coniferæ, almost exactly agreeing with our theoretical assumption, lies at the height of 3800 feet.

In conclusion, we shall glance at the vegetation of the sub-arctic zone, which corresponds with the region of the Coniferæ on the mountains of warmer countries; and we find that the Coniferæ there, viz., Pinus sylvestris, ascend to the height of 1200 feet only, and thus remain in the region of the plain, and no longer ascend into the second region, the vegetation of which is that of the arctic zone.

The region of Abietinæ agrees with the sub-alpine region of Wahlenberg, Schouw, and several other authors.

### 7. *The Region of Rhododendrons.*

The region of Rhododendrons is that which, for the mountains of Europe, has received from other authors the name of the lower alpine region; it corresponds with the vegetation of the arctic zone, viz., from the Arctic Circle to 72° of latitude. This elevated region is destitute of tall trees ; on most of the mountains of the northern hemisphere, only low, bushy species of birch and pine are found in it, and the extremely characteristic genus Rhododendron appears along with dwarf species of willow. This region probably, in general, possesses a mean temperature of from 5° to 7° Cels. On the mountains of the equatorial zone the region of Bejaria, which represents the Rhododendron of the Old World, ought to extend from 11,400 to 13,300 feet in height, and, in fact, a number of observations may be given, which clearly point out this region on various mountains of the torrid zone, though the Bejariæ very seldom appear. Below the equator the Bejariæ of the Andes (chiefly Bejaria æstivans, B. coarctata, and B. grandiflora) clothe the mountains even to the highest Paramos, above the height of 10,000 feet ; however, the Bejariæ, like the Rhododendrons of Switzerland, and particularly Rhododendron ferrugineum in Tyrol, are found in extremely low situations, for instance, on the Silla de Caracas, according to Humboldt,* at the height of 6000 feet ; and in Florida, in latitude 30°, a Bejaria already grows freely on low hills, and Rhododendron lapponicum reaches the plain in the arctic zone. I am inclined to think, that the Bejariæ rather appear in the region of Escalloniæ and Wintera, and that it is in form only that they represent the Rhododendron on the Cordilleras. Bejaria ledifolia on the Silla de Caracas is not above three or four feet high ; its stem divides at the ground into numerous brittle and almost verticillate branches, its leaves are ovate-lanceolate, greyish green on their under surface, and rolled in at the edge. The whole plant is covered with long, clammy hairs, and has a very agreeable resinous scent. Bees frequent its beautiful purple flowers, which, as in all the Rhododendrons, are uncommonly numerous, and are almost an inch in diameter, when fully expanded.

* See A. v. Humboldt, Reise, &c. ii. p. 425.

In this lower alpine region, at the height of 12,700 feet, lies the noble valley of Chuquito round the alpine lake of Titicaca, the vegetation of which I have described at length in my account of my journey to it. In my rapid journey through this rich plateau, and during my short stay in it, I did not find a Bejaria, nor an Escallonia, though I have no doubt that these interesting plants of the Cordilleras grow there. This celebrated plateau of Chuquito is, as I have already often said, perfectly devoid of trees, although there is in it a great profusion of bushes and herbs. The agriculture of the aborigines was confined to the Quinoa and the potato; rye, barley, and oats are now cultivated, yet only the oats and, more rarely, the barley ripen, and the rye is used as fodder. More ample accounts may be found in my Travels, vol. i. p. 403, &c. On the shores of the lake there is often the most luxuriant vegetation; splendid Cassiæ, tall species of Celsia, Gnaphalia, Calceolariæ and Loasæ with extraordinarily large flowers and prickly leaves, occur in great numbers. Discariæ are here tall bushes, and beautiful Cactaceæ, Cerei as well as Pereskiæ, grow on the sides of the hills and are completely covered with flowers. A beautiful turf of numerous new grasses clothes the hills, and a thick forest of rushes encircles the margin of the lake.

On the European mountains, which have already so often been the objects of special phytogeographical investigation, the station of the Rhododendron is universally close beneath the region of alpine herbaceous plants, and I only remark as a further characteristic, that Vaccinium, Andromeda, Ledum palustre, and such little bushes with stiff and glossy leaves appear along with it.

According to Philippi, the Rhododendron is, as it were, replaced on Etna by Astragalus siculus, which is the predominant plant in the corresponding region from 3200 to 7200 feet in height, especially at the latter elevation. This plant forms a thick, half-globular mound about 5 feet in diameter and 2½ in height; a mode of growth which resembles that of several plants in the alpine region of the Cordilleras.

### 8. *The Region of Alpine Plants.*

The region of alpine plants commences on the heights of the mountains at the upper limit of bushes, and thence stretches to the perpetual snow which is the final limit of vegetation. The vegetation of this region corresponds with that of the polar zone, which extends from the northern boundary of vegetation to the limit of bushes and arboreous vegetation, which we have placed in the 72d degree of north latitude. The mean annual temperature of the polar zone lies far below the freezing point of water; that of the region of alpine plants, on the contrary, is far higher, even as much as 3° or 4° above the freezing point, and yet, as we have already shown by examples, p. 17, vegetation is often far richer in the colder countries than on mountain heights, where the temperature of the year stands higher. This is most clearly shown by the culture of the cereals, which extends further towards the pole than towards the corresponding regions of mountains, and this fact is to be explained by the proportionably lower summer temperature on the heights of mountains, of which we spoke more fully in the first part of this book. On those mountains which at great elevations spread into vast plateaux, the mean summer temperature is higher than it generally is at the same altitude on the sides of mountains.

We give the general name of alpine plants to those which grow in the highest region of mountains to the very limit of perpetual snow, and many peculiarities which distinguish them from the plants of the plain have been pointed out. The most general characteristic is, perhaps, their gregariousness, which depends on a singular tenacity of life, on a remarkable development of root, and on a certain inclination to form buds. Almost all alpine plants are perennial; the number of annuals amongst them is exceedingly small, and these are such plants as produce a very large quantity of seeds. The root of these perennial plants which has to bear a rigorous winter, often lasting nine or ten months, is usually very woody, or, like those of bulbous plants, wrapped up in a number of skins; the distribution of these plants, therefore, is regulated chiefly by the mean annual temperature, while the annuals are influenced by the mean summer heat.

Alpine plants are universally celebrated for their proportionably large flowers, which for the most part are of bright and lively colours, and we find confirmation of this praise on the mountains of every zone. On our European mountains the beautiful large-flowered Gentians, the handsome Aretia alpina, the Dryas octopetala with large brilliant white flowers, the lovely Anemones, Primulæ, and the numerous Syngenesia with large yellow flowers, as Arnica montana, Apargia alpina, &c., are well known. On the summits of the Cordilleras of South America, this peculiarity is still more evident; various species of Mimulus, Calceolaria, Calandrinia, Lupinus, and, in particular, several species of Sida with remarkably beautiful and large flowers grow there. We cannot fix on any colour as predominant in the flowers of alpine plants; it has, indeed, been believed, that amongst them white is more frequent than any other colour, but Schouw* has proved that this is not the case on the mountains of Europe; and I can add, that on the heights of the Cordilleras of South America I very rarely met with white flowers, and at several points, where I ascended these mountains almost to the limit of perpetual snow, I did not see a single white flower. Blue, yellow, and violet are the predominant

* Grundzüge, &c. p. 461.

colours among the alpine flowers of the Cordilleras of Peru and Chili.

Alpine plants are usually regarded as those which abound in aromatic, bitter, or resinous substances, and this is confirmed in every zone of the earth. The question now is, whether these substances are a consequence of the station, or whether they are inherent properties of the plants. It is well known that the alpine plants used in medicine are far more powerful when they are gathered in their natural station, than when grown in the gardens of the plain, and this fact is an indisputable proof that the station on the mountain height exercises considerable influence on the production of these medicinal substances. It must not, however, be forgotten that the families and genera most frequent amongst alpine plants are those which universally contain such substances even when growing on the plain. I believe that not a species among alpine plants is known which contains a bitter, aromatic, or resinous substance, if such is not also present in the allied species of the plain; but it is certainly the case that such substances are proportionably much stronger in the alpine species of these genera and families than in those which grow in the plain. The families Compositæ and Umbelliferæ, with the genus Gentiana, are the alpine plants most commonly distinguished by containing a greater quantity of powerful medicinal substances. Near the snow-line on the Cordilleras of South America, the greater number of plants are full of a more or less fragrant, bitter gum, which in the pretty Laretia acaulis *Hook.* (Selinum acaule *Cav.*), is secreted in great quantities, and lies scattered on the surface of the plant. The extraordinarily numerous little shrubby Syngenesia, which, on the Cordilleras of South America, ascend into the region of alpine plants, are exceedingly rich in resinous aromatic substances, and their foliage consists of little, hard, smooth, and glossy leaves, which very rarely show a few hairs. The leaves of these Syngenesia, as well as the whole stem, are generally covered with a resinous secretion, which by no means appears in the same proportion on the species of the same family growing on the plain.

Besides the excessive development of the root and flower, an imperfection of the leaves has been mentioned as a general characteristic of alpine plants; the leaves are said to crumple together, and to become puckered on their upper surface, their green colour is said to disappear in part, and a yellowish hue to appear in place of it, while the leaves at the same time become membranous.* Parrot also thinks that the peculiar character of alpine vegetation consists in this; that the plants during their whole growth are continually striving not to rise high above the ground, and, consequently, form a short and strong, or crooked and prostrate stalk, on which branches, leaves, and flowers are closely pressed on each other.

It is certain that much of what has here been said meets with full confirmation; alpine plants have something so extremely characteristic in their appearance that they may at once be pointed out, even in large collections of dried plants, yet all these characteristic marks belong to those plants also, which grow in the plain in corresponding latitudes, as in the polar and arctic zones. It is, therefore, not the rarefied atmosphere which calls forth the characteristics of alpine vegetation, but it is the operation of the low temperature which prevents the rapid development of leaf-buds; the plants must, therefore, become strong and compact, which is the cause of a greater profusion of flowers. When the plants of the polar zone are, by the peculiarities of coast climate, brought down to the sea shore in the arctic zone, they lose all the peculiarities which once belonged to them as alpine plants; in particular the proportionably large

* See Parrot's Reise nach dem Ararat. Berlin, 1844. 2 Thle.

flowers disappear.* I do not think that we ought to judge of the whole of alpine vegetation from observations on mountains in one and the same latitude. On our northern mountains of the Old World a number of plants are distinguished by wrinkled and very hairy leaves, which are not so bright a green as usual, yet among alpine plants of the same latitude there is no want of such as have thick, fleshy, and smooth leaves. Indeed, plants of this latter sort are those which are found in the greatest numbers on the Chilian Cordilleras, even in the highest regions. The large genus Calandrinia, the genera Alstrœmeria and Oxalis, as well as the Boopidæ show those smooth and more juicy leaves, yet here also there are not wanting plants distinguished by hairy, singularly wrinkled leaves, for example, a number of species of Sida, Calceolaria, and Loasa, even several Syngenesia and also the Nassauviæ. The leaves of these species of Sida and Calceolaria resemble those of our genus Pedicularis, yet several species of these genera which grow in the plain also show Pedicularis-like leaves, just as there are several species of Pedicularis, which belong to the plain of our zone, and yet have as crumpled, though less hairy leaves.

The greater number of the low shrubs which, on the highest parts of the Cordilleras, come as it were in the place of our herbaceous arctic willows, and which for the most part belong to the Syngenesia, have very strong, leathery, and generally smooth leaves, which are mostly so singular in form that without a knowledge of the flowers, one would scarcely recognize in them the leaves of syngenesious plants. The Baccharidæ, the number of whose species is so great, are the most remarkable of these shrubs. I here give the names of a few of the most singular species of this genus which grow in the highest regions of the Cordilleras of Peru, the singular forms of which may generally be guessed from the speci-

* See also Lessing, l. c. p. 291, &c.

fic names, as Baccharis genistelloides *Hook.*, B. phylicæformis *nob.*, B. quadrangularis *nob.*, B. sagittalis *Less.*, &c.

Although the number of alpine herbaceous plants on the various mountains of the globe is so extraordinarily large, yet a remarkable similarity amongst them prevails, at least on the mountains of the same hemisphere, though it is doubtless the case that each of the great mountain chains has its peculiar alpine plants.

The alpine plants, as we have already seen, very exactly agree with the vegetation of the polar zone, and this similarity in the character of the vegetation may be traced from the polar regions to the equator, viz., if we compare, on the mountains of the different zones, the corresponding higher regions with each other. It is difficult to decide in which zone the number of alpine plants is greatest, as this is greatly influenced by the differences of the soil. On the mountains which form extensive plateaus in the region of alpine plants, their number is very great, provided the soil is not too sterile, and on the same grounds I think I may affirm that the greatest quantity of alpine plants, in respect of individuals as well as of numbers of species and genera, belongs to the polar zone. There are, it is true, among alpine plants very few genera which are not also found in the plain, but there are a number of genera which have species either chiefly alpine, or belonging solely to the regions of alpine plants. The following genera are those which furnish the chief alpine herbaceous plants belonging to the northern hemisphere of the Old World and to the mountains of Java (we are acquainted with no other mountain flora for the southern hemisphere); Dryas, Saxifraga, Viola, Phyteuma, Arabis, Epilobium, Draba, Arenaria, Pedicularis, Primula, Androsace, Ramondia, Soldanella, Phaca, Gentiana, Salix, Carex, Astragalus, and some genera of Gramineæ and Compositæ.

It is chiefly of species of these genera that the vegetation on the mountains of Europe and Asia, in the region of alpine plants to the very limit of perpetual snow, is composed. Some of these genera, as Primula, Campanula, and Phyteuma, are more proper to the alpine region of lower latitudes; others again, as Carex, Salix, Arbutus, &c., are more frequent in the alpine regions of higher latitudes, and in the polar zone.

The New World, the vegetation of which differs so much from that of the Old World, also presents considerable differences with respect to alpine plants. Although near the snow line of the Cordilleras many forms appear to which there are plants very similar on the mountains of the Old World, yet in America the greater number of alpine plants are peculiar to the mountains of that continent. The species which occur in the highest regions of the Himalaya mountains belong to the genera Ranunculus, Aconitum, Geranium, Potentilla, Epilobium, Carduus, Senecio, Inula, Cineraria, Myosotis, Primula, Pedicularis, Salvia, Lamium, Origanum, and Polygonum,* and they compose a vegetation which is most decidedly of the same nature as that of the corresponding regions and zones of more northern countries. On the island of Java there are, it is true, no mountains which rise to the snow line, yet the vegetation of the highest regions of this island belongs to the genera Valeriana, Ranunculus, Bellis, Hypericum, Gnaphalium, Swertia, Gentiana, Viola, Potentilla, Centaurea, Spiræa, Carex, Sphagnum, &c.;† it is, however, according to Reinwardt, to be observed, that of all these phænogamous plants on the mountains of Java, there is not a single species which exactly agrees with those of more northern countries; the bog-moss (Sphagnum) alone is said to be the same as that of Northern Europe.

It is a great and sensible defect in our knowledge, that as yet we are not acquainted with the botany of any mountain in the southern hemisphere of the Old World, which rises as high as the limit of perpetual

* See Royle Illustrations, l. c. p. 32.

† See Reinwardt, l. c. p. 13.

snow; we should perceive from it, whether the alpine plants of these mountains show as great differences as the vegetation of the plain in these continents, or whether they agree with the alpine plants of South America.

On comparing the accounts of several travellers who have ascended above the snow-line on the Cordilleras, we find among the plants observed by them in this region of alpine plants, a great number which are very similar to those of our European mountains. As such I name Draba alyssoides, D. aretioides, Cerastium densum, and species of Gentiana, Andromeda, Valeriana, and Lupinus, which Hall found on the peak of Pinchincha,* in the same year in which I ascended to the snow limit of the South American Cordilleras at four different points. I also, in Chili as well as Southern Peru, observed in the region of alpine vegetation many alpine plants which were very similar to ours, as species of Epilobium, Lupinus, Ribes, Viola, Genista, Luzula, Hordeum, Phleum, Plantago, &c.;† but at the same time there were not wanting genera peculiar to the mountains of America. The region of alpine plants on the Cordilleras possesses an extraordinary treasure in the great number of pretty little Umbelliferæ which belong to the Mulineæ, *D. C.* The further south the more frequent become the genera and species of this group of plants, which at last, even in latitude 52° (see p. 206), appear in the plain, where, on account of the peculiarity of coast climate, they find a climate which corresponds with that of the high mountain peaks in Northern Chili and in Peru. In Europe, the Primulaceæ come in the place of the Mulineæ of America; the genus Androsace, and more especially the Aretiæ, on the mountains of Europe frequently present an appearance similar to that of the ge-

* Excursions in the neighbourhood of Quito, &c. Hooker's Journal of Botany. London, 1834, i. p. 338.

† See several parts of my Reise um die Erde, i. p. 315, 348, 349, 451, &c.

nera Fragosa, Bolax, Azorella, Laretia, &c., in America. I have already, in p. 85, spoken at large on the extremely peculiar social growth of these plants, and now refer to that place. The Mulineæ are allied to the singular Boopideæ (Calycereæ *Brown*); they also are peculiar to the highest region of the Cordilleras.

Besides these I name the genera Calandrinia, Espeletia, Oxalis, Acæna, Nierembergia, Alstrœmeria, Culcitium, Chuquiraga, and Sida, which have the greatest share in forming the vegetation of this elevated region of the Cordilleras. I was very much surprised when I saw the magnificent and extremely peculiar vegetation at the base of the volcano of Maipu.* Several species of Oxalis, growing socially, and the rose-coloured flowers of Calandrinia umbellata *R.* et *P.*, C. denticulata *Hook.* and C. biflora *n. sp.*, covered some parts of the country with a red carpet, while noble meadows, composed of Phleum hænkii, (corresponding with Phleum alpinum,) of Vilfa asperifolia *n. sp.*, Deyenzia velutina *n. sp.*, Hordeum comosum, &c., extended even to the snow, and were only occasionally interrupted by large fields covered with thousands of the large, yellow and violet flowers of the Mimulus and Calceolaria, along with which appeared little stunted bushes of Adesmia with orange flowers, and little shrubby Syngenesia.

The region of alpine plants is also not deficient in lichens, and these, even in the most diverse zones of the earth, correspond with the lichens of the polar zone still more than is the case with the phænogamous plants; the Gyrophoræ only have not yet been found on the heights of tropical mountains, but on the mountains of Southern Peru in their place appear large Parmeliæ, which in their form, their shield-like apothecia, and their colour, show quite the habit of the Gyrophoræ. Licidea geographica has been found on the highest parts of the most distant mountains, being generally the last

* Meyen's Reise, i. p. 349.

of the vegetation, if any rocks rise above the ground. Humboldt observed this pretty lichen on the summit of Chimborazo, and Schiede* found it on the Volcan de Orizaba, while it has also been observed by me on various very elevated parts of Southern Peru. The plants which ended the vegetation of the peak of the Volcan de Orizaba belonged, according to Schiede, to the genera Lupinus, Eryngium, Myosotis, Sisymbrium, Draba, Trisetum, Avena, and before all the beautiful Cnicus nivalis was to be remarked.

## II. THE STATISTICS OF PLANTS.

In the beginning of this book (p. 3) I remarked, and proved by examples, that the number of species of plants continually increases as we recede from the poles and approach the equator; want of water and a perfectly barren soil are the only invincible obstacles to this law of nature. Desert tracts at the equator are quite as poor in plants, as they are in our northern regions; but where the greater heat of the tracts lying near the equator is united with a corresponding degree of humidity, the barren soil itself is overcome, and a number of plants appear, which seem to thrive best upon it. But we have also seen that, with the gradual increase of number of species from the polar zone to the equator, the forms of vegetation also become more noble. I have given a sketch of the physiognomy of vegetation from the equator to the polar zones, and in this lies the proof of the latter assertion. The more noble forms of vegetation either appear in the hot zones and are altogether wanting in the colder, as the Palms and Scitamineæ; or they predominate in hot countries, while only a small number are found near the poles, as the Leguminosæ. On account of the less frequent occurrence of the more

* Linnæa, 1829, p. 223.

developed forms of vegetation in colder countries, the number of the less developed forms in them is apparently greater; viz., their number becomes always greater in proportion to the number of the more developed plants, though their absolute number, as well as that of the whole mass of plants, decreases as we recede from the equator. Every catalogue of plants for any country or limited area, which can lay claim to a certain degree of accuracy and completeness, will serve to prove what has now been said; and the method of using these catalogues is founded on the simplest reckoning, by which we find out the number of species of the smaller, as well as of the larger groups of plants, and compare the numbers obtained with each other, or with the whole number of all the species of a country.

I have already shown (page 4), that the whole number of plants is still far from being known with sufficient exactness; but that their number, if we may draw conclusions from the results of recent travels, must be set down as more than 200,000. As yet, however, different countries of the same zone are so unequally known, in respect of the number of species belonging to them, that we can scarcely give with any exactness the number which belong to each of the great zones. It was for a long time thought, that the New World, in proportion, possessed a larger number of species than the Old World, and this opinion was supported by facts; but, judging from the enormous collections of plants which have more recently been sent from some hot countries of the Old World, and from the exceedingly great variety in the luxuriant vegetation of India, and the large islands adjacent to it, which I myself have witnessed, I can no longer assent to this opinion. It is of course necessary, in comparisons of this nature, to choose tracts of equal extent, equal elevation, and containing equal mountain-masses, and above all belonging to the same zone, and of equal fertility, which depends on their possessing an equal quantity of heat, moisture, and mould.

Just as little can there be given, in the present state of knowledge, a proportional number for the plants of the northern and southern hemisphere; in the latter half of the Old World, where the areas of single species are often so extraordinarily limited, there may, in comparison with equal surfaces of the northern hemisphere, be pointed out a greater number of species, yet the greater sterility of many of these countries may again equalize the whole number of species for this zone. The calculations, which might be made with our present materials, could not yield any results approaching the truth.

Another opinion, viz., that islands are poorer in plants than continents, deserves a close examination. L. v. Buch* is the first who has expressed this opinion, which has been repeated by many authors, and has lately received farther support from observations made by M. De Candolle, jun.,† although Schouw has, I think, long ago overturned this opinion by incontestable facts.‡ In his later edition of the Flora§ we have named, Von Buch has stated this opinion somewhat more precisely, probably on account of the objections made to it by Schouw, and I therefore quote the whole passage: "In the small number of species of plants seems expressed the nature of islands, whose number of plants diminishes, the further distant they are from continents, provided they themselves do not extend to a small continent."

The Flora of the Canary Isles, according to Von Buch's account, does not contain more than 377 species of plants, and, in the opinion of this distinguished naturalist, not a fourth of this number would be found

* Allgemeine Uebersicht der Flora der Canarischen Inseln. Berlin, 1819, p. 21.

† Fragment d'un Discours sur la Géographie Botanique prononcé à Genève, lu le 16 Juin, 1834, dans une cérémonie Académique. Bibliothèque Universelle. Mai, 1834.

‡ Grundzüge, p. 493.

§ Contained in the Physikalischen Beschreibung der Canarischen Inseln. Berlin, 1825, p. 130.

to belong to the Azores, were their flora known to us. This supposition may be perfectly correct, but as the Azores are known to be extremely barren islands, we can compare their flora only with equally barren parts of the continent. If the degree of the fertility of the soil, or the quantity of moisture in the atmosphere, be disregarded in such calculations, the result is, in my opinion, of little value. The sandy steppes and well-known deserts in the interior of continents, are just as poor in plants as the most sterile and remotest islands. The distance of the Sandwich Islands from the American continent is three times as great as the distance of the Azores from the coast of Europe, and yet the Sandwich Islands are extraordinarily rich in plants. The lowest region of these lands, which rises little above the highest level of the sea, is, on account of the coralline soil, &c., very barren, and, therefore, very poor in plants; but whenever this plain is left behind, even at the height of 100 and 200 feet, the extremely luxuriant vegetation begins. The number of plants belonging to these islands already described is, it is true, not so extraordinarily large, but the number of those collected there, and now chiefly to be found in the herbaria of the English, may lay claim to be considered so. But I do not think that the group of the Sandwich Islands can be regarded as a small continent; this would be a mere subterfuge. De Candolle, jun., instances the flora of New Zealand, as proving the position laid down by Von Buch, believing that New Zealand does not possess more than 800 species of plants, though it occupies almost the same space as Italy. But I can find nothing in confirmation of this opinion, that New Zealand is so poor in plants; indeed, from the few, but accurate, notices in Cook's voyages, I have conceived a very different idea of the flora of this great island.* A comparison of the vegetation of St. Helena, or of Ascen-

* See p. 196, &c.

sion, with the vegetation of other islands which are situated at equal distances from the nearest Continents —for example, Norfolk Island—will very clearly show how much in all these cases depends on an equal fertility of the soil, in lands or islands of equal size and of the same latitude. Though the island of St. Helena is of considerable size, a large part of it, where the rock consists of a very hard basalt, which resists the influence of the weather, is quite barren. In this island there is a more or less vigorous vegetation in such places only as possess some soil and water, the quantity of which has been increased by artificial means.

I therefore, after allowing for local circumstances, bring the vegetation of islands also under that law of nature, according to which the number of species constantly increases with increasing heat and corresponding humidity. It is again a question, whether vegetation becomes gradually richer in number of individuals also, as it recedes from the pole and approaches the equator, which has certainly been proved to be the case for the number of species. Schouw has touched on this question,* but his opinion is directly opposite to the answer which I would give to it: viz., he thinks that the number of individuals does not increase as the equator is approached, which I, on the contrary, after contemplating the luxuriant vegetation of tropical countries, affirm to be the case. Schouw gives, as the ground for his opinion, that in the torrid zone the plants generally grow to a larger size than in the colder zones, and, consequently, there cannot be so many individuals in an equal space. Little importance, however, can be attached to this reason, for the denseness of tropical forests, and the enormous masses of parasitical plants, which are found on every tree of any importance, certainly do away with the idea of any diminution in the number of individuals, in consequence of their greater

* Grundzüge, p. 391.

size. It is very generally and positively said, that the Cryptogamia decrease in number of species as the latitude becomes lower, and that they must, therefore, be predominant in the colder regions; but neither can I assent to this opinion, and Gaudichaud, who has investigated the botany of so many different parts of the torrid and temperate zones, has also lately expressed himself against this assumption.

In all such comparisons exterior circumstances must of course be attended to, and the vegetation of a dry part of the torrid zone must not, in respect of the number of Cryptogamia, be put in comparison with a damp part of the temperate zone. The Cryptogamia appear in greater numbers in damp places than in dry; and if we examine such damp parts of the torrid zone, we shall often be amazed at the enormous quantity of them. What an immense number of lichens have already been brought from Brazil! In the damp forests of the tropics these lichens also are covered with a still greater number of Jungermanniæ. But a few points only of the torrid zone have been carefully examined in respect of their Cryptogamia; and even in those places such care as we are accustomed to with us, has never yet been employed in their investigation.

If we now accept as proved what has been already intimated,—that, as the temperature on the surface of the globe becomes higher, not only does the number of species and individuals of plants increase, but the forms become more noble,—we shall recognise in this a law, according to which nature has distributed the whole mass of vegetation over the surface of the earth. This so simple result may perhaps be sufficient to oppose every idea of the distribution of organic being by migrations; but there are many other facts which it is impossible to explain on the theory of a migration of plants. Phleum alpinum, Botrychium lunaria, and several other plants, which are quite the same as those growing with us, likewise grow in Terra del Fuego, although they are entirely wanting in all the interjacent zones and regions. By what means can the seeds of these plants have wandered from us to that remotest part of America ? But in Terra del Fuego there prevails a climate similar to ours and to that in the subarctic zone. Why, then, will we not acknowledge what is so obvious, that nature has produced similar, and even the same, forms in these countries so far distant from each other, because not only similar, but often the same conditions are found in these countries ? There is indeed nothing more easy to perceive, in the distribution of organic being over the globe, than the universal law, that nature, in similar circumstances, has always produced similar or perfectly the same creatures.

In the preceding section we have learned that in corresponding zones and in corresponding regions, though in the most remote parts of the earth, vegetation not only shows the greatest similarity in its general features, but also, under similar conditions of climate, frequently a number of similar, or even the same forms ; and this has led us to the indisputable conclusion, that a great many plants cannot have spread from a single station, but that they must have originated in several parts of the earth.

If we take the plants of the region of alpine herbs, which are so often the same on the mountain-tops of the most distant zones, we shall convince ourselves of this with the greatest certainty. The regions of alpine plants on the different mountains are to be considered as islands in a great aërial ocean ; hundreds, nay, often thousands, of miles distant from each other, they yet have many plants in common, and the most of the plants on these islands in the atmospheric sea are at least extraordinarily similar to each other. How can these plants have come from one mountain-summit to another, just where a similar climate prevails, while they are altogether wanting in the plain which lies between these mountains, as well as on the lower parts of them. Nay, we know that a great many of these alpine plants will very rarely grow in warmer plains without particular attention. Such ideas of the migrations of plants must therefore be entirely given up, now that we have such an enormous number of facts relative to the stations of plants. The hypotheses of earlier times, which were invented by the greatest naturalists of their day, may be excused by the small number of observations which had then been collected on this branch of knowledge.

The question, whether nature has created only one individual of each species of plant, or several of them in each circle of its stations, can be answered only by a few conjectures, resting on the uniformity which we everywhere perceive in the creating principle ; facts are not to be made use of here, and all the arguments which we can bring forward against the creation of a single individual of each species, might also be adduced against the origin of the whole human race from a single pair. The question, as to the source and number of the Autochthones, is here, as in the animal creation, very difficult to investigate, and cannot be answered.

The exact observations which have been made, especially in the present century, have irrefragably proved, that nature is still able to create imperfect animals, as well as the lower plants, without eggs or seeds ; only organic matter, water, and air, the absolutely essential conditions of living being, are necessary, in order to produce, with sufficient heat, organic forms. When once these lower forms, in the animal as well as in the vegetable kingdom, have arisen, they reproduce themselves, as observations have sufficiently shown, by eggs or seeds, until they again disappear, when the external conditions are withdrawn by which they were called into existence. Up to what grade of animals and plants this production without a germ can take place, it is still very difficult to say. It has long been established beyond any reasonable doubt, that intestinal worms may be produced without eggs, and this is attended by a diseased condition of the body, of which these creatures are the product. The extremely careful and numerous observations of very recent times, on the appearance of Entozoa in the most secret part of the eye, both in man and in animals, are also too well authenticated to be overturned by ever so ingenious an hypothesis.

This is not the place to defend the doctrine of spontaneous generation, which, I am of opinion, has in no respect been impugned by the recent examinations of the Infusoria; for it has always been allowed that these lower creatures propagate by germs, when once they have arisen. The flying about of little sporules in the air was always the solution given by the opponents of the doctrine, when the production of little fungi in closely shut-up places was spoken of; but not to mention that this assertion rested on no observation, for no one has ever seen these sporules flying about in the air, though they are quite large enough to be seen, such objections may now be completely set aside, for Dutrochet* has made the extremely valuable discovery that the formation of the filaments of fungi can be produced, accelerated or checked by chemical substances. Any one in his apartment may easily make experiments which will convince him of the production of lower organisms without germs. Let him take new rye bread, moisten a piece of it, and put it in a large glass completely closed by a plate of glass or a bell. In three or four days mould will be perceived on the bread, and it will almost always be the same sort of mould which makes its appearance, whether the vessel is left open or is closed up, or placed in a room in which there is not a particle of mould. From the different colours and thicknesses of this mould, it has been divided into several species, of the nullity of which one may

* Observations sur l'origine des moisissures. Annal. des Scienc. Nat. 1834. Tom. i. p. 30–38.

very soon convince himself by a close microscopical examination. If the bread be kept wet by repeated moistenings, the formation of mould goes on uninterruptedly for several months, and then suddenly ceases; the mould which has already been formed, falls in pieces, and the remaining substance of the mouldy bread again becomes visible; this substance may now be exposed to the air of the room for any length of time, in order to catch some of the sporules flying in the air, and, which is certainly very remarkable, no fresh mould will be seen to form upon it. I find in this phenomenon a confirmation of my views concerning the formation of this lower organism, and at the same time regard it as an evidence against the flying about of sporules.

The growth of minute Isariæ on dead flies in autumn, to which I have already (p. 73) drawn attention, must here be recalled to mind.

Although nature can now produce only lower forms without germs, yet she has formerly created in a similar manner the higher plants and animals, which are now propagated only by germs or eggs. But another very important particular yet remains to be discussed, viz., the question whether the very numerous species of plants now existing have existed from the beginning of the present epoch of vegetation, or whether their number has gradually been augmented, as probably some individuals have changed so much by reason of the influences of a different climate and of peculiarities of soil, that they now, as permanent varieties must, appear to us distinct species. It would certainly be easy to explain many things if this latter opinion could be supported by observations; many naturalists would also be very much inclined to admit it, as the great influence which variety of climate and local conditions exercise on the form of plants is universally known, and thus a great number of forms have arisen, which it has been attempted to make into species, though their characters have evidently been caused by the influence of different external circumstances. But the great difficulty, which is not to be overlooked, lies in the recognition of the characters which determine the natural species. "The species," says Link, "is that which is constant in nature, that which is the law in variety, and the aim of natural history is to find out what is constant, and what is the law by which the variety in nature is limited."*

Though it is true, that often the result of a close study of any group of plants is to diminish greatly the number of the species of that group, since it is perceived that this or that species has changed its form in some way or other from the influence of external circumstances, yet we can from this learn nothing more than that those species have been founded on inconstant characters, and that what is inconstant cannot characterise the species.

Oh! if this golden rule were kept, the number of the species of the lower Cryptogamia would certainly be reduced to a third of the present sum.

Let us not, however, go too far in our conjectures, that the number of species of phænogamous plants is too great. Experiments have not yet pointed out a specific change of distinct natural species, and before this has resulted, we must not acknowledge it, else systematic natural history is at an end. I am, nevertheless, of opinion that we are in possession of many valuable observations on certain genera, extremely rich in species, which plainly inform us that almost all the so-called species are nothing more than very slight varieties, which change now to this side, now to that, so that amongst these numerous species there are only a very few immutable forms, which are the natural species. The botanists who have worked out these genera monographically, have, however, passed silently over these observations, and systematic works now accept hundreds of new species, which are said to have been derived from a few forms. To what will this lead?

* Die Urwelt, &c. 2te Aufl. i. p. 280.

We all know how extraordinarily near the different races of man stand, and how they blend into each other; but we also know, that so long as observations have been made, men have always remained the same, whatever climate they have chosen for their abode.

We are here entitled to make various sub-species of man, but not to assume that one race of man has arisen from another, viz., the more perfect, the more beautiful, from the less perfect, the less beautiful. Though the general law seems to be, that nature has first produced the more imperfect, and afterwards the more perfect forms, yet the latter have by no means risen out of the former.

We may also probably learn that, on the whole, the number of phænogamous plants is not too great from the laws which arithmetical botany has pointed out, for if this changing of individuals into permanent varieties went on, all the definite proportions in which the different groups of plants have been distributed would certainly be very soon destroyed. Statistical inquiries into the absolute and relative numbers of species have led to a fixed law, according to which the different groups of plants, viz., families, genera, and species have been distributed in each zone. Families of less perfect plants in the colder zones, increase in number of species, in proportion to the absolute number of species, while the most perfect families become proportionably richer in species towards the equator.

The results of this new science are indeed so extraordinary, that if we know the law according to which the different families of a zone are distributed, we may determine from the exact number of species of a single family in that zone, the whole number of phænogamous plants growing in it, and even indicate the number of species of the other families.

This so important branch of the doctrine of the distribution of plants has arrived at these striking results by infinitely laborious examinations of a great many different floras ; yet this science is still far from having attained the degree of precision which belongs to it, as hitherto very few tracts of land are known, in respect to their number of plants, with the accuracy which is absolutely necessary for these investigations.

This book is not the place to enter upon special inquiries into this subject, but it will be in accordance with its aim, if I here make some remarks on the method which is to be observed in such statistical investigations, and, in conclusion, I shall give statistically the relative proportions of the most important groups of plants, as far as they are at present known with any certainty.

The most important works on the statistics of plants are contained in the treatises of Humboldt* and Beilschmied ;† the other numerous writings of the most learned of botanists, which contain inquiries into the statistical proportions of plants, have been for the most part already named.

When one inquires into the statistics of the flora of a country, he may either compare the absolute numbers of the species of different families with each other, or he may compare these numbers with the whole number of the plants of that country, or, again, he may consider the masses which the species of any family of that country form.

According to Röhling's flora, Germany possesses 2600 Phanerogamæ ; among them we find 328 Glumaceæ, 163 Leguminosæ, &c. These numbers may be compared with each other in two ways, viz., first, the proportion of the Glumaceæ to the Leguminosæ; from which we learn that the number of species of the former is twice as great as of the latter in Germany. Again,

* Sur les Lois que l'on observe dans la Distribution des Formes Végétales. Dict. des Scienc. Nat. T. xviii. p. 422–437.

† Excurs über einige bei Pflanzengeographischen Vergleichen zu beachtende Punkte, &c. Contained in Beilschmied's Pflanzengeographie nach Alexander v. Humboldt's Werken, &c. Breslau, 1831, p. 126.

from these numbers we learn the proportion of the Glumaceæ and Leguminosæ to the whole number of species in Germany, by dividing the whole mass of plants by their numbers. From this it follows, that, in Germany, the Glumaceæ form the 7.9th part of the whole number of species, while the Leguminosæ are only a 16th of the flora.

Inquiries into the distribution of genera over the earth are less fruitful in precise results, for, first, the number of genera depends too much on the will of the observer, and, secondly, the genera do not decrease from the equator to the poles in proportion to the species, for in the coldest zones there are far more genera than belong to an equal number of species at the equator.

The statistical results, drawn from those floras which are accurately known, must be preferred to all uncertain and generally incomplete lists of plants belonging to other countries; the floras of Lapland, Sweden, Germany, England, France, and Switzerland, and also of a few islands, may lay claim to a high degree of completeness, at least as far as respects phænogamous plants; and the results obtained from these floras, as to the relative proportions of the several families, will give us the law of distribution for these zones, as now established by the calculations of Humboldt, Mirbel, Beilschmied, and several others.

The most important rules to be observed in such calculations are briefly the following, to which Beilschmied in particular has drawn attention in the work already named.

First, great care must be taken, that the species in the floras of different countries to be made use of in the calculation, have been founded on a uniform principle; that the number of species are not occasionally so multiplied by monographs of single genera, that the proper relative proportional numbers (which are also called coefficients) are completely lost. The sub-species and varieties, which are contained in one flora and not in the other, must first be reduced to the species to which they belong.

The same must be attended to in reckoning up the families, for different authors have frequently placed this or that genus in different families, whence very considerable errors have arisen, when this has not first been set to rights.

Above all, care must be taken, that the plants of a country, or any area which is to be compared with other floras, belong to the same region of altitude, for if the plants of a country, in which there are mountains 5000 or 6000 feet high, be made use of in a body, very erroneous results will certainly be obtained ; indeed it would be quite as proper to mix the floras of different, remote zones together, though the proportions obtained would not be the same, for, as we shall afterwards see, the quotients of the families do not change in the same ratio with the increasing height as with the latitude.

It is also necessary that for such calculations, floras of equally extensive tracts should be chosen ; for the results of the calculations are not by any means the same for small and large districts, since the areas of different plants vary much in extent.

In the comparison of two floras, the results will be most nearly the same, when both lie in the same zone, or in countries the vegetation of which is characterized by the same form of plants. For example, it is natural that, when the plants of equal tracts, in the sub-tropical zones of the northern and southern hemispheres are compared, the proportionate numbers of the families should often differ, for many families, which are very numerous in the northern hemisphere, have only representatives in the southern hemisphere, as the families of the south have in the north. If, for instance, the relative proportions of the families of New Holland be compared with those of the sub-tropical zone of North America, the most striking differences will be perceived, because perfectly different groups of plants predominate in these countries, so far distant from each other.

Lastly, in statistical comparisons between the floras of two countries, regard must be paid to a certain equality of soil, for much depends on this. When the plants of a wide tract of land with a moory soil, as the Lüneburg heath, are compared with those of a neighbouring sandy country, the result will be strikingly different.

---

The most natural division of the whole mass of vegetation into the larger groups, is still that into Monocotyledones, Dicotyledones, and Acotyledones, and we shall therefore first attempt to ascertain the relative proportions in which these groups appear in certain zones.

With respect to the Acotyledones, which include the Cryptogamia, it has first to be remarked that they must as yet be excluded from such statistical calculations, since their number is at present known for a few places only, with tolerable accuracy ; and in determining the species in some divisions of this group, for example, in the Algæ, Lichens, and some families of Fungi, the views of botanists differ so widely that the number of species given by one author is doubled and trebled by others for the same country. The number of Fungi, Musci, and Algæ in the colder countries is very great wherever there is a sufficient degree of humidity ; yet only in a few floras of single towns have such a number of species been given, that we may consider the cryptogamic flora as in some measure exhausted, for often even in these districts new species are discovered when they are specially sought for.

It is now considered as proved, that in the north, small areas, such as the floras of single towns, show more cryptogamic than phænogamous plants ; this is less the case in the floras of large tracts of land, since the range of the Cryptogamia is often very extensive. We may, however, accept it as an indubitable fact that the total number of Cryptogamia is not nearly so great as that of Phanerogamæ; but it will probably be long before we know even in some measure the Cryptogamia of foreign countries. All results, which can now be obtained by a comparison of Acotyledones with Monocotyledones and Dicotyledones, are so uncertain, that they can scarcely be considered as approaching the truth.

One division only of the Acotyledones, viz., the Ferns, which attract the eye by their beauty, have been collected in foreign countries in the same proportion as flowering plants, and they therefore may safely be made use of in statistical inquiries.

Although we are yet far from having a complete acquaintance with the floras of remote countries in the torrid and temperate zones, we may assume it as probable, that among the plants still unknown to us almost the same relative proportions appear, as amongst those already known, and we may, therefore, draw conclusions for the whole from those we are acquainted with.

The Ferns affect a damp soil, and grow with singular luxuriance in the shade of woods, where they are also very numerous. It may, therefore, be difficult to ascertain the correct proportion of the Ferns to Monocotyledones and Dicotyledones, since their presence, as of the Acotyledones in general, depends much on the humidity of the soil; and in the various zones there are often wide tracts destitute of wood and possessing little water. In the damp forests of South America the number of Ferns is known to be very great, yet we cannot allow that America, when equal areas and equal conditions of climate are considered in comparison, is richer in Ferns than the Old World. The number of these plants in Java and in the damp forests of the Philippines and the East Indies is extraordinarily great, and not inferior to the quantity in America.

Humboldt and Brown give the proportion of Ferns to the whole mass of Phanerogamæ in the torrid zone as 1 : 20; and this proportion has scarcely been changed, though we now possess a greater number of lists of plants from this zone for comparison. At Congo the proportion is almost 1 : 27, and in the plants of New Holland it is 1 : 26. The causes of the number and frequency of Ferns in tropical countries, viz., heat, moisture, and shade, are often found in a still higher degree in islands within the tropics, and the maximum of Ferns is, therefore, to be found there. In Jamaica the proportion of Ferns to phænogamous plants is 1 : 8 ; in Otaheite, according to Banks, 1 : 4, and in St. Helena even 1 : 2.

It has, besides, been a general remark that Ferns are very numerous on islands, in proportion to the corresponding continents. On the Falkland Isles the proportion of them, including the Lycopodia, is 1 : 15 ; in New Zealand, according to Brown, 1 : 6 ; on Norfolk Island, according to Endlicher's Prodromus of 1833, 1 : 3 ; and on Tristan da Cunha, according to Brown, 2 : 3.

The tropical proportion of Ferns, says R. Brown,* on low and open tracts differs much from the instances now given, and it is not improbable that as the maximum of this family falls within the tropics, so the minimum also may be found either within, or a few degrees only beyond, the tropics. It is true that there are some facts which point to a very rapid diminution of the number of Ferns from the torrid to the warmer parts of the temperate zone ; these, however, may be accounted for by the excessive dryness of the soil. It is probable that a further acquaintance with the floras of these countries, small portions of which only are yet known, will not confirm that supposition.

For the temperate zone Humboldt gives the proportion of Ferns to the Phanerogamæ as 1 : 70, which has been obtained from the average of the floras of the colder

* Systematische und geographische Bemerkungen über die Pflanzen in der Nachbarschaft des Congo-Stromes. In R. Brown's Vermischten Schriften. Bd. i. p. 386.

and warmer parts of this zone and of the sub-tropical zone also.

The colder half of the temperate zone is that part of the earth where, at least in the west of the old continent, very few phænogamous plants remain to be discovered; and we can, therefore, place greater reliance on the results obtained from these countries. In France, the proportion of Ferns to the Phanerogamæ is as 1 : 55; in Germany, according to Röhling's Flora, 1 : 45 (1 : 46 according to Wiest, including Switzerland and Istria); in North America, according to Michaux, 1 : 34; in England, 1 : 35; in Scotland, 1 : 31; on the Faroe Islands, 1 : 12.4; and in Iceland, 1 : 18.

In the southern half of the temperate zone the proportion is strikingly smaller, which may be accounted for by the locality, dryness, and want of shade. Thus round Naples the Ferns are only 1-74th of the vascular plants; in Greece, 1-84th; in Portugal, 1-116th; in the Grecian Archipelago, 1-227th; and in Egypt only 1-971st. I cannot, however, think that these are the true proportions for the latitudes of these countries, as in them agriculture has been carried on for thousands of years. We have also lately received accounts of the most eastern countries of the old continent in this latitude, and an immense number of plants, amongst which there are a great many Ferns. In the Canary Isles also we see that the Ferns are as 1 : 14 to the whole mass of vascular plants. We have, however, already shown that the proportion of Ferns is as small in many parts of the torrid zone. Brown mentions that the islands in the Bay of Carpentaria possess more than 200 flowering plants, and no more than three Ferns. I found the Ferns in almost the same proportion in Southern Peru; in the plain of this country they were almost entirely wanting, but on the mountains I found them in the same small proportion to the Phanerogamæ.

From these statements it may, I think, be concluded, that the proportion of the Ferns to phænogamous plants is least in the middle of the temperate zone, and that it becomes larger both towards the equator and towards the poles. It is at the same time very remarkable, that these plants reach their absolute maximum in the torrid, and their absolute minimum in the arctic zone. At North Cape only four Ferns have been found, yet their proportion to the Phanerogamæ is 1 : 7 there, and in Greenland 1 : 10. No Ferns have yet been discovered in the most northern parts of the arctic zone, as is shown by the catalogues of plants for Melville's Island and Spitzbergen. In Baffin's Bay Lycopodium selago only has yet been found.

For the Monocotyledones and Dicotyledones, on the contrary, the law, according to which their species are distributed in the different zones from the equator to the poles, is now determined with considerable certainty, when large tracts of land are taken into account. In small districts, even when lying in the middle of a large country in which the general law is confirmed, the proportions are quite different, for the station of the Monocotyledones depends much more on local conditions, than that of the Dicotyledones; their greater frequency or more vigorous growth is always connected with greater humidity of the soil, and their number of species also diminishes, the drier the air and soil in which they grow. These local conditions must be carefully attended to, in the consideration of the relative numbers of the Monocotyledones in different countries, for by them alone can be explained the great irregularities, which may be here and there remarked.

The Monocotyledones increase in number of species, in proportion to the Dicotyledones, in cold countries; it may be said, because they being less perfect plants occur in greater frequency at a distance from the torrid zone; but the fact may be accounted for by the decrease of the more perfect plants towards the poles, by which the quotients of the Monocotyledones become larger as the distance from the equator increases.

By many laborious reckonings Humboldt determined, that in the torrid zone the Monocotyledones are as 1 : 6 to the Dicotyledones, while in the temperate zone the proportion is 1 : 4, and in the arctic zone 1 : 3. All the observations which have since been made quite confirm these statements, and perfect them by the addition of some intermediate numbers. Von Martius has, for example, reckoned up the plants brought by Wallich from the East Indies.* The proportion of Monocotyledones in this great collection of 7643 species is found to be almost 1 : 6½. In the northern part of New Holland, according to Brown, the proportion of them is 1 : 3½ ; and in the sub-tropical zone of this country, for instance round Port Jackson, it is 1 : 3. The proportion of Monocotyledones to Dicotyledones for the arctic zone has been still more exactly given in recent works; on Melville's Island, it is 1 : 2½ ; on the Faroe Isles, according to Trevelyan's Flora,† 1 : 2½ ; on the Falkland Isles, 1 : 2, according to D'Urville's Flore des Malouines; and in Lapland as well as Iceland the proportion is even somewhat higher.

By these statements is clearly shown the regular increase of Monocotyledonous plants in proportion to Dicotyledonous as the latitude becomes higher, and we now pass on to the change in the proportions of these groups of plants as the elevation increases. A very valuable work by Osw. Heer‡ has given us the desired information relative to this for the Swiss mountains. In Switzerland the proportion of Monocotyledones to Dicotyledones is nearly 1 : 4.9, according to Ringier's calculations ; they, however, diminish very much in number on the higher parts of the mountains, where

* Flora of 1834, p. 1.

† The Edinb. New Phil. Journ., Octob. 1834–Jan. 1835, p. 154–164.

‡ The relative proportions of Monocotyledonous and Dicotyledonous plants on the Alps of the eastern part of Switzerland, compared with those in other zones and regions. See Mittheilungen aus dem Gebiete der theoretischen Erdkunde. Heft i. Zürich, 1834, p. 99.

the proportion is 1 : 5½, and at still greater elevations 1 : 6, 1 : 7, and on calcareous soil, even 1 : 9.

Observations on the Rhætian Alps give the following results. The proportion of Monocotyledones to Dicotyledones, from the height of 5000 to 6000 feet, is 1 : 5; from 6000 to 7000 feet 1 : 5¼; and from 7000 to 8000 feet 1 : 5 9-14ths. On the mountain chain from St. Gothard to Bernina the proportion diminishes in almost the same degree; Heer found it to be 1 : 4 13-14ths, from 5000 to 6000 feet; 1 : 5¼ from 6000 to 7000 feet; and from 7000 to 8000 feet, 1 : 5 3-5ths; but here there is the greatest difference between the lists of plants of particular points, which, however, may be explained by the unequal supplies of water. In marshy places the proportion of Monocotyledonous to Dicotyledonous plants is 1 : 3, while in dry places close at hand it is 1 : 6. On the dry side of the Andula chain, Heer found the proportion to be 1 : 7, and on the eastern side of St. Bernhard 1 : 9.

Ramond* found only 10 Monocotyledones and 61 Dicotyledones on the summit of the Pic de Midi de Baynes; the proportion there is consequently 1 : 6. On comparing this vegetation in the region of alpine plants with the corresponding vegetation in the polar zone, for example, that of Melville's Island, we find the most astonishing difference in the relative proportions of Monocotyledonous and Dicotyledonous plants in these two regions, for on Melville's Island the proportion of these groups is as 1 : 2½. The dampness of the soil of this island is evidently the reason, that the number of Monocotyledones as well as of Cryptogamia is so much greater there, than in the corresponding region of the Pic de Midi.

In these results it is singularly striking, that the usual proportion of Monocotyledones to Dicotyledones on the heights of the Swiss mountains, viz., 1 : 6, is

* Mém. du Muséum, vol. xiii. p. 217.

exactly the same as that which has been established for tropical countries. In damp parts of these mountains, however, the proportion is 1 : 3, and the question now is whether the number of Monocotyledones increases so much, in comparison with the Dicotyledones, in damp parts of tropical countries that the proportion is in them also 1 : 3. This question may, I think, be answered in the affirmative, although special numerical data for this are still wanting. An investigation of the causes, why the proportion of Monocotyledonous to Dicotyledonous plants within the tropics is the same as that on the lower heights of the Swiss mountains, would perhaps be important. I think that the extensive range of the Monocotyledones, in general, may be one of the chief causes, for hitherto in statistical inquiries it has always been necessary to take very large tracts of land into account; for the torrid zone, and in them, on account of their extensive areas, the relative number of species of Monocotyledones must naturally diminish, in proportion to the Monocotyledones of a small district. When we are in possession of a flora of each part of the torrid zone, we shall probably find the full explanation.

By these researches Heer has likewise proved, that the vegetation on the different sides of a mountain does not, as has often been affirmed, differ when the conditions of soil are similar.

Now that the laws, according to which the Monocotyledones are proportioned to the Dicotyledones in the different zones, have been pointed out, we proceed to show the proportions of some of the chief families of Dicotyledonous plants to other families, and to the absolute number of the whole mass of plants.

Humboldt, in the treatise already mentioned,* has unfolded the laws according to which the principal families appear in the different zones, and I can the rather refer to the results given in that celebrated work, as

* Dict. des Scienc. Nat. T. xviii. p. 433, &c.

since that time few floras of single countries have appeared, which make any considerable alterations in these, but on the contrary they are more and more confirmed.

A perfect agreement and an explanation of every deviation from the general law cannot of course be obtained, until the floras of every zone are exactly known.

At present many systematic works are in progress, which, when completed, will furnish a great mass of materials for new statistical investigations, and fill up many gaps now very apparent.

The following statement of the relative proportions of the chief families of plants in the principal zones is given by Humboldt. The Junceæ, Cyperaceæ, and Gramineæ increase in proportion to all the species of Phanerogamæ as the latitude becomes higher, for their proportions in the different zones are as follows :—

| | Torrid Zone. | Temperate Zone. | Arctic Zone. |
|---|---|---|---|
| Junceæ, ... | 1 : 400 | 1 : 90 | 1 : 25 |
| Cyperaceæ, | 1 : 22 (1 : 50 in America) | 1 : 20 | 1 : 9 |
| Gramineæ, | 1 : 14 | 1 : 12 | 1 : 10 |

The Glumaceæ, viz., the three above-named families, are to the whole number of Phanerogamæ in the torrid zone as 1 : 11 ; in the temperate zone as 1 : 8, and in the arctic zone as 1 : 4.

The relative number of species of the four following families, viz., the Rubiaceæ, Leguminosæ, Euphorbiaceæ, and Malvaceæ, on the contrary, decreases as the distance from the equator becomes greater. Their proportions are the following :—

| | Torrid Zone. | Temperate Zone. | Arctic Zone. |
|---|---|---|---|
| Rubiaceæ, | 1 : 14 (1 : 25 in America) | 1 : 60 | 1 : 80 |
| Leguminosæ, | 1 : 10 | 1 : 18 | 1 : 35 |
| Euphorbiaceæ, | 1 : 32 | 1 : 80 | 1 : 500 |
| Malvaceæ, | 1 : 35 | 1 : 200 | entirely wanting |

The families Cruciferæ, Umbelliferæ, and Compositæ are distributed in a different manner, for their quotients are highest in the temperate zone, and diminish towards the equator and towards the poles. Their proportions are the following :—

| | Torrid Zone. | Temperate Zone. | Arctic Zone. |
|---|---|---|---|
| Cruciferæ, | 1 : 800 | 1 18 (1 : 60 in America) | 1 : 24 |
| Umbelliferæ, | 1 : 500 | 1 : 40 | 1 : 60 |
| Compositæ, | 1 : 18 (1 : 12 in America). | 1 : 8 (1 : 6 in America). | 1 : 13 |

We have already seen that the mode of distribution of the Ferns is the opposite to that of these families, for the Ferns are least frequent in the temperate zone, and relatively increase towards the equator and the pole.

After the appearance of these calculations of Humboldt, Schouw also specially treated of the geographical distribution of some of these families, and the proportions he gives occasionally differ from the former. The grasses, for example, according to Schouw's reckoning,* form 1-10th or 1-12th of the whole mass of phænogamous plants, while, according to the above account, they are only 1-14th. The explanation of this difference may perhaps be, that Schouw has made more use of special floras, in which the quotients of monocotyledonous families are for the most part larger than when large tracts of country are taken into account, since the areas of these plants are generally extensive, and they therefore count equally in small districts and in large countries, while in the latter there are a number of additional phænogamous plants.

In the warmer part of the temperate zone Schouw finds the proportion of grasses to be from 1-12th to 1-14th, and in the colder half of this zone 1-10th. In the arctic zone he considers the grasses to form 1-8th, for in Kamschatka, Iceland, Greenland, and the Loffoden Isles, they are 1-7th, 1-8th, or, at least, 1-9th, and in Melville's Island

* Gründzüge, p. 288.

they are in the proportion of 1 : 4.7. In the Florula of the Loffodens, which Lessing has given us,* the proportion is only 1 : 8 ; it is probably, however, really much higher, for the number of Monocotyledones in this list seems to be too small. The Florula contains 162 Phanerogamæ (without Ferns), and 127 Dicotyledones ; the proportion of Monocotyledones is, therefore, 1 : 3.6, a proportion which is quite foreign to this zone. If, however, this striking deviation for the Loffoden Isles should be confirmed, it would be interesting to investigate the causes of it.

Schouw thinks it must be admitted that the relative proportion of grasses to the Phanerogamæ decreases with the increasing altitude; several calculations seem to me, however, to confirm the opinion that the relative proportion of grasses becomes higher with the elevation as with the latitude. I have reckoned for the different regions the whole mass of mountain plants in the Flora of France which De Candolle has given with the exact height of their stations,† and I have found that in the region from 700 to 1400 metres the proportion of grasses is 1 : 28.3 ; from 1400 to 2100 metres, 1 : 23.8 ; from 2100 to 2800 metres, 1 : 26 ; and from 2800 to 3500 metres, 1 : 15. The correct proportion for the plain cannot be given, as these plants are not specially distinguished. For the whole Flora of France the relative number of the grasses to the Phanerogamæ is 1 : 14. The large collection of mountain plants brought by Humboldt from America, seems also to show an increase of grasses with the higher elevation, and I have observed it on the southern parts of the Cordilleras whenever the soil possessed any moisture.

The Cyperaceæ are quite as widely diffused as the Gramineæ ; we have already seen that Cyperus and Carex, the principal genera of this family, are the opposite of each other in their geographical distribution. The genus Cyperus attains its maximum in the torrid zone,

* Reise durch Norwegen nach den Loffoden. Berlin, 1831.

† Mém. de la Soc. d'Arcueil, iii. p. 262.

and the Carices are most numerous near the arctic circle, for in Lapland, Iceland, Greenland, and Kamschatka they always form 1-9th or 1-10th of the whole flora ; in higher latitudes, as on Melville's Island, they become less frequent, for they are there only 1-16th. The decrease of the Cyperaceæ from the sub-polar zone to the tropics is extraordinarily regular. In Denmark they are 1-15th ; in England, 1-17th ; in Germany, 1-20th ; in France, 1-26th ; in Greece, 1-50th, and so on.

The proportion of Cyperaceæ in the torrid zone cannot, with our present materials, be given so exactly ; it seems that their number there increases again, which may also agree with the fact that the species of Cyperus are at the maximum in the torrid zone. The proportion of Cyperaceæ in the torrid zone may, perhaps, be from 1-15th to 1-18th.

The increase of the Cyperaceæ with the altitude is certainly not universal ; their presence is only too often connected with a damp, marshy soil, which is found on most mountains of great height.

Schouw has also specially treated of the geographical distribution of several other families, as the Compositæ, Leguminosæ, Cruciferæ, Cacteæ, Proteaceæ, and the Palmæ, to which I shall only refer, as strikingly different results have become apparent from late systematic works. Besides I think that the statistical data which I have arranged here as shortly as possible, will give an idea of the state of this branch of botanical geography. More special researches, which are attended with a constant repetition of the facts already so often used, are beyond the scope of this outline of botanical geography, but they shall be particularly attended to in my lectures on this science.

In the section before the last I pointed out at length the parallelism between the zones and regions corresponding with each other in similarity and equality of climate, in respect of the general appearance of their vegetation. It might from this be expected that the statistical results in the corresponding zones and regions would likewise agree; no investigations concerning this, however, have yet been made which can lay claim to the necessary degree of accuracy. Plants belonging to those families which depend much on the peculiarity of the soil, will, indeed, in this comparison of the vegetation of certain zones with the corresponding regions, show great differences, but in others this will not be the case. The relative proportion of Monocotyledones, for example, increases with the higher latitude, but it greatly diminishes in the corresponding heights, because in them the atmosphere and soil are much less humid.

Unfortunately the floras of particular countries are not yet adapted to statistical investigations of this nature, viz., in which each region is reckoned separately, for it is absolutely necessary for this that the lowest as well as the highest station of every plant should be given. The altitudes of the vertical range of the mountain plants in France which De Candolle has given,* are at present the only data which can here be used, but even here the data for the plants of the plain are wanting, and they would doubtless be very incorrect if they were supplied from the Flora of France.

In the following Table I have compared the mountain plants of France in the different regions, and expressed by proportional numbers the size of the families. As those plants have for the most part been collected below the 45th degree of latitude, consequently within the warmer part of the temperate zone, I have been obliged to admit five regions for the flora of France, the lowest of which falls in the warmer temperate zone, therefore the second answers to the colder temperate zone, the third to the sub-arctic, the fourth to the arctic, and the fifth to the polar zone. The heights at which these plants are found go as high as 3500 metres; consequently, for this flora, a height of 700 metres answers to a region.

* Mém. de la Soc. d'Arcueil. Tom. iii.

It is true, we before accepted 1900 or 2000 feet as the vertical extent of each region, yet here the warmer climate which belongs to the west coast of a continent may account for the difference. Hitherto in statistical reckonings of plants the separation of the flora into different regions, and the separate reckoning of each region have been neglected, and that because circumstantial data relative to the extent of the vertical range of the plants were almost entirely wanting. But it is easy to see that the results obtained in this way could not be exact, for not only do the proportions of the families in the plain and on the heights very seldom agree, but for the most part great differences are to be found in them ; it is, therefore, evident that no determination of the proportional size of individual families can be attained, so long as the plants of the mountain flora are reckoned along with those of the plain of the same latitude. Those botanists who have already worked at mountain floras, or who shall in future work at them, are therefore urged to give the whole extent of the vertical range of each plant in their floras.

I am well aware that there are some other works in which the vertical range of different plants is exactly given, but these works seem to me too incomplete to furnish certain results for the large tracts they embrace. The number of plants will probably be doubled or trebled when these countries are carefully examined.

The results of reckonings which are given in the following Table can only be regarded as imperfect, since the materials that could be made use of are so. I am, however, convinced that when they are completed, it will be found that the quotient of each family agrees with that in the corresponding zone. It is, of course, understood that every plant whose range extends through different regions is counted in each, in order to obtain the absolute number of plants for each region.

*Table of the Proportionate Numbers of some of the Principal Families belonging to France, arranged according to the different Regions of the Country:—*

| Regions. | From the plain to 700 metres. | From 700 to 1400 metres. | From 1400 to 2100 metres. | From 2100 to 2800 metres. | From 2800 to 3500 metres. | For France, from the plain to the height of 3500 metres. |
|---|---|---|---|---|---|---|
| These regions correspond to the following zones. | The warmer temperate zone. | The colder temperate zone. | The sub-arctic zone. | The arctic zone. | The polar zone. | |
| Total number of Phanerogamæ :— | | 653 | 650 | 269 | 79 | 3540 |
| The following are the proportions of the different families :— | In consequence of the incompleteness of the data this region must remain unreckoned. | | | | | |
| Monocotyledones | | 1 : 4.9 | 1 : 5.1 | 1 : 6.7 | 1 : 6.1 | 1 : 4 |
| Gramineæ | | 1 : 28.3 | 1 : 23.8 | 1 : 26 | 1 : 15 | 1 : 14 |
| Cyperaceæ | | 1 : 19 | 1 : 20.3 | 1 : 29.8 | 1 : 26 | 1 : 26 |
| Junceæ | | 1 : 72 | 1 : 65 | 1 : 44.9 | 1 : 26 | 1 : 106 |
| Glumaceæ | | 1 : 9.9 | 1 : 9.4 | 1 : 10.7 | 1 : 7 | 1 : 8.6 |
| Liliaceæ | | 1 : 34.3 | 1 : 36 | 1 : 67 | wanting. | 1 : 95 |
| Orchideæ | | 1 : 54.4 | 1 : 65 | 1 : 89.6 | wanting. | 1 : 69.4 |
| Coniferæ | | 1 : 93 | 1 : 92.8 | 1 : 269 | 1 : 79 | 1 : 208 |
| Amentaceæ | | 1 : 81.6 | 1 : 59 | 1 : 53.8 | 1 : 79 | 1 : 104 |
| Primulaceæ | | 1 : 65 | 1 : 40.6 | 1 : 24.4 | 1 : 9.8 | 1 : 86.3 |
| Labiatæ | | 1 : 32.6 | 1 : 34 | 1 : 134 | 1 : 79 | 1 : 26.2 |
| Rhinanthaceæ | | 1 : 29.7 | 1 : 23.6 | 1 : 20.6 | 1 : 26.3 | 1 : 26.2 |
| Gentianeæ | | 1 : 43.7 | 1 : 38 | 1 : 38 | 1 : 26.3 | 1 : 118 |
| Ericinæ | | 1 : 50 | 1 : 54 | 1 : 38 | 1 : 39.5 | 1 : 136 |
| Campanulaceæ | | 1 : 34.3 | 1 : 29.5 | 1 : 269 | wanting. | 1 : 95 |
| Compositæ | | 1 : 7.5 | 1 : 6 | 1 : 10.3 | 1 : 11.2 | 1 : 8 |
| Rubiaceæ | | 1 : 54 | 1 : 61 | 1 : 89.6 | 1 : 26.3 | 1 : 72.2 |
| Umbelliferæ | | 1 : 20.4 | 1 : 26 | 1 : 67 | 1 : 79 | 1 : 26.8 |
| Saxifragæ | | 1 : 32.6 | 1 : 20.9 | 1 : 11.6 | 1 : 7.9 | 1 : 93 |
| Rosaceæ | | 1 : 17.2 | 1 : 18 | 1 : 15 | 1 : 19.7 | 1 : 29.5 |
| Leguminosæ | | 1 : 15.9 | 1 : 20.9 | 1 : 20.7 | 1 : 39.5 | 1 : 10.2 |
| Caryophyllaceæ | | 1 : 21.7 | 1 : 16.6 | 1 : 12.8 | 1 : 11.2 | 1 : 23.2 |
| Cruciferæ | | 1 : 19.8 | 1 : 19.6 | 1 : 20.6 | 1 : 13 | 1 : 18.2 |
| Ranunculaceæ | | 1 : 28.3 | 1 : 23.6 | 1 : 20.6 | 1 : 39.5 | 1 : 29.2 |

From this Table it will be seen how extraordinarily the quotients of the families differ for the different regions, and when compared to those which are drawn from the species of the whole country. It will, therefore, be the better perceived how necessary it is to reckon each region separately. For example, if a country with high mountains is compared with a flat country in the same latitude in respect of their number of plants, the proportional numbers of individual families will by no means agree, or if they do the agreement has been occasioned by accidental circumstances.

In the Table the plants of the alpine region are probably the most complete, and when we compare them with the plants of Melville's Island, and with those which Ramond has collected in the corresponding region of the Pic de Midi de Bagnes,* we find considerable agreement in the numerical data, though there are not wanting differences which future observations will probably either remove or explain.

| | Region of Alpine Plants in France. | Summit of the Pic de Midi de Bagnes. | Melville's Island. |
|---|---|---|---|
| Gramineæ | 1 : 15 | 1 : 10.1 | 1 : 4.7 |
| Cyperaceæ | 1 : 26 | 1 : 25.3 | 1 : 16.7 |
| Compositæ | 1 : 11.2 | 1 : 5.4 | 1 : 13.4 |
| Saxifragæ | 1 : 7.9 | 1 : 17.7 | 1 : 6.7 |
| Rosaceæ | 1 : 19.7 | 1 : 17.7 | 1 : 16.7 |
| Leguminosæ | 1 : 39.5 | 1 : 17.7 | 1 : 32.5 |
| Ranunculaceæ | 1 : 39.5 | wanting. | 1 : 13.4 |
| Caryophylleæ | 1 : 11.2 | 1 : 11.9 | 1 : 13.4 |
| Cruciferæ | 1 : 13 | 1 : 11.9 | 1 : 4.9 |
| Campanulaceæ | —— | 1 : 71 | 1 : 6.7 |

From the prevailing differences in this Table we may conclude that no great importance can be attributed to the results of single countries, however exact they may be; only the mean of the greatest possible number of observations can be acknowledged as a law which may the nearest approach the truth.

* Mém. du Mus. vol. xiii. p. 217.

# SUPPLEMENT.

## THE HISTORY OF CULTIVATED PLANTS,

CONTAINING ENQUIRIES CONCERNING THE NATIVE COUNTRY, THE DISTRIBUTION, THE CULTURE, AND THE USES OF THE PRINCIPAL CULTIVATED PLANTS, WHICH SERVE AS WELL FOR THE FOOD AS FOR THE COMFORT, THE LUXURY, AND THE TRADE OF NATIONS, AND ARE THE FOUNDATION OF THEIR PROSPERITY.

### I. THE CULTURE OF THE CEREALS.

WE begin by considering the culture of the cereals, which we understand by agriculture in the restricted sense. Agriculture precedes all civilization ; with it is connected rest, peace, and domestic happiness, of which the nomade knows nothing. In order to pursue it nations must take possession of certain lands ; and when this has been done and their existence is firmly established, improvement of manners and other culture may gradually advance.

The people who carry on agriculture are no longer always inclined to bloody war ; they fight only to defend their country, from which they derive their support.

The principal cereals are wheat, spelt, rye, barley, and oats, for Europe and the adjacent parts of Asia ; rice, and several kinds of millet* throughout the south-east of

* Panicum miliaceum, P. italicum, P. frumentaceum, et Eleusine coracana.

Asia ; maize, in the New World, and Sorghum vulgare or Indian millet, Eleusine coracana and Poa abyssinica in Africa.

We are ignorant under which of these cereals the first civilization of man has arisen, but it is probable that the people of Eastern Asia, who cultivate rice, have been the first to resolve on fixed settlements, and, as we shall afterwards see, rice also appears to grow wild in these countries.

Improvement of manners in the west has been attended by the culture of the wheat, yet it cannot be determined where it was first cultivated. Agriculture, without doubt, came from Egypt to Greece,* whence it has gradually spread its beneficent influence over all Europe. We hear it constantly asked where the cereals which we now cultivate are found in their wild state, as if their cultivation must necessarily have proceeded thence. Recent observations, however, strongly oppose this assumption. Von Martius, as we shall by and bye more fully see, has found rice growing wild in the interior of South America ; nay, he has seen the inhabitants of that country gather it in, and yet they have never engaged in the culture of this valuable plant, but still live in their original condition. It is often determined by accidental causes that the people of different countries cultivate sometimes one, sometimes another plant for food, when they might have likewise cultivated other even more advantageous plants.

The native country of our cereals is generally said to be unknown, but in this respect too little dependence has been placed on the observations of very trustworthy travellers. Michaux† and Olivier‡ have informed us that spelt, barley, and wheat grow wild in Persia round

* See the learned researches in Link's Urwelt und das Alterthum. Berlin, 1834. 2te Auflage, p. 400.

† Encyclop. méthod. Art. Botanique. T. i. p. 211.

‡ Voyage dans l'empire Ottoman, l'Egypte, et la Perse. Paris, 1807. 4to. Vol. iii. p. 460.

Hamadan, in Mesopotamia, and on the banks of Euphrates. Were we not so little acquainted with these countries, which were the cradle of western civilization, we should in all probability possess still more exact information relative to the station of the cereals. Link* thinks that they may have become wild in those countries which have so long been under culture, yet it may be with equal reason objected that this is extremely doubtful, for in our part of the world, at least, these plants have never become wild, nor have I seen or heard of any instance of maize or rice becoming so in tropical countries. It is well known with us, that where such cultivated plants are accidentally sown beyond the limits of the ploughed land, they grow at most one year only, and afterwards altogether disappear. I cannot, therefore, assent to the opinion of those who maintain that our cereals have no longer a native country, although it is true that for some of them it is still unknown.

### 1. *The Wheat (Triticum sativum L.).*

Of our cereals wheat is the one which requires most heat. It seems to succeed best on the borders of the sub-tropical zone, where it yields an extraordinarily large crop. The culture of the wheat is at present widely extended ; it is carried on in every quarter of the globe. In Europe the wheat ascends above 62° north latitude ; indeed Schouw† gives 64° as the polar limit of its culture on the western side of the Scandinavian peninsula, but at the same time he remarks that the culture of wheat first begins to be of importance below 60° north latitude. From the meteorological observations in these latitudes, we must conclude that a mean heat of at least 4° Cels. is necessary for the growth of wheat, and during three or four months the mean summer heat must rise above 13° Cels. By this one must be guided in

* Die Urwelt, &c. I. p. 403.

† Europa. Kopenhagen, 1833, p. 9.

attempting to introduce wheat into elevated mountain plains. Wheat does not bear tropical heat well ; in countries within the tropics it first occurs at altitudes which in climate correspond with the sub-tropical and temperate zones.

The low elevations at which Humboldt met with wheat in America are remarkable,* viz., at Victoria, near Caraccas, at the height of 1600 feet, and at Las Quatro Villas in the island of Cuba, which lies on the borders of the torrid zone, at still lower altitudes ; nay, in the Isle of France wheat is grown almost close to the seashore. Similar instances have been observed in the Island of Luçon, but there the mean heat is very much lowered by the prevailing monsoons. Besides, in several tropical countries, wheat and the other northern cereals are sown in winter often in the very places where tropical fruits are grown during the wet summer months. I have seen this in the neighbourhood of Canton, and Royle† mentions it for India, where in winter the vegetation has a perfectly European aspect, and many species of true European genera make their appearance.

In the middle of the temperate zone of Europe, viz., in France, wheat is cultivated to the height of 5400 feet only. In Mexico its culture first begins at the height of 2500 or 3000 feet ; nay, between Vera Cruz and Acapulco, according to Humboldt, fields of wheat are first met with at the height of 3600 feet, and ascend above 9000 feet. On the plateau of Southern Peru, wheat fields of extraordinary productiveness lie at the altitude of 8000 feet ; and at Cangallo, at the foot of the volcano of Arequipa, wheat succeeds well, even as high as 10,000 feet. Round the lake of Titicaca, at a height of 12,700 feet, where there reigns a perpetual spring, but where sufficient summer heat is wanting, wheat and rye do not ripen. I observed that the temperature on the shores

* De distributione geogr. plantarum, p. 161.

† Illustrat. of the Indian Botan. &c. Fasc. i. p. 10, &c.

of that lake in summer at six A.M., was no more than 6° R., and at mid-day, with a somewhat clouded sky, it did not rise above 12° R. This is the reason that wheat does not ripen on this plateau, where the winter is extremely mild.

We do not yet exactly know to what height the culture of the wheat ascends within the tropics, but it probably goes higher on the plateau of Tacora than on the Himalaya mountains, where there are no such great and uninterrupted plains.

Wheat is extremely productive in Chili and in the republic of Rio de la Plata, so that the export of Chilian wheat is of great importance. Not only are immense quantities sent to Peru, but even round Cape Horn to Rio Janeiro, and the Chilian wheat is of excellent quality. It is cultivated in all parts of Chili, where there is sufficient water, from the sea to a height of 5200 feet; yet it is scarcely credible that at the present time North American flour is sold in the market of Valparaiso, and the bakers are obliged to buy it as it is cheaper than flour made in the country, because there are no roads in the interior, and wages are exceedingly high from want of sufficient hands.

I here pass over the mode of cultivating wheat and its uses as already known, but I have still some remarks to make on the various degrees of productiveness of this grain in different countries, in order to show how much more productive the soil would be in a better climate than with us, if the inhabitants would bestow the labour on it. In our cold countries each wheat plant generally sends up a single stalk which bears only one ear, and, therefore, the average produce of the seed is not more than five or six fold.* In Hungary, Croatia, and Sclavonia the average produce is from eight to ten fold; in La Plata, twelve fold; and in the north of Mexico, seventeen fold; and in the equatorial parts of that country, twenty-

* See here Baron A. v. Humboldt's Neu-Spanien, iii. p. 60.

four, and in plentiful years even thirty-five fold. Humboldt mentions an instance of extraordinary productiveness which was observed in Mexico,* viz., of wheat plants sending up forty, sixty, or even seventy stalks, the ears of which were almost equally well filled, and contained from 100 to 120 grains each. Yet how much greater is the produce of the maize! On the plateau of Mexico, where the wheat crop is on an average eighteen or twenty fold, the maize, to which we shall afterwards revert, produces 200 grains.

Besides wheat, spelt† is frequently grown in southern countries; it was known to the Greeks and Romans, indeed, with the latter, according to Pliny,‡ it was the oldest cereal, and was called far, ador, and also adoreum.§

The other grains we cultivate, as barley, rye, and oats, are of importance in the colder countries only. They resist the cold better than wheat, and are, therefore, the only cereals which can be grown in high latitudes. The culture of rye prevails in the sub-arctic zone; that of oats and barley in the arctic, and in the larger part of the sub-arctic zone of the eastern countries of the continent. On the Scandinavian peninsula, barley extends to 70° N. latitude, rye to 65° and 67°, and oats to 65° and 62½°, while the culture of wheat, even to a trifling extent, goes no higher than 62° or 64°, and is first of importance below 60°.

In the same proportion the latter cereals are grown at much higher elevations than the wheat. In the south of Lapland, for example, in 67° N. latitude, where there is not a trace of the culture of wheat, barley ascends to the height of 800 feet above the sea.|| The limit of the culture of corn on the Alps of Tyrol, is at the height of 3800 feet, in the Tyrolese mountains 4500, and on Monte Rosa of 5880 feet. In France, according to De Candolle,

* L. c. iii. p. 52.

† Triticum Spelta et var.

‡ Hist. Nat. Lib. xviii. cap 8.

§ See Link, die Urwelt, &c. p. 406.

|| Schouw's Europa, p. 10.

rye is cultivated at the height of 6600 feet, and in southern countries at still more considerable altitudes, where the highest temperature of the day seldom rises above 14° Cels. In the plateau of Peru barley and rye seldom ripen above the height of 10,000 feet. I have seen ripe oats at the lake of Titicaca, at the height of 12,700 feet; but rye, wheat, and in general barley also, are used only as green fodder, though the latter still here and there ripens even at this elevation. Barley is cultivated in Peru for fodder even at the height of 13,800 feet; for example, Rivero has observed it at the Alto de Jacaibamba.* Gerard found barley, buck-wheat, and rape at the height of 13,000 feet in the western part of the Himalayas, where Kunawar is situated, but it is not said whether the barley ripens there. In Chili, in the latitude between Quillota and Valparaiso, barley is at present cultivated at the height of 5200 feet.

The native country of these latter cereals, viz. rye and oats, is unknown.† Barley is probably indigenous to the north of Africa, for Diodorus places its native country in Egypt, where wine was made from it at an early period.

Barley was used by the Hebrews, Greeks, and Romans. Rye seems, from Link's recent investigations, to have been likewise known to the ancients, but no trace of oats is to be found in the oldest records. Horses, at the time of the Trojan war, were fed with barley instead of oats; in later authors, for example Galen, we find the first mention of the use of oats. Oats are now chiefly used for feeding cattle, but the old Germans ate oatmeal porridge, and at the present day oat-bread is frequently eaten in Ireland, Scotland, Norway, and Sweden.

* Memorial de ciencias nat. Lima, 1828, i. p. 102.

† See Link's learned investigation of this subject in his Urwelt, u. s. w. Bd. i. p. 407, 2te Ausgabe.

## 2. *The Rice (Oryza sativa L.).*

Rice is probably that grain which supports the greatest number of persons. Formerly it was the generally received opinion, that rice belonged to the Old World only, and though it is no longer disputed that it grows wild in the interior of South America also, as on the Rio Negro and in Pará, where it was found by Von Martius,* and is acknowledged as wild by Nees von Esenbeck,† yet this plant was not an object of agriculture in America previous to the arrival of Europeans. Von Martius mentions, that on the Rio Iraria, a branch of the Rio Madeira, the wild rice grew as thickly as if it had been artificially sown, and that the Indians gathered in rich crops of it, by guiding little boats amongst the ripe stalks and beating out the grain into them.‡

It is a remarkable circumstance, which can only be accounted for by the stupidity of the Indians, that this valuable grain should not long ago have become an object of agriculture in a place where it grows wild in such quantities; nay, that it has remained hidden from our knowledge until a very recent period. I may here observe, that the so-called wild rice, on which the aborigines of Canada live during winter, is a totally different plant, viz., Zizania aquatica.

Rice is most generally cultivated in the eastern and southern parts of Asia, and there it forms the universal food, but it is also a common article of food in the north of Africa, in Egypt, Nubia, Persia, Arabia, Asia Minor, Greece, Italy, and in the southern parts of Portugal, Spain, and France. The culture of rice has passed with the Europeans into America, and it is very exten-

* Reise nach Brasilien, iii. p. 309.

† Flora brasil. Vol. ii. pars i. p. 318, 560.

‡ Von Martius, however, is now of opinion, that this wild American rice is a different species from the Asiatic.

sively cultivated there in the tropical and sub-tropical zones, and in still higher latitudes. In the Southern States of North America the culture of rice has become so general that it has long been the common food. In the West Indies, Venezuela, and the Brazils, this grain is also a favourite object of culture, and it may justly supersede the maize there; the negro slaves in America even prefer rice to the manioca. It may be seen from the interesting accounts in Von Martius' Travels, what an extraordinary quantity of rice is now raised in Brazil. The province of Maranhâo alone produces 560,000 or 600,000 Alqueires* yearly, and a steam engine to clean the rice, has been erected in it.

In India and China, where rice is the chief food, famine and death are the inevitable consequences of a failure of the rice crop, and this is not unfrequent. Not only too little or too much rain, but insects also, cause a total failure of the harvest and famine, wherever the people give themselves up to the culture of a single kind of food. In China, the land no longer produces sufficient food for the excessive population, notwithstanding all the industry with which agriculture is carried on, and this country therefore consumes all the superfluous produce of the fertile islands of the Indian Archipelago. But when the rice crop fails in China, the greatest fleets could scarcely bring so much of the grain as would be required to prevent a famine.

In tropical countries, the chief seats of rice culture, there are a great many varieties of this plant, one of which is better adapted for certain soils than another. Above all, two principal varieties must be mentioned, one of which is grown on hills, the other in marshy, or very wet places. The former is known by the name of mountain rice; many botanists still doubt whether it can grow in a dry soil, not artificially irrigated. But

* Four Alqueires are a Fanega, and 100 Fanegas are equal to 100.696 Berlin bushels.

this is certainly doubted without reason, for Marsden* mentions the culture of mountain rice in Sumatra, where it is called Ladang, so fully and circumstantially, that one must be convinced of it. This variety is also cultivated with great success in Java and Brazil, apparently in very damp soils. I may here conveniently notice the culture of the dry Tarro (Caladium esculentum), which is grown on the Society and Sandwich Islands in wet ground, even at a considerable elevation, although the other variety of this plant stands always under water.

The common or marsh rice (Sawuhr in the Malay) is generally cultivated in the following manner: it is sowed either in the cleaned mud of a natural marsh, or, which is more usual, in hollows prepared for it, which are dug two or three feet deep, and can be laid under water. In the south of China these rice-grounds occupy the whole of the level ground, and ascend high up the mountains; they are either supplied with water which comes down from the mountains, or the water is pumped from one field to the one lying above it, and in this way, in this land of wonders, water is carried above the height of 1000 feet.

The rice-ground is called Pihring in the Malay. The grain is first sown in it very thickly in little heaps. When the young plants are two or three inches high, the tops are broken off, that each plant may form several side shoots. In many parts of China the plants are several times transplanted in order to obtain a more abundant crop. In Sumatra they are transplanted on the 40th day after sowing, when the middle shoots have long been broken off. After this is done, the skill of the cultivator consists in the exact proportioning of the water, which he admits into the Sawuhr or rice-ground, for the same water must not remain long on it, and when the rice begins to flower, all the water must be

* The Hist. of Sumatra. London, 1811, p. 67, &c.

withdrawn. The harvest commences from three to four months after the transplanting; the ears are either cut closely off and the stalks left to decay in the earth, or they are cut with the stalks and bound into small sheafs.

Mountain rice, or Ladang, is sown on elevated ground and succeeds best in soil which has been obtained by rooting out and burning down forests. This is everywhere done in Sumatra, Java, Luçon, and the Brazils, where there is much wood and a scanty population. But as it is impossible to burn the woods of tropical countries in their fresh state, the branches and tops of the trees growing in a place which it is intended to sow afterwards, are usually hewed off at the beginning of the dry season and left lying about until they become dry, and then the whole are set on fire. The fire is often maintained for months, until all is burnt to the ground, and the soil is manured by the remaining ashes in a way which cannot be so easily done in other places. If the wet season commences during the time of burning, the fire is extinguished, and the whole work must be put off until the next dry season. The mountain rice is sown at the commencement of the wet season, which in the northern hemisphere takes place in April and May, in the southern in September and October. Holes are first made with a pointed instrument, and another person throws a few seeds in each, leaving the hole to fill up of itself. The harvest follows about five months after seed-time. The mountain rice is reaped in the same manner as the marsh rice; in Brazil, however, Von Martius mentions that the stalks are trodden down, in consequence of which they shoot up again and yield a second crop after one or two months. The modes of separating the rice from the ear are in different countries as various as those of threshing wheat. In Sumatra the Malays trample out the ears, while they support themselves by a bamboo pole.

Rice in the husk is called Päddih in the Malay (Paddee in England), Palay in the Tagalli, and the separation of the grain from the husk is a very difficult task. Machines are employed for this purpose in countries which are more civilized, but the poor Indian has this labour every day, before he can eat his rice. Among the inhabitants of the Philippine Islands, if the palay is not ground the day or night before, there is nothing to eat the following day. But we must not altogether wonder at the indolence of these people ; they are even in this instance to be excused, for the palay, on account of its hard shell, keeps much better than shelled rice.

The palay or päddih is ground in large mortars with heavy pestles of hard wood, and this is generally the employment of the female part of the family, and occupies them during a third of the night. The cleaned rice measures about half the quantity of palay ; the husk is easily removed by throwing the rice about, for the grains are very heavy.

The yield of the rice crop is very different according to the moisture of different places. Mountain rice, in soil newly taken in by burning the wood, yields from 60 to 80 fold; but in manured ground, where the rice is grown every year, the husbandman must be contented with 40 fold. Marsh rice, on the contrary, produces 100 or 120 fold ; I have, however, seen very fertile places, for example at the Laguna de Bay in the island of Luçon, where the crop was only 70 fold. But on the Philippines there are also places where the marsh rice, by being several times transplanted, yields 400 fold. The mountain rice, though less productive, is better, and is also more esteemed, as it keeps longer than the marsh rice.

In those countries which are the chief seats of the culture of the rice, and where it is the most common food, the modes of cooking it are infinitely various. When boiled in clear water, it stands in the place of bread to the inhabitants of Eastern Asia. A great many dishes are prepared from the flour, and in China a strong spirit is obtained from the grain. The wine of the Chinese, called Samdschu, is distilled from it, and resembles a strong arrack. Although this liquor is extremely intoxicating, it is always drunk boiling hot at Chinese tables.*

Rice is very cheap, when the harvest is abundant; it is said that in Manilla the average price of a Cavan which holds 137 Spanish pounds is eight reals; therefore for about a penny of our money there would be obtained more than three pounds of cleaned rice. But there are seasons and certain places, where it is only half so dear, and again times when it is thrice the price.

### 3. *The Maize (Zea Mays L.).*

Maize is indigenous to the New World only; it was there, before the arrival of Europeans, the principal cereal,† and it still is so in the tropical parts of this continent. Maize thrives best in the hottest and dampest tropical climates; there are some places where it brings forth 800 fold; in less fertile lands the produce is 300 or 400 fold, and 100 fold is looked on as a poor crop of this grain in tropical countries. Maize is less productive in the temperate zone; the average yield in California, between latitudes 33° and 38°, is not greater than 70 fold. In yet colder countries, the yield is still smaller, and our cereals gradually supersede the maize; this is the case in Chili, where maize is grown as a vegetable only, and wheat is used for bread. We do not know the exact polar limits of maize culture in the New World, yet so much is certain, that they lie in the 40th parallel; even in the southern hemisphere, where, particularly in Chili, from many causes the climate is much

* See Meyen's Reise, ii. p. 392.

† See the oldest Spanish writers on America, who have given a clear account of this.

lower in proportion to the latitude, the maize is cultivated as low as 40°. On the western coast of Europe maize is grown in 45½° north latitude, on the Rhine to 49°, and in our country even in 52° large and abundant crops of maize are raised in gardens, yet with us there is little taste for this fine grain, and therefore its culture is neglected. Maize is only grown to adorn our gardens, and the rich produce is given to cattle. In Germany maize is most extensively cultivated in the fertile valley of the Rhine, known by the name of the Bergstrasse, but this district is also the warmest in all Germany.

The culture of maize has spread rapidly over the old continent, and has reached India, China, and Japan by some way of which there remains no tradition. The Malays in Sumatra and the natives of the Philippines cultivate maize, but with them it is by no means so common a food as rice. Maize is said to have been brought to Japan 1200 years ago; Siebold has seen a writing which is said to contain the statement, that maize at that period was carried by the sea to the shores of Japan, and that it has ever since been cultivated there. But even though there should have been no mistake in the interpretation of that Japanese writing (which, indeed, Klaproth, the person best acquainted with that eastern language, affirmed after he had read a copy of that writing), yet the most weighty objections may be made to that statement, so that no grounds remain on which to believe it.

The currents in the Pacific Ocean flow uninterruptedly from the American coast to the Marian Isles; a fast-sailing ship takes two months for this distance; how then is it possible, that maize, the grains of which, as of all the other cereals, are so easily injured by moisture, can have remained so long a time in salt water of so high a temperature without being destroyed? It would besides be very strange, if it were true that maize had been carried by currents from America to Japan, without having reached and become indigenous in the South Sea Islands, particularly the Sandwich and Society Isles. I cannot, therefore, be convinced of the growth of maize in Japan for so long a time, but I think it may be assumed, that it has been introduced there by the Portuguese.

At the present day maize is cultivated in all the countries of the tropical and temperate zones which European civilization has reached; it cannot, however, supersede the culture of the earlier cultivated cereals.

Although maize is a plant which succeeds best in the hottest climate, it ascends to an incredible height on the mountains of America. Humboldt mentions that vast maize-fields are to be found on the plateau of Mexico, at a height of 8680 feet; and in Peru, on the road between Lima and Pasco, maize is cultivated as high as 3824 metres, almost 12,000 feet; nay, at the time of the Incas, it was grown by artificial means on the island of Titicaca, in the great lake of the same name. On that island, at a height of 12,800 English feet, was the celebrated temple of the sun; in it the Incas offered some of the maize, grown in the island, to the deity, and the rest was carried by the virgins devoted to the service of the sun to the other cloisters and temples of the kingdom, whence it was distributed amongst the people. The people believed that whoever preserved a single grain of this maize, would never want bread during his whole life.

Like our cereals, maize shows several varieties, some of which are distinguished by the size of the grain, others by early ripening. Yet of all the cereals, which are grown by various nations, none except the rice is so unequal in its produce. Humboldt says,* on the same soil, the produce of a grain will vary from 40 to 200 or 300 fold, according to the variation of the moisture and mean temperature of the year.

* L. c. page 37.

When the harvest is good, maize is more profitable to the colonist than wheat, and we may say that the culture of the maize has the advantages and disadvantages of the culture of the vine.

The modes of using maize are very various among the different peoples of America ; the Peruvians and Mexicans prepared several kinds of bread from it. The former people had one kind which they used in sacrifice and called Canen, another, which was their ordinary bread, named Canta, and a third sort which they used at their feasts. At the present day, in the different countries of America in which maize is cultivated, there are a great many ways of cooking it ; the whole head is very often brought to table boiled and is eaten with salt, and it then tastes much like our shelled barley.

This is not the place to describe the modes of cooking the plants used for food, but it may be conceived how general the use of maize is in the tropical parts of America from the fact, that at the beginning of this century, in Mexico alone, 800 millions of kilogrammes, above 1,600,000,000 lbs. were consumed by a population of not more than 5,000,000. But the consumption of this grain is so enormous, because in places which are destitute of grass the mules are fed on it. There is therefore the greatest distress in this country when the maize-crop fails.

Several fermented liquors are likewise prepared from maize, which were known in Peru at the time of the Incas, under the name of Chicha. The chicha, which is the common beverage, is like our white beer, or, still more, the weak beverage which is known in many parts of Germany by the name of Broihan. Other kinds of chicha taste like cider, and become very strong by age. I have tasted cider, which was found in an ancient tomb, and which must have been at least 300 years old, and found it to be like alcohol. On the sides of the Cordilleras there is everywhere abundance of chicha de mays, while on the plateau other liquors are prepared, for example, pulque in Mexico, and in Peru a chicha de quinoa.

The stalk of the maize is extremely saccharine, and not only is there a honey-like syrup prepared from it, but the stalks are crushed in the same way as our sugar-cane, and a well-flavoured brandy is obtained, which in Mexico is called pulque de Mahio or pulque de Tlaolli.

Several varieties of maize, which are distinguished by extraordinary height and great beauty of leaf, are grown in our gardens and are a valuable addition to our ornamental plants. In South America the varieties are endless, and in hot and fertile parts, as at some points of Northern Chili, maize plants not unfrequently attain the height of 10 and 15 feet.

*Remark.*—A great deal has been written concerning maize, and in many states it has been attempted to cultivate this beautiful cereal and introduce it as a common article of food. One of the greatest works on this subject is that of Parmentier (Le Maïs ou Blé de Turque, à Paris, 1812, 8vo, 1 vol.) in which are cited the old Spanish authors who have written about maize (see pages 14–19 in it). F. de Neufchateau's Supplément au Memoire de M. Parmentier sur le Maïs, Paris, 1817, must also be mentioned.

---

Besides the cereals already named, I must notice the several kinds of millet, which are generally used in the southern and eastern parts of the Old World, even in the sub-tropical part of China and Japan.

Turkey millet or Negro-corn* is grown in all the torrid parts of Africa, in the south of Europe, especially in Portugal, in the Levant, and in the East Indies. It is a plant of hot countries, but its limits are not yet determined. In the East Indies, where it is much grown, and indeed is often the common food, especially where

* Sorghum vulgare *Willd.*

the rice is not cultivated, Turkey millet ascends to very considerable heights.

The native country of this plant is unknown, yet, as the name implies, it seems to have come from Africa, and it is there of as great importance as wheat with us.

The numerous species of millet with small grains, which are cultivated over all Europe, in the East Indies, China, and Japan, and on the islands of the Indian Archipelago, as Panicum miliaceum, P. germanicum, P. frumentaceum, P. miliare, and P. italicum, are extremely well tasted, but they are a general article of food in some parts of India only.

Now that we have become acquainted with the principal cereals and their range of distribution, this, perhaps, is not an improper place to put the question, How man has been induced to cultivate the grasses, which have so often small seeds, in order to protect himself most surely against famine? This would be extremely difficult to answer, if these nutritious grasses in their wild state grew singly and scattered, but the case is the reverse.

When speaking of the native country of the rice, we became acquainted with the wild rice fields, which Von Martius found on the Rio Madeira, from which the inhabitants of that country gathered nearly as good a crop as others from their cultivated fields. We shall take a similar example from our native country. Glyceria fluitans* grows wild with us on the margin of standing water, as well as on very wet meadows; round Berlin, where this plant grows singly, no one thinks of the well-tasted seeds which it bears in its spike; but farther east, in East Prussia, Masuria, and on the Lower Vistula, it grows in such quantities that its seeds are gathered with great profit without the plant having been previously sown. From the seeds of this grass are made several fine kinds of groats. Where grasses with nutri-

* Festuca fluitans *L.*

tious seeds thus grow wild in vast quantities, it must very soon have occurred to man, to imitate such natural corn fields, and to remove them to those places which seemed to him best adapted either to protect them from the weather or against enemies. Such was the origin of fixed dwellings and agriculture.

All the nations who have reached any degree of civilization have highly valued agriculture, and regarded it as the foundation of prosperity ; they have, therefore, held its inventor or introducer as sacred, and recognized in him a divinity. In the Chinese empire an annual festival is kept, on which the emperor guides the plough with his own hand in his garden at the north gate of Pekin, while in every province of the vast empire, the high officials perform this ceremony in place of the emperor, thus to show in what estimation agriculture should be held.

If we, in conclusion, consider the relative produce of the cereals in order to see which of them gives the largest crop, with the least sacrifice, it follows from the preceding statements, that maize stands far above all the others, then rice, and after it the other cereals.

### 4. *The Quinoa (Chenopodium Quinoa W.).*

I shall notice the culture of a few other plants, immediately after the cereals, as they are cultivated in an exactly similar manner.

The station of the quinoa is, it is true, extremely limited, but for those countries in which it is grown, it is, next to the potato, the best gift which nature has bestowed on man. Over all the plateau of Southern Peru, above the height at which rye and barley still ripen, the Chenopodium Quinoa *W.* is a principal object of agriculture, and on the plateau of Chuquito, almost 13,000 feet high, are vast fields quite covered with this plant, which, however, do not give the landscape the charm of our beautiful green corn-fields. On good soil this plant attains the height of three or four feet, and bears an immense quantity of seeds, which unfortunately, for a long time, feed an innumerable flock of birds, like sparrows, for this plant has the disadvantage, that all the seeds do not ripen at the same time.

The leaves of the quinoa are very commonly eaten as a vegetable, and much resemble those of our Chenopodium viride, which is eaten as spinage by the poor of our country, and a variety of the quinoa, as well as of the latter plant, with red leaves, is not uncommon.

The quinoa is still cultivated in Southern Chili, but, before the introduction of our grasses, it must doubtless have been a more general food, and that not only in Chili, but in Peru, where it is now superseded by our cereals, wherever the climate permits their culture. The variety which, according to Molina, is called Dahue by the Indians of Chili, and which has ash-grey leaves and white seeds, is the one commonly cultivated round the lake of Titicaca.

The little, mealy and very oily seeds of the quinoa are a very pleasant and nutritious food, and they, along with the potato, are the common food of the poor inhabitants of the plateau of Southern Peru. There are many different ways of cooking this grain. Sometimes the seeds are crushed between stones, and boiled as soup or frumenty ; sometimes the meal is roasted and then forms the chocolate of the plateau ; sometimes Chicha de quinoa is made from it.

On the table-lands of the Himalayas, in Southern Asia, a plant very similar to the quinoa, viz., Amaranthus fariniferus, is cultivated in the same manner ; and the uses of, as well as the modes of using, the two plants are also similar.

### 5. *Buck-wheat (Polygonum Fagopyrum et spec. var.).*

Buck-wheat is cultivated with great advantage in many parts of Northern Europe, as it prefers the very worst soils. We have received buck-wheat from the interior of Asia, and, from Beckmann's investigations into the subject, it appears that it was not known in Europe before the beginning of the 16th century. Its native country is still not exactly known; it seems, however, that it must be sought for in the north-west of the Chinese Empire. Another kind of buck-wheat, Polygonum tataricum, which has a preference for a sandy soil, and grows wild in Southern Siberia, beyond the Sea of Baikal and in Yenisei, has long been known. In the countries where this plant is wild, the seeds are gathered, like those of Glyceria fluitans with us, and the wild rice with the Americans.

In the elevated mountain lands of Southern Asia, the group of the genus Polygonum, which furnishes buck-wheat, seems to be peculiarly at home, for there, in the more or less elevated plateaux, a great many species are cultivated, and are frequently the chief food of the inhabitants of those countries.

## II. THE CULTURE OF THE PRINCIPAL TUBEROUS ROOTS.

### 1. *The Potato (Solanum Tuberosum L.).*

From the consideration of the cereals of the Old and New Worlds, we pass to the potato, which is a gift of America to the Old World. It is true that prosperity and civilization have been developed on the eastern continent without an acquaintance with the potato, yet the universal diffusion of this plant amongst us has produced a complete revolution in our system of agriculture, and has furnished us with the most powerful means of preventing the general famines from which the inhabitants of Europe formerly so often suffered, and in whose train followed the most dreadful pestilences. The exigencies of the poor are completely met by the cultivation of the potato, for since its introduction there has never been an instance of its failure and that of our cereals at the same time. The crop of the potato is generally more productive when the grain crops are deficient; but now we see the greatest misery amongst the peasantry and the poor when the potato fails, a case which is indeed rare, but which, however, happened in many dry districts in the summer of 1834. So great is the influence of the culture of this root on our social relations, and this must increase more and more with the constant increase of the population, and at the same time of the number of the poor. In unhappy Ireland the potato and oaten bread are the common food, and, when the former fails, thousands must die of famine. But even with us how important is it; not to mention that we eat it almost daily, and that in many provinces the rye bread is mixed with potatoes, the preparation of starch, sago, brandy, wine, and even sugar from it is the source of the support of millions. Nay meat, milk, butter, and cheese can be kept at low prices only by the culture of the potato.

The artificial distribution of the potato furnishes material for instructive reflections, which this is not the place to pursue specially. It is very astonishing to see how a plant which is indigenous in the cold regions of the South American Cordilleras, should, in a short time and in an incredibly rapid manner, have become the general food in whole quarters of the globe. Over all Europe, from Hammerfest in Lapland, about 71° north latitude, on Iceland, and in the Faroe Isles, the potato is cultivated, and its culture has been introduced to the lower plateaux of India, into China, Japan, the South Sea Islands, New Holland, and New Zealand. The potato has been extensively grown in Saxony since 1717; in Scotland since 1728, and in Prussia since 1738.*

* See Beckmann's Grundzüge der deutschen Landwirthschaft, 1806.—J. Banks, An attempt to ascertain the time of the introduction of potatoes, 1808.—Lambert's Descript. of the gen. Pinus, &c., Sec. Ed. ii. App. p. 11, where Montevideo, Lima, and other places are pointed out as the native country of the potato.

It is well known with what reluctance the country people formerly engaged in the culture of the potato; nay, Frederic the Great was obliged to compel the Pomeranians to accept this immense benefit.

Nothing is yet positively known concerning the extent of the native country of the potato; it is certain that this plant was cultivated in the colder regions of the South American Cordilleras before the discovery of America, but it is equally certain that it was unknown to the Mexicans. At the present day the potato, Papa in the old Peruvian language, forms the chief food on the plateau of Peru; and on the shores of the Lake of Titicaca, as at the time of the Incas, this root is planted with even greater care than in our country. The potato is grown in Chili also, and is there called Pogni, to distinguish it from Maglia, the wild potato, which produces only small and bitter tubers. If the potato had migrated from Chili to Peru, it would probably have retained its Chilian name; but this conjecture is no longer necessary, for it grows wild in both countries. I myself have found it at two different places, on the Cordilleras of these countries, and Ruiz and Pavon mention the mountain of Chancay as a station where the potato is to be found wild.

As I have already remarked, it is quite certain that the potato was not cultivated in Mexico before the arrival of Europeans, and it is also believed that it is not indigenous on the mountains of Mexico. Schiede,* indeed, found a potato wild on the volcano of Orizaba, which was sent to us, and has been planted here, but many doubts have arisen whether this potato is really the Solanum tuberosum, and it rather seems certain that it is another species.

Although the range of the potato was thus interrupted by its absence in Mexico, before the emigration of Europeans to America, yet there are various sources which seem to prove the cultivation or rather the presence of

* Linnæa von 1829, p. 227.

this plant in some parts of North America, and in all probability we have received the potato directly thence.

The colonists who arrived in Virginia in 1584, found the potato there,* and ships which returned from the Bay of Albemarle in 1586, brought the first tubers to Ireland ;† therefore the statement that Sir Francis Drake introduced this root into Europe seems to be unfounded. In the account of the remarkable voyage of the English buccaneer, there is no mention of it, and when Drake, after his return to England, (when he entered the Thames with silken sails, the plunder of the Spanish galleons from Manilla,) was honoured by a visit to his ship from Queen Elizabeth, all the kinds of food and fruits which that voyager had brought home with him were put upon the table. In the account of that feast all the dishes are named, but the potato is not mentioned. Thus the name of the man who brought this blessing to Europe has perished. Had it been misery, or war, with bloody battles, all the histories of that period would have been full of it. We need not wonder that the potato was not brought to Europe by the Spaniards as speedily as maize and the sweet potato, for this plant was cultivated on the western coasts of South America, and the voyage round Cape Horn then took up too long a time, and was too unfrequent to send the potato by this way to Europe.

I pass by the mode of using the potato as already known ; there are some peculiarities in this among the mountain inhabitants of South America, its native country. Amongst the many varieties which are grown in America, a small, very sweet potato is chiefly used for roasting in the embers. In the towns of Puno and Chuquito, on the shores of Lake Titicaca, these roasted potatoes may be got at any time of the day fresh from the fire, just like the roasted chesnuts in the south of Europe.

* A. v. Humboldt's Neu-Spanien, iii. p. 75.

† Beckmann, Grundzüge, &c. p. 289.

In those countries there is a very good method of preserving the roots for future use ; the potatoes are cut into slices, and dried until pretty hard. These slices are very convenient on journeys.

## 2. *The Arum or Aron.*

The roots of several species of Arum are cultivated with extraordinary care in the hottest part of the torrid zone, and they are there even a still more general article of food than the potatoes or bread with us. The Arum is grown in the most distant countries of both continents ; we found the Arum macrorrhizon and Caladium esculentum in the Sandwich Islands, and both species were also cultivated in the Friendly Islands. The Arum macrorrhizon is chiefly grown in the East Indies and China, and also Arum colocasia, which is said to have been brought from Africa. The former species and Caladium acre *Brown* are also grown in the tropical parts of New Holland ; while Caladium esculentum is found on the Indian Islands, in the West Indies, and at several points of the continent of America.

Few other agricultural plants require so high a degree of heat as the species of Arum with large farinaceous tubers. Europe does not possess one of them. The Sandwich Islands, where these plants succeed remarkably well, indeed, lie on the border of the tropics, and enjoy a very agreeable climate, without the excessive heat of other tropical countries ; yet, as we saw in an earlier part of the book, the mean heat of Hawaii amounts to 19,12° R., and for more than five months of the year the mean temperature is above 20° R.

All the roots of the Arums have an acrid, somewhat poisonous principle, which, however, is so slightly connected with the nutritive ingredients that it is lost in the drying, or by roasting and baking, and the root is then perfectly innocuous.

The culture of the Arum takes place in the same zone

in which the banana, the sugar-cane, and the cocoa-palm are grown, but the banana and sugar-cane extend much farther beyond the tropics.

No other agricultural plants can give the landscape so agreeable a tone as the Arum fields, surrounded by the plantain and sugar-cane, whose various shades of green contrast so pleasantly with each other.

On the Sandwich Islands the Arum is called Taro, and the fields in which it is planted taro fields. These fields are generally quadrangular pieces of ground, about forty-five or fifty feet square. They are dug out two or three feet deep, and so situated that a running stream can be turned into them. These hollows generally lie like terraces, one above another, so that the water can be carried from the higher to the lower ones, and the borders of them, which at the same time separate the possessions of the different proprietors from each other, are generally used as foot-paths; at least this is the case in the richly cultivated districts.

The hollows of the taro fields are so deep that the leaves of the plants project but little above their level; the plants are set rather wider than the potato with us, about the distance of the cabbages in our fields from each other. Just as with us those agricultural plants which are forced to an excessive development of root, do not generally bear fruit, so we rarely see any of the taro plants in flower, and these grow wild in the vicinity of old taro fields, deep in the water like our Acorus calamus.

The tuber of the Taro plant attains the size of a child's head, and when boiled or baked in hot earth, it has a great resemblance to the sweet potato; it is, however, more delicate in flavour, and probably more nutritious. A variety of the Arum macrorrhizon is grown in dry land, and even at heights above 800 and 1000 feet. This plant, the tubers of which are never so large nor so well flavoured as those of the wet taro, must, however, be kept excessively wet; for this purpose it is usual to surround each plant with a little hollow in order to collect the more moisture about the root.

In Oahu the culture of the banana ceases at the same height as that of the taro; above the height of 800 feet neither the wet taro nor the banana is found.

The modes of preparing the taro are, as has been said, very various. It is commonly eaten like bread, with or without salt, after having been boiled or baked. The tubers are also cut in slices and fried in lard. But the most common mode of all is to mash it after being boiled, into a thick frumenty. More water is then poured to this, and the whole mass is allowed to ferment, which generally begins within twenty-four hours. This fermented semi-fluid frumenty is called Poe, and it is the favourite food of the Sandwich Islanders, who often swallow incredible quantities of it. As the use of the spoon is unknown in those parts where the taro is cultivated, this frumenty is eaten with the fingers.

The young leaves of the taro plant are used as well as the tubers, but not so generally, as they require a quantity of lard. When a pig is baked in the earth its belly is filled with these leaves, which are then considered an exceedingly good vegetable.

The taro and some bananas, a cocoa-nut, or a baked bread-fruit are the common diet of the natives of the South Sea Islands. The flesh of the pig and the dog is still confined to the richer class, wherever Christianity has been introduced by the missionaries without first laying the foundations of the prosperity and civilization of the people. The sweet potato, the yam, and some other roots are less generally used than the taro in the South Sea Islands.

### 3. *The Manioc or Mandiocca.*

The root of the Manioc is one of the most important articles of food in the tropical parts of America, and it seems very certain that this plant is indigenous in the New and not in the Old World; at least the very current opinion that it was brought from Guinea to America, is entirely without proof. The manioca plant grows in that zone in which the banana ripens, but the latter ascends much higher up the mountains than the former. According to Humboldt, the manioca on the mountains of Mexico does not ascend higher than 600 or 800 metres, while the banana reaches much greater altitudes.

Two varieties of the manioc plant are cultivated by the inhabitants of America; the one is called by the Spaniards Juca dulce,* the other Juca amarga. Botanists formerly considered these two plants as mere varieties, and gave them the name of Jatropha manihot; however Pohl, who for a long time travelled in Brazil, thinks that they are systematically different species, and calls the bitter manioc plant, Manihot utilissima, the sweet one Manihot aipi. The root of the latter plant is perfectly innocuous, while that of the other species is a violent poison, if the poisonous juice is not most carefully separated from the flour, which is effected by mere expression of the grated root.

From the flour of the manioca is made bread, which is called Cazavi and Cassava (Pan de tierra caliente by the Spaniards) and is extremely nutritious and palatable. This nutritiousness is ascribed to the sugar and a clammy substance; the latter has some resemblance to the Caoutchouc, which is common to all the plants of the family Tithymaloideæ. The Cassava bread is generally in the shape of discs, 18 or 20 inches in diameter, and three millemetres thick, which are called Turtas. One pound of this bread is sufficient for the daily food of a native of America. The fine starch of the manioc root is also frequently used as flour, and this, which is known in Europe under the name of Tapioca, forms in several countries a considerable branch of commerce. Manihot-sago, which is also prepared from this starch, is likewise an article of commerce. The flour of the grated, dried and smoked manioc is indestructible, which is of

* Juca is the name of this plant in the language of Haïti.

the greatest importance in tropical countries, and it is therefore particularly useful in journeys.

The culture of the manioca requires more industry and patience than that of the banana. It thrives best in dry and elevated situations;* in damp low lands the root is of an extraordinary size, and soon rots if not gathered in at the exactly proper time.

The plant is propagated by shoots. The roots ripen at very different periods according to the different varieties and the degrees of heat. There is a variety in Brazil, which produces large roots in six or eight months; in Mexico nine months usually elapse before the crop is gathered in, but there are also varieties, whose roots cannot be dug for fifteen or even eighteen months.

Besides the flour of the manioca root, the expressed juice of Juca amarga is used; it contains the poisonous substance before mentioned, which, however, is dissipated by long boiling. The thickened juice is of a brown colour, and forms a kind of souy, which resembles a thick broth.

Too much cannot be said in praise of the valuable manioca plant. The Indians who cultivate it are compensated by it for the rice and the other cereals of the Old World. It is true that its culture is not so soon repaid as that of the other agricultural plants, and therefore a people, who resolve on the culture of such a plant, which does not bear edible roots for eight or even eighteen months, must necessarily possess some degree of civilization.

### 4. *The Batata or Camota (Convolvulus Batatas L. and Ipomœa tuberosa L.).*

The Batata is almost universally called Camotes in the Spanish colonies, from the Aztec word Cacamotic;†

* See Spix and Martius' Reise, ii. p. 875.

† See Alex. v. Humboldt, Ueber Neu-Spanien, iii. p. 81.

it is a plant belonging to the New World, and, very probably, to the South Sea Islands also. On the Sandwich Islands the culture of the camota was widely spread long before the arrival of the Spaniards and English. This plant requires a very high temperature, and is cultivated in every part of the torrid zone; but as it is only an annual, it can also be grown beyond the tropics, wherever the heat of the summer is equal to that within them. The sweet potato is even cultivated in New Zealand.

The camota bears tubers, which are very like those of the potato, but have a sweeter taste, so that it is generally called the sweet potato. It succeeds best in a hot, but dry climate, where the tubers grow to two, three, or four times the size of the fist, are mealy and of the most agreeable flavour, so that they are preferable to the potato, particularly when roasted in hot ashes. In the valley of Arequipa, almost at the height of 8000 feet, I found the most excellent camotas, far superior to the potato. But the batata is very different in a hot and damp climate, as in the East Indies and in the south of China, where the summer is the rainy season. The camota here is a tuberous root, which when boiled is soft, waxy, and of an unpleasant sweet taste; it is better even in the South Sea Islands.

Two varieties of the camota are cultivated, one with a yellow, the other with a white tuber. Its culture is quite the same as that of the potato; its tubers, however, as an article of food are nowhere of the same importance as the potato and the cereals with us, and the manioca and maize in tropical America. The batata is still to be seen in the south of Europe.

I have already mentioned that the tubers of two different plants are called batatas, viz., of Ipomœa tuberosa *L.* which is cultivated in the West Indies, and of Convolvulus batatas *L.*; the area of each of these plants is not yet so well known as is desirable.

### 5. *The Igname or Yam (Dioscorea alata L.)*

Another very nutritious root, which often attains an enormous size, is universally known under the name of Yam. The appellations Igname (on the coast of Paria) and Axes are American, the latter, as Humboldt informs us, in the Haytian language. The yam is cultivated not only over the whole tropical zone of the Old, as well as the New World, and in the South Sea Islands, but even still farther south, for Cook* found it in New Zealand. Towards the north, I do not think that so high a station for this plant is known. This nutritious root is grown in Java, Manilla, Sumatra, China, and all parts of the torrid zone. In a hot and damp climate it often weighs as much as 30 or 40 lbs.; but in point of flavour it is far inferior to the Batata. In Cochin China Finlayson† observed yams, which were 9½ feet in circumference and weighed 474 lbs. As in this instance the root was very fibrous, a kind of sago was made from the starch contained in it.

---

We have yet to mention some other tuberous roots, which are used as food in different parts of the globe. The Oca (Oxalis tuberosa *Mol.*) is grown on the Cordilleras of Mexico, Peru, and Chili; between 11° and 12° south latitude, its culture ascends above 8000 feet, and in Mexico also, it is said to be grown in the coldest regions along with the potato and the quinoa.

In China a Sagittaria sagittata is cultivated, whose tubers are as large as the fist and are much eaten. Nelumbium speciosum is grown in China, Japan, over great part of tropical Asia, and even in the eastern part of Africa. This root is generally to be found in the Chinese markets. In Upper Peru the Aracacha (Conium maculatum H. B. K.) is cultivated to the height of some thousand feet.

* Erste Reise. Berlin, 1774, ii. p. 33.

† The Mission to Siam and Hue, &c. London, 1826, p. 272.

The tubers of Tacca pinnatifida are eaten in the Society Islands, the Moluccas, and several other Indian islands. They are very sharp and bitter, but become somewhat milder by culture, and are rendered innocuous by the mode of preparing them. The grated tubers are squeezed in the same way as the roots of the Manihot, so that only the farina remains, from which is made a kind of bread of the same nature as the sago bread of the Moluccas, which it is said to surpass in flavour.

The acrid tuber of Dracontium polyphyllum is eaten in the Society Islands, particularly when there is a scarcity of the bread-fruit. The bulb of Hæmodorum spicatum supplies the place of bread to the aborigines of King George's Sound: when roasted it is mealy, but always continues somewhat acrid.

## III. THE CULTURE OF THE PRINCIPAL TREES, WHOSE FRUITS ARE THE GENERAL FOOD OF NATIONS.

### 1. *The Bread-fruit (Artocarpus incisa F.).*

The Bread-fruit is one of the most important plants to the inhabitants of the torrid zone, and alone is sufficient to furnish an agreeable and highly nutritious food, on which man can support himself. The native country of this useful tree is very extensive, but lies in the torrid zone only. On the islands of the Indian Archipelago and the island-groups in the South Sea which lie within the tropics, the bread-fruit is found. It has, however, never been observed in the wild state, but the whole species has passed into the cultivated state,* and it is therefore probable that man settled wherever he found a bread-fruit tree. Even yet the favourite situation of the fragile Indian huts is under its shady branches.

* The two Forsters were of opinion that the Artocarpus integrifolia or Jacca, which grows in the East Indies, was the parent plant of the cultivated bread-fruit; however, so many objections may be urged against this opinion, that it is very improbable.

The general appearance of the bread-fruit is beautiful, and none of our forest trees can compare with it. It attains the height of 40 feet, and its large and thick crown is adorned with the most beautiful green foliage. The leaves are about a foot and a half long, and between 10 and 11 inches broad ; they are divided into finger-like lobes. It is the fruit which furnishes the agreeable food. It is almost round and frequently attains a considerable size ; it is pulpy, enclosed in a harder rind, and generally contains seeds, which are somewhat larger than the seeds of the horse-chesnut. The hexangular pannels on the surface of the bread-fruit point out the single fruits of which the whole mass is composed.

The bread-fruit tree bears fruits abundantly, and they cover the tree for eight or nine months without interruption and ripen successively. The tree is without fruit for three months only, and during that time the Indians live on the preserved fruit. The bread-fruit is plucked before being perfectly ripe, while the rind is still green, but the pith snow-white and of a porous and mealy texture. It is then peeled, wrapped in leaves and baked on hot stones, for it cannot be eaten raw. When roasted or baked it tastes like wheaten-bread, sometimes rather sweeter. It is best on the Friendly and Marquesas Islands. When it is quite ripe, the pith is pulpy and of a yellow colour, and it can then be eaten uncooked ; but it has still a disagreeable flavour. G. Forster, to whom we owe a little monograph of the bread-fruit tree, describes the different modes of preparation by which the fruit is rendered edible. For example, the unripe fruits, after the removal of the rind, are laid in a paved pit and covered with a heap of leaves and stones until they become sour and ferment. The paste, says Forster,* tastes then like the black Westphalian bread, when not thoroughly baked. From the store

* L. c. p. 20, On the Bread-fruit. 1787.

in the pit the quantity necessary for daily use is taken, and made into lumps about the size of the fist, which are rolled up in leaves and baked between hot stones. These lumps of bread keep for weeks, and are a very good provision in journeys. The Indians live on these stores during the three or four months that the tree bears no fruit.

This valuable tree produces so abundant a crop, that three trees are quite sufficient to maintain a man for eight months. The great discoverer Cook gives this tree the highest praise in a few words; he says, "Whoever has planted ten bread-fruit trees has fulfilled his duty to his own and succeeding generations as completely and amply, as an inhabitant of our rude clime, who, throughout his whole life, has ploughed during the rigour of winter, reaped in the heat of summer, and not only provided his present household with bread, but painfully saved some money for his children."

The bread-fruit is propagated partly by young suckers from the root, which are laid bare of soil and their surface notched. From these notches sprout a number of shoots, which are cut away with a portion of the root and put into the ground.

On many islands of the Indian Archipelago the chesnut-like seeds of the bread-fruit are regarded as a principal article of food. They are rendered palatable by roasting, but in most cases the seeds entirely disappear in the cultivated fruit. By culture a number of varieties of this tree have arisen,* which are chiefly distinguished by the shape of the fruit, and by the presence or absence of the seeds.

The trunk of the bread-fruit tree is also useful. The wood is soft and light, and is easily worked into little canoes and various household utensils. From the fibres of the young tree stuffs are made similar to those made from the paper mulberry tree. For this purpose the

* See Forster, De Plantis Esculentis. Berolini, 1786.

young shoots are planted closely together, that they may grow perfectly straight and yield the longer fibres.

### 2. *The Plantain or Banana (Musæ spec. var.) Platano in the Spanish.*

This is one of the commonest and most nutritious fruits of the tropics. The flavour of the common sorts is generally not so agreeable to strangers as one would expect from the descriptions of travellers. One must first be somewhat accustomed to it, and then he finds the plantain exceedingly sweet and pleasant. The varieties of this fruit in different countries are certainly numberless; on the Philippines alone more than 70 are cultivated, which have peculiar names. But there are also several distinct species of the genus Musa, which produce excellent fruits, six or seven of which are systematically distinct; and the different species also require different climates. According to Humboldt,* who has also given us the best and fullest information concerning the culture of the plantain, in his well-known work on Mexico, the Camburi (Musa sapientum *L.*) grows in the plain of the tropics, with a temperature of from 19° to 21° Cels., and extends to 30° and even 35° of latitude; on the mountains its culture ascends to the height of 900 toises, while the Platano Harton or Arton (Musa paradisiaca *L.*), even at the equator, ascends no higher than 500 toises, and requires a mean heat of between 23° and 28°. The fruit of Musa regia *Rumph.* is called Dominico in the Spanish colonies. The common banana, with all its varieties, which is grown in the tropical parts of Asia and Africa, even on the western side of the latter continent, appears, after many investigations, likewise to belong to Musa sapientum.

There have not been wanting writers who have con-

* De Distributione Geogr. Plant. p. 156.

sidered the banana to be that apple, which, in the garden of Eden, was the cause of such woe, and, therefore, it has long borne the name of the Apple of Paradise.

Where is the native country of the plantain, whether it is indigenous in the Old or New World only, or in both hemispheres, are questions which cannot at present be answered with absolute certainty, but with great probability. That the plantain is indigenous in the tropical zone of the Old World has been clearly proved; it grows wild in the forests of Ceylon, and it has been universally found in the South Sea Islands, and even at the present day is found on them in a wild state.* In the forests of the Sandwich Islands a wild plantain grows abundantly, which, in every thing except the fruit, resembles the cultivated one. It is almost as certain that America possessed the banana before the Europeans arrived there; at least in several countries a tradition prevails, that the varieties Arton and Dominico were cultivated before the arrival of the Spaniards, and Humboldt found the plantain and manioca cultivated among all the Indians in the remote lands on the Orinoco, which certainly had no communication with Europeans. But it is rendered still more certain by the account of Garcilasso de la Vega,† who names clearly and plainly all the kinds of food which were commonly used at the time of the Incas, and amongst them the plantain is mentioned for the torrid and temperate zones of Peru.

In all the Spanish colonies of the Old and New Worlds the plantations of the banana are called Platanar (Banarin). The culture of these plantations, fortunately for the Indians, is extremely easy, for when the fruit is ripe the old trunk has merely to be cut down, to permit the new shoots to grow freely. One of the shoots is generally about two-thirds of the height of the old plant, and it bears fruit again in the space of three months.

* See Sawers in Mem. of the Wernerian Society. Edinburgh, vol. iv. p. 403.

† Coment. reales de los Incas, i. p. 282.

When the shoots are planted, fruit cannot be expected until the 10th or 11th month. The average produce of a plantain is about 30 or 40 lbs. of fruit, but not unfrequently between 60 and 80 lbs., and as the Indian may reckon on four crops in the year, a single tree yields at least more than 100 lbs. of fruit in one year. There is, therefore, scarcely any other plant which produces such a quantity of fruit on an equally small space.

The fruit of the Musæ is soft, more or less saccharine, and of agreeable scent and taste. Seeds are generally wanting, and there are even wild species, in whose fruits none have been found ; but in India, Cochin China, Java, and Luçon* there is a constant variety (the Platano de Pepita), the fruit of which contains an immense number of large seeds, and therefore is less esteemed for eating.†

Finlayson‡ has expressly mentioned the occurrence of the cultivated Musa with perfect seeds ; on the island of Ubi, on the coast of the Eastern Peninsula, he found a wild Musa, the fruit of which was full of seeds, and he considered it identical with Musa sapientum, only the fruit had very little edible pulp in comparison with the cultivated banana. The Platano de Pepita, which I found in the island of Luçon, is a permanent variety, which is propagated by shoots, and though it contains a great number of seeds, the pulpy substance of the fruit is exceedingly well-flavoured.

The modes of using the banana are infinitely various ; the ripe fruit is usually eaten raw, after the thick skin is removed, which is very easily done. It is also often eaten roasted in the skin, and is excellent when fried in butter. It cannot be denied that the banana is one of

* Meyen's Reise, &c. ii. p. 214.

† Compare here Forster de Plantis Esculentis, p. 38, who names this variety Musa granulosa.

‡ Journal of the Voyage to Siam. Lond. 1826, p. 86.

the most nutritious fruits, though a person can eat a great quantity of it.

The beautiful plantain trees, which are the peculiar ornament of the country houses in tropical countries, are useful in several other respects to the inhabitants. The leaf serves the Indians for a tablecloth and a plate; before each meal he plucks the necessary leaves. When the Indian bakes an animal in the ground, he first wraps it up in plantain leaves. He also uses this beautiful leaf as a protection against the sun, and for keeping off troublesome insects.

The plantain is still more valuable on account of the tenacity of its fibres, which are used for making hemp, flax, and still finer threads. The hemp of Manilla, the Avaca of the Tagalli, is an important article of commerce to the inhabitants of the Philippines, for already ship-loads of it have been sent to Europe, and the cordage of Manilla, with which every ship there is rigged, is of remarkably excellent quality.

*Remark.*—As the stem of the plantain-tree is softer than that of the hemp, the preparation of this hemp is much easier and quicker than that of European hemp. The fibres, which lie in the outer layers of the stem, are the coarsest and are used for making ropes; those of the inner layers are much finer, and from them is made several stuffs which form the dress of the inhabitants of the Philippines. The natives of the Sandwich Islands make their excellent fishing-lines of these fibres. In Luçon the finest Avaca thread is woven with silk, and the stuff thus made is extremely valuable; it resembles that of the Pinna and is equally costly. The stem of the plantain is seven or eight feet high, and the fibres, which run through it without interruption, and which furnish the Avaca after being prepared by rotting, are all of equal length, and thus are superior to the best European hemp. The Avaca is also much stronger than an equal bulk of our hemp.

The culture of the plantain is, therefore, most important to all the inhabitants of tropical countries, and as this plant, which furnishes their chief support, may be said to grow of itself, without requiring any labour from the Indians, it has been thought that it is this fruit which permits, or at least promotes, the indolence of the Indians. I do not think that this view is correct: a man whose circle of ideas is confined, feels no need of employment, but if the poor Indian were educated, he would certainly show as great activity as we generally see amongst Europeans.

*Remark.*—During the printing of this book the fourth volume of Ritter's Erdkunde has appeared, in which the celebrated author has investigated the geographical distribution of several Indian agricultural plants, as well as the influence of their culture on man, with his usual learning. In this work there is ample information concerning the following agricultural plants, which I have likewise treated of in this book, viz., the banana, the pepper, the date-tree, the cocoa-nut, the areca palm, and the wine palm; and the great geographer has brought to light a number of most interesting observations, with which the public could not easily have become acquainted in the scattered literature concerning India.

### 3. *The Olive (Olea Europæa L.).*

The olive is one of the most valuable plants to mankind, notwithstanding that its culture is very limited; it has extended somewhat more widely since the discovery of America. Southern Europe, from 44° to 36° N. latitude, is the proper seat of the culture of the European olive; it requires a mean heat of from 14.5° to 19° Cels., but chiefly depends on the severity of the winter. In countries where the average winter temperature is below 5.5° Cels., the olive will not succeed without protection, which cannot be afforded to large plantations. In Europe it is cultivated to 44½° N. latitude; above this we find only a few plantations in well sheltered situations. Even in the peninsula of Southern Europe the olive thrives exceedingly well in the coast districts, where the winter is mild, but not so well on the table-lands, though of no great altitude. On account of the milder winter of coast climate, the olive is still grown in the Crimea, but the fruit yields an inferior oil. It is also met with in the lower part of the valley of the Rhone, and on the southern side of the Cevennes. With a slight protection the olive tree can go much further north, as is shown by the little tree which grows in the open air in the botanic garden at Bonn.

The reason that the zone of the culture of the olive extends so little to the east and south may be, that in these countries there are other plants, whose seeds yield as good oil as that of the olive, for example the Camellia oleifera in China and Japan, Camellia drupifera in Cochin China, and Thea oleosa in China, as well as the numerous palms and species of Ricinus, whose seeds likewise produce oil. The Castor-oil is made by long boiling and pressing the seeds of the Ricinus; its use is more general than we probably believe. In India, China, Africa, and even in the Australian colonies, castor-oil is employed in cooking.

In the East Indies the oil expressed from the seeds of Sesamum orientale and Raphanus (Brassica) orientalis, is as commonly used as olive-oil in the south of Europe, and may even be considered there as one of the chief articles of food.

The olive can bear a considerably higher mean temperature than it receives in Southern Europe, and it seems then to grow more luxuriantly than in our colder climate. At present olive trees are rare in the Canary Isles; they, however, thrive there like the willows of our climate.* It is true that in these beautiful islands the olive first grows in the region of European agricul-

* See L. von Buch, Beschreibung der Canarischen Inseln, p. 122, &c.

tural plants, from 1200 to 2500 feet, with a mean heat of 17.6° Cels., but this is the fault of man only.

The olive is not a native of the New World, where it now, in several places at least, grows with the greatest vigour. It is the Andalusian olive which was introduced into Mexico by Cortes. At the beginning of this century there was one of the most beautiful olive plantations on the plateau of Mexico at the height of 1168 toises,* where, however, the climate is generally as fine as at Naples; even in January and February the mean daily heat is between 13° and 14° Cels. At present the olive is grown in many parts of Mexico,† and it is hoped that soon this branch of agriculture will supply the whole demand of the country.

In the preceding century the olive tree was planted in New California in the neighbourhood of San Diego. It succeeds extremely well on the western coast of Peru, when in latitude 15° and 17°, it appears even on the heights of the coast, and grows to the size of our apple tree. The olives of the coast district of Arica, of Tacna, Islay, and Cumana are excellent, and the consumption of them is enormous. Roasted olives are carried about the streets of Arica and Islay, and are sent in little baskets, woven of palm leaves or rushes, to the higher plains, where, for example in the market of Arequipa, they are sold daily in immense quantities. The olive of the Old World greatly enriches the hot, dry countries of the west coast of South America, for it is incredible in what sterile soil this tree is found on the Peruvian coast; in the neighbourhood of a little spring grows a whole grove of olives close to a plantation of aloes and water melons. Although the olive was brought to Peru only a few centuries ago, I have seen there very thick trunks, which, considering the slow growth of this tree, indicate a very great age. As examples of the great

* See A. v. Humboldt, l. c. i. p. 56, and iii. p. 93.

† Becher, Mexico. Berlin, 1834, p. 142.

age of the olive, and the enormous thickness which it attains, must be mentioned the trees which grow on the Mount of Olives near Jerusalem, and are probably those which stood there at the time of Christ. These olives are eight in number; they are at least six metres in circumference,* and between nine and ten in height. The olive thrives over the whole of Chili, particularly in Coquimbo, but also at St. Jago, in latitude 33°, though elevated almost 2000 feet above the sea.

### 4. *The Cocoa-palm (Cocos nucifera L.).*

The palm has ever been the queen of the woods, not only on account of the beauty of its form, but of its extraordinary utility. There are few, perhaps no palms, which are not in some way useful to man. I cannot here mention them all, but only those which are to be regarded as a principal food of nations, or which, from some other useful qualities, exercise considerable influence on the prosperity of man. No other family of plants possesses such enormous powers of producing fruit; the Alfonsia in South America,† about six feet in height, bears about 200,000 flowers in a single flower-sheath, and often more than 600,000 at one time, which, it is true, do not all come to maturity.

Before all must be named the Cocoa, whose true country lies in the Old World and the South Sea Islands; it has probably migrated to America,‡ and is grown there in great numbers, on the West Indian Islands, and in Brazil.

The cocoa is an inhabitant of the coast, and only a few instances are known in which it has been observed in the interior of a country; this tree, however, may be

* Bové, Relation d'un Voyage Botan. en Egypte, &c.—Ann. des Sciences Nat. 1834, T. i.

† A. v. Humboldt's Reise, &c. Buch xi. p. 52.

‡ The currents in both the oceans flow in a direction very proper for the migration of the cocoa from the Old World to America.

successfully removed from its natural station by the fostering hand of man, and may be cultivated under other conditions of locality. Humboldt found the cocoa in the steppes of Venezuela, and Duke Paul Wilhelm of Würtemberg has very lately observed this palm in the island of Cuba, growing with singular luxuriance far from the sea shore; in India also, we find the culture of the cocoa far in the interior; however, it does not always succeed.*

Few of the South Sea Islands are without the cocoa; Easter Island, the most easterly of them, is one on which it has not been found. The islands in the Chinese, Malay, and Javanese Seas, as well as all those lying in the torrid part of the Indian Ocean, are more or less endowed with this valuable palm, but it is nowhere of greater importance to the support of the inhabitants, than in the numerous islets of the Laccadive and Maldive groups. The damp coasts of India, and particularly the rich island of Ceylon, possess by far the greatest number of cocoa-palms; they grow there in millions, and this branch of agriculture is no longer carried on for the support of the inhabitants, but its produce is an object of profitable trade. In the South Sea and the Indian Ocean the groups of cocoa-palms everywhere greet the approaching voyager, and under their scanty shade he first sees the straggling huts of the natives; but in the East Indies, where peculiar circumstances have occasioned the extensive culture of the cocoa, villages and large towns are situated amidst groves of it, and the whole coast of Malabar is shaded with innumerable cocoa-palms. In the south of Ceylon there is a forest of the cocoa-palm which stretches along the sea-shore for 26 English miles, is several leagues broad,† and contains about eleven millions of full grown palms. When the Dutch were masters of Ceylon, 6000 casks of arrack,

* Hamilton, Descrip. of Hind. ii. p. 210.

† See Transactions of the Royal Asiat. Society of Gr. Brit. vol. i. p. 546.

3,000,000 lbs. of cordage made from cocoa-fibres, and an immense quantity of oil were obtained from the forest every year. The cocoa-palm supplies the universal food of the people in this island as well as on the Maldives and Laccadives.

The narratives of travellers have taught us much of the utility of the cocoa-palm, and some works have been entirely devoted to it. Like all other fruit trees, it, as well as many other palms, is improved by careful culture. Its growth is rapid, and it frequently yields about thirty nuts in the sixth year. A full grown cocoa-palm often bears 200 or 300 nuts, and is nearly 100 years old. The fresh, ripe nut is full of a fluid as clear as water, and of a sweetish taste, which is called the milk. This cocoa-milk is esteemed as a cooling and extremely agreeable drink, and is often enthusiastically praised by travellers. The kernel afterwards forms in the nut, and the fluid gradually disappears, and in this state the nuts are brought to us by commerce. The kernel consists of a hard, white substance, which has somewhat the taste of the sweet almond, but is much less agreeable, and in a raw state it can be but little recommended as an article of food, on account of its extreme solidity.

The sugar-cane is cultivated in all the countries within the tropics in which the cocoa grows, and it is, therefore, very easy for the people to preserve the fruit of this palm. The kernel boiled in sugar forms a delicious sweetmeat, which is universally eaten in the Spanish colonies in tropical countries.

The well-known and highly esteemed cocoa-nut oil which was formerly called Oleum Calappi,* or Oleum Palmæ, is obtained by boiling the kernel for a long time in water and then pressing it. In some countries the kernels are first allowed to rot, the mass is then boiled, and a thick oil is thus obtained from it. It is well known what an immense quantity of cocoa-nut oil

* Kulapa is the Malay name of the cocoa-palm.

is at present used in our manufactories; it is said to be particularly excellent for making gas for illumination. When fresh, the oil is used in cooking, and many rude people anoint their bodies with it to prevent too rapid perspiration. From cocoa milk also an oil is obtained by boiling, which is particularly excellent and is used instead of butter. But from the cocoa milk, after it has fermented, is also obtained by distillation a very strong and fine spirit, a kind of arrack, which is much esteemed and rarely comes to us in trade.

The use of the hard cocoa-nut shell is well-known in our country. Its extraordinary hardness and the beautiful polish it takes, make it suitable for various articles, such as the knobs of canes and the tips of pipes, and in India and China beautiful vessels made out of it and mounted with silver and gold, are not unfrequently met with. The natives of the tropical countries, where the cocoa grows, use the shells for drinking cups, and from the fibres which form the thick pericarpium are made cordage, ropes, mats, brushes, and such articles, which have become known to us by commerce. For foot-mats the sheath of the leaves is mostly used. The leaves are less frequently used for plaiting than for thatching; those of several other palms, however, for example in the interior of Brazil according to Martius, furnish a substance which supplies the place of flax to the Indians. Bennett* has added to his travels an appendix treating of the cocoa palm, in which are described the plaited articles, and the method of making arrack from this palm in Ceylon.

The fibrous covering of the cocoa nut is also very good fuel, which becomes heated like charcoal, and in the island of Luçon is employed in burning pottery.†

The trunk of the cocoa palm has only a thin layer of wood, so that it cannot very well be worked for

* Wanderings in New South Wales, Batavia, Pedir Coast, Singapore, and China. Lond. 1834, ii. App. p. 295–342.

† Meyen's Reise, ii. p. 246.

building purposes, but the whole trunks, when used as pillars, are very durable and form the supports in the larger buildings of the Indians. The pith is praised as an excellent manure; for other purposes it is useless.

The cabbage of the cocoa palm is highly esteemed, and with reason. It is formed of the young shoots which contain the young leaves not yet unfolded and still of a marrow-like substance, and it is often of extraordinary size, even weighing above twenty pounds.

The palm wine is also known, which is obtained by fermentation from the sap of the cocoa-palm. A spadix pretty far advanced is cut off, and the wound enlarged every day; the sap which flows from this wound is received in vessels of bamboo-reeds, and becomes wine by fermentation. After eight days, however, it becomes as sour as vinegar, so that it is generally made into arrack before that time.

In order to collect the sap of the cocoa palm, the Indians lay poles from tree to tree, and thus they can easily go round their plantations several times a day to pour out the sap from the bamboo reed, which they had fastened to each tree before the sap began to flow. The next morning the sap is changed into palm wine, but the following evening it turns sour if the vessels are not closely stopped up.

By mixing the raw sap with lime is obtained the palm sugar, which is much used, and was brought to Europe, though rarely, in remote antiquity.

The inhabitants of the countries where the cocoa palm abounds, are aware that the wine obtained from it is much more pleasant than that of the wine from other palms, but they are also taught by experience that the crop of fruit is extraordinarily lessened by drawing off the nourishment of the plant. But when the wine is the object, it is obtained from a vigorous tree throughout the whole year; yet the strongest tree dies if the sap is withdrawn for several successive years. The cocoa palm is generally taken advantage of for the wine for a certain period, and then left to bear fruit again, which it ripens during the whole year, and therefore flowers; unripe and ripe nuts are almost always to be found on the same tree.

I have here mentioned the modes of using the various parts of the cocoa palm, and these are sufficient to convince every one how extraordinarily useful a plant it is. It must not be thought, however, that the cocoa nut is the sole support of those people who cultivate it, for wherever it grows, there are also found other still more important alimentary plants, such as the rice and the plantain in India, the arum, plantain, batata, and yam in the South Sea Islands, and the maize, manioca, &c., in America. Even in the Sandwich Islands, where the cocoa is called the tree of life, it is of very subordinate value, for these islands abound in other and much better fruits. Rumpfius and Von Martius in their writings mention the extraordinarily salutary powers attributed to cocoa nut oil.

According to Humboldt the northern limit of the cocoa palm lies below 28° of latitude, and it ascends to the height of 700 toises above the plain.

Like the cocoa nut tree, many other palms are made use of in various ways; in each, however, some particular part of the tree is more especially used, of which the following most remarkable examples are given.

### 5. *The Date (Phœnix dactylifera L.)*.

The date is indigenous in the north of Africa, in Egypt, Nubia, Syria, and Arabia Felix; it does not go farther east than the mouth of the Indus, yet it is found artificially planted in India also. The date trees which grow round Batavia, were brought from Persia.

The date palm requires a sandy and well-watered soil, and therefore it is found in the great deserts of Africa only near springs. It here not only affords food to the traveller, but its leaves are excellent fodder for the beasts of burden. Thus the date tree extends through the whole of Africa to the Atlantic, and is even indigenous in the Canary Isles; it is, however, wanting to the south of Senegal, and also in the southern hemisphere. It no longer appears in the oasis of Darfur, between 13° and 15° N. latitude; the zone in which it grows well, in general, extends from 19° to 35° N. latitude. The date also passes into Europe and is grown on walls in Italy, in latitude 44°, with a mean temperature of 13° or 14° Cels. Osbeck found it growing at a convent near Cadiz as high as the building. According to Link* it flowers freely in the south of Europe, as in Sicily, the Morea, and the south of Spain, and also bears fruit, though it is not sweet. In Spain, in the plain of Elche alone, the date tree is cultivated for the sake of its sweet fruit; in other more northern countries it is planted for the sake of its leaves. In Sicily the date still grows at the height of 1700 feet, viz., at Adernò and Trecastagne on Etna,† but it probably does not bear fruit there. The date tree has been introduced into America, and it is said that it thrives well in the West Indies, and is grown on the western coast of South America, even at Copiapó, in latitude 27°, but I myself have not seen it in this country; and I doubt whether the date tree can bear good fruit in Northern Chili, where the temperature from peculiar circumstances is so low.

The date palm is an example of extraordinary fruitfulness, and its fruit forms the chief support of the inhabitants of the barren parts of Arabia and Egypt. It is a diœcious plant, and where the male tree is wanting, the people are obliged to bring the pollen to fertilize the female flowers, which would otherwise fall off. The inhabitants of the deserts of Arabia and Africa have therefore been long acquainted with the difference in the sex of the plants. The Arabs even keep the pollen from one year to another, in case the male flowers

* Die Urwelt. I. p. 347.

† See Philippi on the Vegetation of Etna. Linnæa, vol. vii. p. 731.

should fail the succeeding season. Theophrastus was acquainted with this artificial mode of fertilizing the date palm. The fruit of the date has been very much changed by culture, and there are now several distinct varieties of it. Bové saw even a white variety in Arabia Felix.*

From the sap of the date-palm wine also is made. The young leaves or heart of the plant is eaten as cabbage ; and the uses of the leaves and pith are the same as of those of the cocoa nut.

### 6. *The Chilian Palm (Molinæa micrococos Bert.).*

The Chilian palm is not, as Molina supposed, a species of Cocos, but it forms a peculiar genus to which Bertero has given the name of Molinæa micrococos, in honour of the historian and naturalist Molina.† This palm, of which there is said to have been immense forests in Chili in earlier times, but which is now rare, is the most southern in America, where it descends to about 35° south latitude, and there finds so low a temperature that in winter snow often lies for several hours. The Molinæa also grows naturally in the island of Juan Fernandez, and it is still planted in the 37th degree of latitude.

This palm is extraordinarily fruitful and each spadix bears above 1000 nuts. The kernel is much sought after for making sweetmeats, and is even an article of export to Peru. A very well-tasted oil is also obtained from it.

The leaves, the young shoots, the sheaths, &c., are used for the same purposes as those of the cocoa palm.

### 7. *The Mauritius Palm (Mauritia flexuosa L.).*

The Fan palm at the mouth of the Orinoco is likewise

* Ann. des Scienc. Nat. 1834, i.

† El Mercurio Chileno. Santiago, 1828.

a remarkable alimentary plant, for, as Humboldt tells us,* it alone supports the unconquered tribe of the Guarauni. In the rainy season, when the delta of the Orinoco is inundated, the Guarauni live, like apes, on the trees of this palm, from trunk to trunk of which they stretch mats woven of the leaf stalks of the Mauritia. After they have covered these suspended floors with clay they can kindle fires on them.

The Mauritia is a social palm, which grows in marshy places and on the margin of standing water, as well as near rivers. The whole north of South America, east of the Cordilleras, seems to be gifted with this noble species; more or less extensive forests of it are found from the mouth of the Orinoco to the Amazon river, through the whole of Guiana, Surinam, and the north of Brazil, and also in numerous places along the Amazon river as far as the eastern declivities of the Cordilleras. The grey and smooth trunks of the Mauritia are said to grow so socially in the north of Brazil, that they, a hundred feet high, were ranged, as Von Martius says, close to each other like the palisades of a giant's stronghold.

The leaves are fan-shaped, and their fibres furnish the materials for various fabrics, such as mats and ropes, with which the Guarauni construct their super-terrene habitations on the tops of the palms or on felled trees. The Otomakas, in the delta of the Orinoco, are also acquainted with the method of making a fly-net, which is likewise woven of the fibres of the Mauritia leaves. The fruit of the Mauritia, which is red and scaled like fir-cones, hangs in immense clusters from the top of the palm, and has the flavour of ripe apples.†

From the sap of the Mauritius palm, the Guarauni make their sweet and intoxicating palm wine; and the pith of the trunk, before the male palm pushes forth its flower-sheaths, yields a sago-like flour, which, like the

* Ansichten der Natur. I. p. 26.

† Alexander von Humboldt, Reise, &c. p. 8. Buch ix. cap. xxv.

manioca, is made into large thin cakes, and is the chief subsistence of the people, so that this palm is sometimes called the sago palm of South America.

### 8. *The Sago Palm.*

Sago, which we receive from several tropical countries, is prepared from various palms, many of which may, perhaps, be yet unknown to us. Those which chiefly yield it, are Sagus Rumphii, Cycas circinalis, C. revoluta, Corypha umbraculifera, Caryota urens, and Phœnix farinifera. The native country of Cycas circinalis is far extended; it is found from Japan to Siam, and grows on all the Indian islands, as on Java, Sumatra, Borneo, Macassar, and Ceram, where great forests of it cover the morasses.

As is generally known, sago is prepared from the pith of the trunk; however, the pith is not fit for making it at all seasons, and that period must be selected when the palm has formed its spadix, but not yet opened it. At a later stage, either no sago is obtained or only a very bad woody kind. After the palm has been felled, the pith is taken out and rubbed down in water into extremely small particles, which will pass through a sieve.

In those countries, where sago is prepared, a palatable bread* is made from it, which is baked in square hollow stones, which are first heated to a proper degree. When new the sago bread is soft, but it afterwards becomes as hard as stone, for the farina of the sago is jellied by the effects of heat into a transparent, extremely brittle mass. Sago is prepared on almost every island of India; many sorts are very bad, but others are extremely fine and distinguished by the purest whiteness.

* The word Sago in the language of the Papuas means bread, and as bread in those parts is made of the pith of palms, the name of the artificial product has been transferred to the pith from which it is prepared. This word was first used by Pigafetta; he witnessed the preparation of sago in the Moluccas.

This is enough to show of what utility these palms are, whose pith yields a very delicate and nutritious bread. They are plants, however, which rarely pass beyond the tropics. They grow wild in the tropical forests, and the Indian, who is too indolent to cultivate other plants, procures sufficient food from them with ease.

In the East Indies the preparation of sago has become an important branch of agriculture, and is at the same time very profitable. The Indian sago tree, Sagus Rumphii or Metroxylon sagus *Roxb.*, yields so large a quantity of nutritious matter, that it far surpasses in this respect every other agricultural plant. A single trunk in its 15th year, sometimes yields 600 lbs. of sago; and an English acre of land (40 roods long and 4 broad) can support 435 palms, which yield an annual produce of 8000 lbs. of sago.*

### 9. *The Guinea Oil Palm (Elais guineensis L.).*

The large quantity of palm oil, which is at present used in our manufactures, is chiefly derived from the Elais guineensis *Linn.;* while the well-known cocoa-nut oil is obtained from Cocos nucifera, and that which comes from America chiefly from Cocos butyracea. The oil of Elais guineensis lies in the fleshy coating of the seeds, and it is obtained by mere expression; while that of the cocoa-nut is procured by boiling, or by previous putrefaction and boiling of the nuts.

This palm oil is both white and yellowish, of the consistency of butter, and has a smell of violets and a delicate and pleasant flavour. It has become a considerable object of commerce, since so much is consumed in our manufactures, and now increases the prosperity of the inhabitants of the tropics.

* See Crawford's Hist. of the Ind. Archip. i. p. 337 and 393.

## 10. *The Wine Palm.*

As we have already seen that some palms are distinguished by the peculiar usefulness of some particular part, one, for example, yielding a remarkably useful fruit, another a nutritious pith, others excellent oil, &c., so there are other palms of the tropical zone, which are generally grown for the sake of the palm-wine only. I name here Phœnix (Elate) sylvestris on the coasts of Malabar and on the lower plains of India; the Nipa palm in the Philippines and Java; Cocos butyracea (the wine palm) in South America; and above all, the Borassus flabelliformis, although many others yield wine, but not in so great quantities nor of so good a quality. The wine from the Borassus is obtained in the same way as that from the sap of the cocoa, which is particularly esteemed. The flower sheath, before it is fully formed, is either nicked near the top, and the nick daily renewed, or the whole shoot is cut off and the wound widened every day. The sap of several of these wine palms also yields a large quantity of sugar, and in this case it is generally distilled to arrack.

A different method is followed in preparing the wine from the wine or king-palm (Cocos butyracea) of South America, a circumstantial account of which Humboldt has given us.* After the trunk, which tapers but little towards the top, is felled, a hollow eighteen inches long, eight broad, and six deep, is made in the woody part of it, below the crown of leaves and flowers, and after three days a yellowish-white, clear sap, of a sweetish and vinous taste, collects in it. For eighteen or twenty days this palm wine, which, as it were, ferments in the trunk after it is felled, is collected every day, and it is a remarkable circumstance that this mass of nutritive sap is secreted long after the felling of the tree. A palm generally yields eighteen flasks of sap,

* Reise in die Æquinoctial-Gegenden, vi. 2, 1832, p. 55.

and the produce is said to be richer if the leaf-stalks are burned.

In the East Indies the culture of the Fan-palm, Palmyra or Brab of the English (Borassus flabelliformis), is of great importance, for the quantity of palm wine consumed there is enormous. Unfortunately this noble palm grows very slowly, and does not yield the wine until it is thirty or forty years old. The wine of Phœnix sylvestris is not so highly esteemed, and in India is chiefly used by the poor, while the rich drink the wine of the Palmyra.

This is, perhaps, enough to show that there are a great number of species in the family of palms, which produce a quantity of the best articles of food for man; but we must not share in the common opinion that these are as abundant and as easily to be procured, as the climate in which the palms grow is delightful. It is certain that the culture of the palms is inferior to the regular and nearly certain crop of our cereals. Only men as contented and as easily satisfied as the Indians could long support themselves on the produce of the palms; and though it is certainly true that the wild Indian, who roams amidst the mountain-forests, is chiefly supported by the wild palms, yet it must not be forgotten, that many days may pass in which he has nothing to eat.

I should exceed the prescribed limits of this manual, were I to enumerate and give the geographical distribution of all the principal fruits, which in different parts of the globe are not only luxuries, but real articles of food. Even in those countries where the most beautiful tropical fruits grow in the greatest abundance, they are seldom used as a common food, but rather as luxuries or to improve the flavour of the ordinary diet. Indeed, amongst almost all those people who cultivate the delightful fruits of the tropics, we find a higher or lower degree of civilization; the rude Indian does not concern himself about them, if they do not afford him his ordinary food. The juicy orange, the delicious mango, the pine apple, the Anonas, and many others of these noble fruits are seldom found amongst uncultivated tribes; I therefore pass them over and only notice the uses of a few other fruits, which, for certain periods, more or less form the support of particular nations, or are in some way directly connected with their mode of life, and which, on account of their peculiar chemical composition, can be kept for some time. Amongst the fruits to be here particularly noticed, there are many which become of great importance to particular nations on account of certain local circumstances, while in other places, where there is a greater abundance of better fruits, they seem to be of little consequence and are often entirely overlooked.

I mention the following:

### 11. *The Water-Nut (Trapæ spec. var.).*

The water-nut of European lakes has very large seeds, which are rich in starch and oil, but although they are found in very large quantities in our lakes, little use is made of them. On the contrary, in India and China, where, on account of the dense population, even the most fertile soil cannot always produce sufficient food, the fruit of the various species of water-nut, which grow in enormous quantities in the lakes of those countries, are commonly eaten. In India it is the Trapa bispinosa *Roxb.*, and in China the Trapa bicornis *Linn.*, which produce farinaceous and oily seeds without any particular flavour. The fruit of the water-nut is exposed for sale, as a food for the poorer classes, in all the markets of Southern China, in large towns such as Canton and Macao, as well as in all the villages which I visited. The produce of this fruit must be extraordinarily great on the heights of the Himalayas, where, on the plateau of Cashmere for instance, as Moorcroft,* and before

* Notices of the Nature, Production, and Agriculture of Cashmere. Journ. of the Roy. Geogr. Soc. of London. Vol. ii. p. 253, &c.

him G. Forster,* mentions, such a quantity is consumed that this trade, which the state has monopolized, brings in a clear revenue of £12,000 sterling. This plant grows in prodigious quantities in the lakes and ponds of Cashmere, and a great number of people are employed during the greatest part of the year in collecting the nuts.

---

The ever verdant land of Cashmere possesses several other plants which are there common articles of food, while with us and in other countries, where there is not so excessive a population and an abundance of other fruits, they are eaten only occasionally and chiefly by the richer classes only. The walnut is an example of this, for it seems to be very extensively cultivated in the vale of Cashmere, so that, according to Moorcroft, four different species are met with there, which are probably only varieties produced by cultivation. It is well known that the walnut contains as large a proportion of very well-flavoured oil as the olive, and great quantities of the nuts are consumed in Cashmere in the preparation of the oil which is then exported to Thibet; the oil of the sesame is, however, preferred to it. As the wood of the walnut-tree of Cashmere is also highly esteemed, this tree is extensively planted on this plateau, and the profit of this branch of commerce must be very considerable, since the government at least shares in it with the proprietors.

In European forests, which are, properly speaking, poor in edible fruits, there are several trees whose fruits in earlier times, before agriculture had arisen, were the ordinary food of the rude people, and are still of more or less importance to the inhabitants of particular districts; as

* Voyage du Beng. à Pét. Paris, 1802. Tom. i. p. 318.

### 12. *The Chesnut (Castanea vesca Gaertn.).*

The chesnut grows over the whole south of Europe, and finds its true home in the warmer part of the temperate zone. In Asia, this chesnut is found in Western Grusia and on the higher mountains of Caucasus, where it was discovered by Bieberstein; and, in all probability, our European chesnut is indigenous in the north of China, although the fruit which is brought to us in commerce from Canton differs somewhat in shape from the common nuts. Link,* who is most fully acquainted with the South European flora, has given with great precision the distribution of the chesnut in the south of Europe. According to his researches, it grows in the northern parts of Greece; in the middle part, it ascends the hills, and in the southern, it is found only at very considerable heights.† In Italy it is the same; the chesnut forms the forests on the mountains of Piedmont, and is one of the chief means of support to the inhabitants of the valley of the Waldenses and the adjacent districts; it rises gradually higher, and at last forms a well-known wood on Etna. In the warmer parts of Switzerland, and the south of the Tyrol, it is likewise a common forest tree, which greets the traveller from the north on the southern descent of the Simplon, about 1400 or 1500 feet below the greatest height of that famous pass. Chesnuts are chiefly used as food amongst the Cevennes and in Limousin. Link found the high mountains of Spain and Portugal often entirely covered with chesnut trees, or, as on the Sierra de Marao, they formed a belt below the colder summit. But this area of the chesnut has been very much enlarged by art, towards the north, as well as towards the south. In Germany, woods of the true chesnut are not unfrequent; they extend high up the

* Die Urwelt, u. s. w. 2te Auflage, 1834, i. p. 355.

† The south of Greece, however, has a climate which is that of the subtropical zone.

Rhine; and they grow very well, though not in such numbers, on the Harz mountains, and even round Berlin and Potsdam, so that more extensive plantations of this beautiful and useful tree ought to be attempted, particularly in places sheltered towards the north; they would probably be more profitable than the bad wine of these districts.

In conclusion, I draw attention to Link's learned antiquarian researches concerning this subject, contained in his work on the Ancient World, vol. i. p. 356, &c., in which it is shown that the oldest Greek naturalists held this valuable fruit in proper estimation, and have described it under the name of the Oak of Jupiter (*διὸς βάλανος*).

Several other trees and shrubs in our European forests produce similar nutritious and palatable fruits, which, however, are of less importance than the chesnut in the economy of man. It is known, that in the south of Europe oaks with edible fruit occur; this species is Quercus ægilops *L.*, a lofty and beautiful evergreen tree belonging to Greece. The fruit, however, is not particularly agreeable, and therefore the Greeks, as Link observes, leave it to the swine, when they have any better food. In Albania, also, this oak forms vast forests. A second oak with edible fruit is Quercus ballote *Desf.*, which was first discovered in northern Africa. Link, however, has found that there are forests of this tree in the south of Portugal and the adjacent parts of Spain, and that the fruit is frequently eaten in these countries, and is even sold along with chesnuts at the gates of Madrid.

Besides these edible fruits of our woods, we have to mention the hazel-nut (Corylus avellana) and the pine seeds (Pinus pinea *L.* and P. cembra *L.*). The edible pine is a native of Southern Europe, and we find its fruit even exposed for sale in the market of Upper Italy. The Siberian stone-pine is indigenous in Switzerland, the Tyrol, and Siberia; and in these countries, where there is not a great abundance of other articles of food, its fruit is commonly eaten. In the high lands of the East Indies, also, are found many pines with edible fruit. The hazel-nut is, on the contrary, a much more nutritive and palatable fruit, which is proportionably richer in oil than the walnut and the kernel of the olive. In the north of Europe, even far above the arctic circle, the hazel-nut plays an important part in the domestic economy of the peasantry, and large plantations of this bush would probably be profitable, were it not that the kernel easily becomes rancid, from the great quantity of oil it contains.

In South America, the Araucaria, on the Cordilleras of Southern Chili, is of great importance to the rude Indian. This most magnificent tree of all the Coniferæ, which I formerly described (p. 132), ripens a large quantity of well-flavoured seeds, which are twice the size of almonds, and are a favourite food of the natives of the South Chilian Cordilleras. The most northern of the Araucaria forests occur in the latitude of Conception : therefore, in parts where the settlements of the whites are unimportant. But the Araucani are they who enjoy the fruit of this noble tree, and, as amongst other rude tribes, the time of gathering it is with them a time of universal rejoicings. The abundance is so great that many of those Indians, without carrying on any branch of agriculture, use these nuts as their only vegetable food.

The Catappa (Terminalia catappa) is a tropical tree, resembling a lime, and bearing well-tasted edible nuts, which occurs on many islands of the Indian Archipelago, particularly the Moluccas. The fruit bears some resemblance to a walnut, but is smooth; it is at first red, but becomes black when ripe, and contains one or two almond-like kernels. The Catappa-fruit would probably play a more important part in the house-keeping of the Indians, if there was not a superabundance of many other far more productive plants. The Juvias, in the forests of equatorial South America, seem to be of more importance.

### 13. *The Brazilian Nut or Juvia (the fruit of Bertholletia excelsa Humb. et Bonpland).*

The Juvias which are brought to us from Brazil, under the name of Brazil nuts, &c., are the fruit of a magnificent forest-tree, which grows in the interior of South America, and belongs to the family of Myrtaceæ. Humboldt, before whose journey to America this tree, with its pleasant and exceedingly useful fruit, was almost unknown to us, travelled over the country in which it is found, and has given us an ample and extremely interesting account of it.*

The Juvia tree has a trunk two or three feet in diameter, and 100 or 120 feet in height. The long boughs of the Bertholletia, says Baron v. Humboldt, spread far apart; they are almost bare underneath, but towards the extremities are set with thick bunches of leaves. This distribution of the leaves, which are half leathery, somewhat silvery on the under side, and about two feet long, bends the branches towards the ground, as is the case with the branches of the palm. This splendid tree flowers in the 15th year, generally at the end of March or the beginning of April, and the fruit is ripe in the latter part of May. The large fruit is 12 or 13 inches in diameter and globular; it has a very hard woody shell, half an inch thick, and contains from 15 to 22 of such nuts as come to us in trade. As the nuts, when ripe, lie loosely within the large shell, the fruit makes an extraordinary noise when it falls from the tree.

The flavour of the large almond-like seeds is very pleasant so long as they are fresh, but on account of the quantity of oil contained in them, they soon become

* See his Reise in die Æquinoctial-Gegenden, iv. p. 466, &c. Buch viii. cap. xxiv.

rancid ; however, they often taste very well with us, even after so long a voyage.

The Bertholletia excelsa, which produces this exceedingly nutritious fruit, seems to possess a pretty extensive station in the interior of South America, though it does not extend far on either side of the equator ; it has been observed in the forests at the mouth of the Amazon river, as well as in the Cerros de Guayanna in 3° of latitude. Humboldt and Bonpland first found this noble tree at the mouth of the Cassiquiare, and large woods of it are to be found in the boundless forest ocean, which waves round the shores of the Rio Negro, as well as the whole system of streams which connects the river Amazon with the Orinoco. According to the accounts given to Humboldt the juvia and cacao trees are extremely common above Gehette and Chiguire.

At the commencement of the time when the juvia nuts are ripe, the Indians of the Upper Orinoco proceed to the forests, where these trees occur in great numbers, and there collect and carry home immense quantities of the precious fruit. The return from this harvest is then celebrated by a feast at which the rudest excesses are committed, and which Humboldt has very vividly described.*

The planting of this tree, which yields so pleasant and nutritive a food in other tropical countries, cannot be too strongly recommended to those governments which possess extensive tropical colonies. The kernel of the cocoa nut is a very harsh fruit in comparison with the beautiful almond of the Bertholletia excelsa.

* L. c. page 463.

## IV. THE PRINCIPAL AGRICULTURAL PLANTS WHICH ARE MORE OR LESS USED AS LUXURIES.

### 1. *The Areca Palm.*

The inhabitants of the East Indies and the neighbouring South Sea Islands, by chewing the areca nut, which is so universally known under the name betel-nut, procure themselves an enjoyment quite similar to that which the Peruvians obtain by chewing the coca. The palm, which produces the betel-nut, is the Areca catechu *Linn.;* it is one of the most beautiful forms we have seen in India, and is a plant of the hottest countries within the tropics, though at the same time it prefers a damp climate. On all the coasts of the East Indies and the adjacent South Sea Islands, the Philippines, the Carolinas, the Marian, and Society Islands, where chewing the betel is a more or less common practice, the areca palm is planted near the dwellings, where it, in company with the banana, forms a beautiful object. In the town of Manilla, we find the betel palms planted in regular rows before the houses; and plantains, Anona squamosa, Averrhoa Belimbi, and other tropical trees, grow between them. Pigafetta found the practice of chewing the betel in the Philippines as common as it is at present.

In consequence of the enormous consumption of betel in the Indian peninsulas and in China, the betel-nut forms a very extensive branch of traffic ; the export of it to China from the Eastern Peninsula and Sumatra is very large. The trade in betel and pepper, between China and Sumatra, is carried on by large armed ships belonging to our northern nations, among whom North Americans are not wanting. The betel is procured by exchange for powder and arms, and although this trade is attended by great danger, for the Malays attempt to surprise every ship, it is in recompense extremely lucrative in the market of Canton. Single ships sometimes carry as much as 10,000 pikels (133⅓ lbs. English) from Sumatra, and this island furnishes yearly at least between 40,000 and 60,000 pikels, which are shipped from May until August. Sumatra, together with the adjacent countries, indeed, exports 80,000 or 90,000 pikels annually, which for the most part go to China. The ships are laden with fresh areca nuts without further packing, and in consequence of being heaped together they produce a high degree of heat during the voyage.

The areca palm, although probably wild only on the Sunda Isles and the adjacent Philippines, is most extensively cultivated not only in Sumatra but in India. In Ceylon, and especially throughout Malabar, and still higher up the coast, there are vast plantations of this beautiful palm, and their produce is of great importance, for as every one chews betel, the consumption of areca nuts in India is incredibly great.

In preparing the betel-rolls the areca nut is cut into long narrow pieces, and rolled up in leaves of the betel pepper, which are striped on one side with moist chalk. In Luçon, we find in every corner of the house a little box or dish, in which are kept the betel-rolls (Buyos) prepared for the day's consumption, and a buyo is offered to every one who enters, just as a pinch of snuff or a pipe is with us. Travellers and those who are obliged to work in the open air carry the buyos for the day in little boxes or bags, just as the Peruvians do the coca rolls. The preparation of the betel falls on the female members of the family, and during the forenoon they may be almost always seen lying on the ground and making buyos. The little chest which is used for this purpose contains some areca nuts, some leaves of the betel pepper, a strong knife for breaking the nut, and a little dish of moistened chalk, which is laid on the leaves by means of a wooden spatula. The use of the betel is very largely indulged in; in the countries formerly named, for example the Philippines, every one must chew betel. Every one who can afford it puts a fresh buyo in his mouth every hour, which he can chew and suck for a half an hour at least.

We who are not accustomed to it cannot judge how great the enjoyment of the betel is; those nations speak of it with enthusiasm, and in the Philippines the labourer is paid in betel rolls, just as in Upper Peru with coca. The use of the betel may, perhaps, not have an injurious influence on the health, and we see here an example of the power of custom. The chewing of it is one of the most disgusting practices which can be found amongst a people; and its use is scarcely persisted in for a few years before the teeth become red, the gums even are coloured a dark brown, a perpetual salivation is caused, and the spittle at last is of a reddish brown colour. The Tagali maidens regard it as a proof of the uprightness of the intentions, and strength of the affection, of their lovers, if they take the buyo out of their mouths.

As the betel nut is always chewed with the leaves of the betel pepper,* the culture of this plant is likewise of great importance to the countryman in the neighbourhood of large towns, for every day incredible quantities of fresh betel pepper leaves are sold in the markets. We see piles three and four feet high of these large, beautiful, heart-shaped leaves, carried about in baskets; yet every person who possesses a little piece of land, usually grows the leaves he requires. The plantations of betel pepper are laid out like our bean fields, but the plants stand further apart, and their beautiful leaf gives the whole field a bright green colour, such as belongs to few other plants.

The betel pepper requires good soil and also a low situation and much water; the plantations are surrounded by a ditch and a wall, on which is made a hedge of various plants, for example, in India of Euphorbia Tiru-

* Piper Betle *L.*

calli, Arundo tibialis *Roxb.*, and several others. If the plantation does not possess sufficient water, it must be watered for six months ; it is divided into regular beds, surrounded by enclosed canals, but between the wall and the tilled ground a vacant space twenty feet broad is left. In the middle of each bed are made holes for the shoots, which are always planted a foot and a half from each other. In each hole are set two shoots three feet long, in such a manner that the centre of each is fixed in the ground, and consequently the ends stand out on each side, and afterwards bud out. For the first eighteen months these shoots are allowed to climb up poles ; during this period they require much water, often twice a day. Between the shoots are planted young trees of Æschinomene grandiflora, Guilandina Moringa, or Erythrina indica, &c., which grow very rapidly, and at a later period serve as supports to the betel plants ; for after eighteen months, these plants are taken off the poles, the lower part of the stems is laid in the ground for about three feet, and the stem directed so that it can climb up the trees. In the second year the stalk is again laid in the ground for three feet of its length, and this is done every year. The leaves may be plucked in the fourth year, and during the next six or seven years, after which the plants die and must be replaced by new ones.

Another custom, similar to the use of the betel, is the chewing of the Terra japonica, or the Succus catechu, also called Caschu, which is likewise very general in the East Indies. The greater part of this catechu is made from the nuts of the Areca catechu by repeated boilings and repeated evaporations to perfect dryness. But in many parts of India, particularly in the north, at the foot of the Himalayas, the catechu is prepared from the wood of the Mimosa catechu, which grows wild there. This tree is also found in Ava.

In order to make the catechu the trees are felled, the outer white wood is removed, and the inner cut into small pieces, and three times boiled. The extracts are then poured together, inspissated, and dried in the sun, in which it must lie for seven days. This catechu, called caschu in India, greatly resembles the Gambir extract.

## 2. *The Gambir Extract.*

An article of luxury, similar to the betel, is now, in some parts of India, received from year to year with greater favour ; I mean the Gambir extract.

The Nauclea Gambir and N. aculeata *Linn.*, shrubs from five to seven feet in height, are the plants whose leaves yield by boiling this favourite substance which is chewed by the Malays in Sumatra, but especially by the Dutch colonists in India. A salutary effect is attributed to this substance, so that it is enjoyed under the pretence of being an aid to digestion, but it is with this as with our customs of the same nature. The true tobacco smoker knows to ascribe to the use of tobacco the most beneficial influence on his health.

At present the gambir plant is principally cultivated in the Dutch colonies in India, and in Java, especially in the charming island Bintang,* in Sumatra,† Malacca, at Singapore, and probably on many other islands of these seas. The Dutch government has prohibited the import of the gambir extract, and thus encouraged this new branch of industry in its own possessions. On the Dutch island Bintang, where there is the station Rhio, there were, in the year 1832, 6000 gambir plantations, the larger of which contained from 80,000 to 100,000 trees, and the smaller at least 3000 or 4000 ; we may infer from this what influence this branch of culture has on the physiognomy of the vegetation of that island. At the different places where the gambir plant is cultivated, there seems to be a slight difference in the

* See Bennett's Wanderings. London, 1834. II. p. 183, &c.

† See Anderson's Miss. to the East Coast of Sumatra. Lond. 1826.

method of gathering the leaves. The gambir plant is covered with leaves during ten months; and in Malacca these are plucked off four times a-year, but on Bintang, where the best gambir extract is made, they are gathered only twice a year. If the leaves have not been fully developed, the extract is bad, and what is worst of all, the plantations are soon ruined. When the gambir bush is three years old, they begin to make use of the leaves, and if they are plucked only once in six months, the bushes last for 25 or 30 years.

After the leaves have been either plucked directly from the bush, or stripped from the cut shoots, they are rapidly boiled for five or six hours in large iron kettles, separated from the liquor, and either boiled again or thrown away. The liquor is then inspissated to a thick extract, which is poured into long moulds. After the mass has become somewhat hardened in them, it is cut into pieces and dried in the sun, and thus this extract comes in trade in the shape of hard dry pieces, of a blackish brown colour; it has even been sent to England, in order to try it as a mordent in dyeing.

The taste of this kind of catechu is at first sweetish, with a pleasant aromatic flavour; afterwards it becomes somewhat astringent and bitter.

At Rhio, in Bintang, this gambir extract is said to be rendered almost colourless by being repeatedly dissolved and purified; this white extract, however, does not come into trade. The quantity produced on the island of Benang amounted, in the year 1829, to 31,000 pikels, 133⅓ lbs. each; in 1830, to 35,000; 1831, to 47,000; 1832, to 63,000; 1833, to 70,000; and this article was then worth eight rupees the pikel, at which price the government bought it, to re-sell it at a much higher rate. Accordingly the exportation of gambir extract from Benang amounted to 360,000 rupees in the year 1833, and this new branch of culture is extending every year. Nay, at Singapore, where 150 gambir plantations were planted in 1833, 20,000 pikels are made every year by the Chinese.

This article, like every other, is more or less esteemed according to its excellence; it is said that on the whole, the gambir extract of Benang and the coast of Bengal is the best. It is worse the more granulous it is, which is often attributed to it having been badly evaporated.

### 3. *The Culture of Opium.*

The use of opium is as general in the east as the indulgence in spirituous liquors amongst us; the mode of using it, however, varies amongst different nations. The Turks, as is well known, eat the opium; the Chinese and Malays smoke it, and swallow the smoke. The use of opium, like all other luxuries, is adopted by, and spreads over whole nations, with wonderful rapidity. The inhabitants of the eastern parts of Asia were, not a very long time ago, seized with this passion for the use of the opium, and now this new luxury is spreading with the most amazing speed, to which no law, no punishment, in fact nothing can be opposed. We shall in the sequel see to what an incredible height the use of opium has risen in China, a country in which it is strictly prohibited.

We do not here speak of the culture of the small quantity of opium which, under the name of Turkish opium, is brought to us from the east for medicinal purposes, but of the culture of the immense quantity in India for which so many millions are spent. The Turkish opium is stronger than the Indian, and is on that account preferred to the latter in medicine. But the inhabitants of Eastern Asia who smoke the opium, prefer the Indian to the Turkish, so that the latter bears only half the value. I have not had an opportunity of comparing the effect of the Indian and of the Turkish opium in a medicinal respect, yet it seems to me as if the Indian opium had a less rapid effect on the system of the blood, and thus deserved to be preferred to the Turkish. Although the Indian opium is very dear, the inhabitants of the East Indies show no particular liking to this branch of agriculture, as it is exceedingly laborious. The merchants, therefore, travel about and advance large sums to the country people, in order to induce them to pursue it; in consideration of this, the farmer then sells a pound of the opium for 15s. to the advancer of the capital.*

The opium plant† is cultivated not only for the sake of the opium, but also for the seeds, which yield the very useful poppy oil, and probably the plant was first grown for the sake of the seeds. The opium plant requires the best soil, and also so much constant attention that the clear profit of its culture is not so great as that of the sugar-cane or tobacco.‡

The ground on which the opium plant is to be grown is first prepared as if for the culture of rice, and is then divided into large squares, quite similar to the rice grounds, which can be irrigated by little canals. The ground is prepared in September and October, and the seed is sown in November, after which the ground must be watered every fourth day. In six or seven days the young plants are two inches high, and they are then thinned out, so that the remaining plants stand four inches apart. After twenty days, when the plants are six inches high, the weeds must be pulled up, and the ground must receive some manure. In two and a half months the plant is ready for the preparation of the opium, and in three months the seeds are also ripe. During the time of making the opium, which lasts two or three weeks, a great number of persons are employed in the opium fields in making several incisions in the outer side of the capsules either by thorns or fine pointed

* See Buchanan. A Journey from Madras through Mysore, Canana, and Malabar. Lond. 1807. T. i. p. 295, &c.

† Papaver somniferum *L.*

‡ Tennant's Indian Recreations, consisting chiefly of strictures on the domestic and rural economy of the Mahommedans and Hindoos. Edinburgh, 1803.

needles,* so that the white juice, which is present in such abundance in the vessels close beneath the outer skin of the capsule, may flow out. The next morning they return, and with a shell scrape off, from the wound, the hardened juice, which has assumed a yellowish-brown hue, and in this manner obtain the opium. These incisions of the capsule are generally thrice repeated, and each time the wished-for sap is obtained, which then hardens in the air.

The sap is then subjected to some operations under the inspection of the merchant who has bought it, in order to prevent adulteration. It is first dried in the sun to evaporate the water from it, but this is compensated by poppy oil in order to prevent the resin being completely dried up.† The opium is then formed into flat cakes, about four inches in diameter, which are wrapped in poppy leaves; finally, when it is sufficiently dry, it is packed in chests with the chaff of the poppy seed. In this state it is brought to the Chinese market; each chest contains 133⅓ lbs. English, or 100 catti, which vary in price according to the abundance of the supply or the prospects of the speculation. The different sorts also vary in price according to the countries in which they have been cultivated; for example, in December 1831, the prices were the following :‡—

| | | |
|---|---|---|
| Patua opium, | 935–945 | Span. piastres the chest. |
| Benares, | 940 | ... ... ... |
| Malwa, | 655–660 | ... ... |
| Damann, | 655 | ... ... ... |
| Turkish, | 555–560 | ... ... |

The finest opium, therefore, was worth more than 1400 Prussian dollars for 127.6 Berlin lbs., or 133⅓ lbs. English. But the farmer receives according to contract only 660 or 670 Prussian dollars.

* In Persia the incision is made with an instrument having five edges. (See Kæmpheri Amoenit. exot. Fasc. iii. Lemgoviæ, 1712, p. 643.) The first wound gives the lachryma opii, which is yellower in colour.

† See Tennant, l. c. p. 300. ‡ Meyen's Reise. II. p. 299.

The clear profit of the opium culture is, therefore, of less importance to the planter, and is said to amount to no more than from twenty to thirty rupees for the acre, which yields only 30, or at most 40, lbs. of opium. The produce is, however, extremely uncertain, depending on the weather, the number of insects, &c. ; one year it may be very small, and the next very great.

The poppy oil obtained from an acre of land is said to be worth only two or three rupees.

In conclusion, we shall add a few words on the extent of this remarkable trade in opium. In the market of Canton alone, since the year 1828, more than eighteen or nineteen millions of dollars' worth of this article have been imported to China ; it is at the same time known that along the whole Chinese coast to Corea, a very considerable smuggling trade in this article exists ; so that the whole sum of money annually drained from the Chinese empire for opium must be several millions of dollars more, leaving out of consideration that opium which is sent direct from India to the northern parts of China by Chinese and Siamese ships. Through Canton alone, from 1818 to 1831, above fourteen millions of lbs. of opium have been brought to China, which have cost the sum of 115,672,339 piastres.

The quantity of opium which is consumed by the Malays of the Indian Archipelago, in Cochin China, and Siam, as well as in India and Persia, is so immense that if we could obtain an exact statement of it, we should find the consumption of this article, so injurious to the health, quite incredible. Burnes has observed,* that in some countries it is even given to the horses to incite them to greater exertions. A Cutchee horseman, says Burnes, shares very honourably his store of opium with his horse, which then makes an incredible stretch, although wearied out before.

* Narrative of a Visit to the Court of Sinda, p. 230.

### 4. *The Tobacco (Nicotianæ spec. var.).*

The aborigines of Hayti smoked the American tobacco when the Spaniards discovered the island; and towards the end of the 16th, or in the beginning of the 17th century, this custom passed over to the nations of Europe. It has long been the opinion that the use of tobacco, as well as its culture, was peculiar to the people of America, but this is now proved to be incorrect by our present more exact acquaintance with China and India. The consumption of tobacco in the Chinese empire is of immense extent, and the practice seems to be of great antiquity, for on very old sculptures I have observed the very same tobacco pipes which are still used. Besides, we now know the plant which furnishes the Chinese tobacco; it is even said to grow wild in the East Indies. It is certain that this tobacco plant of Eastern Asia is quite different from the American species.

The genus Nicotiana, generally speaking, belongs to the warmer zones, yet a few species of it have a very extensive area and a great power of resisting the influence of climate, for they can be grown under the equator and in the temperate zone, even far above 55° north latitude, where the mean summer heat is equal to 15.87° Cels.

It is well known what immense quantities of this noble herb are cultivated even in our native country; as far as regards quantity, it produces as much here as within the tropics, but the quality is very different. The southern polar limit for the culture of tobacco is not exactly known, but it seems to extend to the 40th degree of latitude, for in South America tobacco is cultivated at Conception, and in New Zealand enough is grown for the consumption there.

Havannah is celebrated on account of its tobacco, and this island alone, at the time of the former Spanish system of trade, produced 350,000 arrobas of 25 lbs. each, consequently the immense quantity of 895 millions of lbs., 128,000 arrobas of which were sent to Spain.* The tobacco of Caraccas, and of the present Venezuela in general, has likewise attained great celebrity, and is at present largely exported. The culture of tobacco is also vigorously prosecuted in the Philippines, and the produce is little inferior to that of Havannah. The cigars of Manilla are in great repute throughout India, and are also highly esteemed with us, though they very seldom come into the market. They are easily known from being cut at both ends, and in trade they are always in bundles of 32 cigars. In the Philippines, as well as in the island of Cuba, the culture of tobacco is monopolised, and officials are seen wandering over the whole country, often in great numbers, in order to prevent the illicit culture of this herb. The cigar manufactory at Manilla at present employs 1500 men, and above 3000 women, natives of Luçon. These women sit in long rows under sheds, roll the chopped tobacco in selected moistened leaves cut into triangles, and finally fasten them on both sides with gum. Manilla, it is true, in the year 1829 exported only 4591 arrobas of cigars,† but the consumption in the country itself must be enormous, for there every one smokes.

It is unnecessary to remark with what eagerness every people has adopted the practice of smoking tobacco; even the rudest hordes may be conciliated by means of it, and, indeed, there are few nations which, in the absence of tobacco, have not some other similar sort of relaxation.

### 5. *The Coca of the Peruvians.*

The leaves of the coca plant‡ are to the Peruvians what the opium is to the Turks, the betel to the inhabi-

* See V. Humboldt. Ueber Neu-Spanien, iii. p. 177.

† Meyen's Reise, ii. p. 376.

‡ Erythroxylum Coca *Linn.*

tants of the East Indies, and tobacco to other nations. The native habitat of the coca-plant is probably on the eastern side of the Peruvian Cordilleras; as yet, however, it has not been found in a wild state.

One or two days' journey from La Paz, in the same district in which the Cinchona forests first occur, the coca-plant is also cultivated; and La Paz, the capital of Bolivia, carries on the chief trade in it.

Poeppig,* who, during his stay at Huallaga, on the eastern side of the Peruvian Cordilleras, was for several months in parts where the coca is grown, has given us full information concerning this branch of Peruvian agriculture. Just as it is extremely difficult for a drunkard or a real tobacco smoker to give up his favourite indulgence, so it is difficult for a coquero to refrain from the use of the coca. We also learn from these Travels, that the use of coca is as universal on the eastern side of the Cordilleras of Northern Peru, as on the table land in the south, and the consequences of it are said to be very bad in the warmer and damp districts. In the cold and elevated parts of the plateau of Chuquito, where the use of coca is certainly very general, and that not only amongst the Indians, but amongst the mixed races, as well as the whites, little is to be observed of all the dreadful diseases, which are said to arise from its use. In the villages and towns, around the basin of Chuquito, Indians, negroes, whites, and persons of mixed blood, of extreme age are seen going about, and now as before using the coca. The women of these parts, who, as of mixed blood, are known under the name of Zambitas, are exceedingly corpulent, and chew coca as generally as betel is used in India, without exhibiting any signs of the dreadful consequences. By the mixture of burnt lime, which is much more general in Northern Peru, than in the South, the teeth are dyed in a disgusting manner, but they do not suffer any injury, which can

* Reise in Chile, Peru, &c. ii. p. 210, &c.

easily be seen to be the case on the islands of India, where the betel, mixed with lime, is one of the ordinary enjoyments of life. It is certainly the case, that the excessive use of coca, on account of its volatile principle, which produces an effect similar to that of opium, weakens the organs of digestion and gradually over-excites the nervous system, and a number of diseases thus arise, which, however, are far from being so dangerous as those which are the consequence of using opium. A death by emaciation, in consequence of the use of the coca, which has lately been affirmed to have occurred, seems to me to be something very inexplicable.

Von Martius has given us some account of the manner of cultivating this plant.* He found plantations of the Erythroxylum coca on the River Amazon at Ego, and he supposes that the plant has been introduced there, because there they possess tobacco also, and use it much more frequently than coca. Von Martius saw there bushes three feet high which were planted in rows, three feet from each other. The leaves were dried in ovens, then reduced to powder in mortars, mixed with the ashes of the leaves of Cecropia palmata, and kept in layers of grass until required for use. In Peru, however, the coca is used in a different way ; the leaves are there chewed, like the tobacco leaves with us. The Peruvians carry the coca about with them in little bags, made of woollen stuff or the skins of young animals. The leaves are shaped like young cherry leaves, and have a pleasant bitter, and a stringent taste, and a fine ethereal odour. The Peruvian chews these leaves as often as he can, nay, almost the whole day ; their effect is, on the whole, stimulating, but afterwards, it appears to me, somewhat stupefying, resembling the stupefaction consequent on the use of opium. This excitement renders the Indian labourer, who is of an extremely melancholy disposition, cheerful, and keeps him

* Reise nach Brasilien, p. 1169.

from becoming weary; on arduous journeys the coca assuages hunger for several days, and warms against cold; in short, in that country are attributed to coca all the effects, for which tobacco is praised with us.

The Indians chew the coca leaves either by themselves, or with clay or lime, which they first mix with the leaves, and work the mass into little balls to be used gradually. They keep each ball in the mouth as long as they perceive a bitter and strong taste; when this taste ceases, they throw the ball away and take another.

It is remarkable, that the use of the coca, though so highly esteemed within its area, has so very limited a range.

The coca plant, which is said to resemble the black thorn, is cultivated on the River Huallaga, where Poeppig remained for a long time, on warm and damp slopes, from about the height of 2000 to 5000 feet, where there is still no night frost, which is injurious to the plants. The coca which is grown in hot districts is said to be inferior in strength. The berries of the coca plant are sown on a piece of ground, made arable by burning at the end of the rainy season. Holes about 9 inches square and 18 inches deep, are dug in this according to a certain rule, and in each hole is thrown a handful of seed which is left lying uncovered. About 100 plants spring up in each of these pits, and they are allowed to stand in them for 15 or 18 months, when they are transplanted, and the young bushes set in regular parallel rows. Irrigation, weeding, and stirring up the soil are the labours necessary to the success of these young plantations, in the spaces of which maize is sown in the first year. The first crop is obtained after from three to five years, according to the soil, and every 13 or 14 months afterwards; in large plantations, however, the leaves are said to be gathered during the whole year. The leaves are reckoned ripe and fit for drying when they have become rigid; size and colour are not regarded. They are stripped off and dried by the heat of the sun, because it is said, probably in consequence of prejudices, that artificial heat diminishes their strength. Beside the dwelling-house on each cocale (that is, the hacienda, where the coca is cultivated), there is a sort of floor on which the leaves are exposed to the sun's rays. On the eastern side of Northern Peru the dried leaves are packed in large woollen sacks, and each bale (tercio) when fresh weighs 80 lbs., though it loses considerably in weight by lying long. In Upper Peru, where the llamas are bred in such numbers, the coca, as well as the Peruvian bark, and almost all the other products of the country, is packed in llama hides, and these bales of coca (cestos) which are sent over the whole of Southern Peru, weigh from 20 to 30 Spanish lbs. each.*

According to a work on this subject, which appeared in La Paz and was consulted by Poeppig, about 40,000 of these cestos, containing on an average an arroba, are annually brought to market on the eastern slope of Upper Peru, the present Bolivia, for in Upper Peru proper no coca is grown; and as one is worth between six and seven piastres at La Paz, the value of this branch of culture for Bolivia is between 2,400,000 and 2,800,000 piastres. We learn also from this work, that about 40,000 arrobas of coca are raised in the provinces Arequipa, Moquegua, and Arica, in the lower part of the western side of the Cordilleras of Southern Peru; however, the districts in which the coca is cultivated there are not mentioned, and I am quite ignorant of any such. In my journey through the extremely dry provinces, Arica and Arequipa, I nowhere heard of the cultivation of the coca, which, it is known, requires a damper climate; and in the markets of Arica, Arequipa, Tacna, and Islay, I saw only the coca of La Paz, in bales of llama skins, before which sat Zambitas with weights

* See Meyen's Reise, ii. p. 16.

and scales, selling these dried leaves. The coca consumed in Peru seems far to exceed in value the sum of 4,500,000 Prussian dollars, of which, according to Poeppig's statement, Huanuco produces the amount of 90,000 piastres, Jauga 40,000, and Truxilla 20,000 piastres. A large quantity of coca, of which, however, we have no estimate, is likewise produced in the fertile province of Cuzco.

### 6. *The Vine (Vitis vinifera L.).*

The distribution of the vine over the globe is of particular importance to mankind; the use of wine or beer, as the ordinary drink, produces so very different an effect on the people, that the influence of the culture of the vine on the nation is not to be overlooked. Before we give an account of its distribution and culture, it will be necessary to give some information respecting the native country of the vine (Vitis vinifera *L.*). Unfortunately, its native station, like that of several of the cereals and most other agricultural plants, is far from being known in its whole extent. We already know several places where the vine grows wild, for example, the Neapolitan territory, where is found a small sweet grape, which yields very good wine; and in Portugal, where there grows a small and very sour grape, which is not at all esteemed.* The vine of Northern Africa yields, without much culture, the finest grapes; it may, therefore, have been introduced thence to the southern countries of Europe. In France and Germany also, the vine occurs wild in the forests, for example, in the vast woods on the Rhine between Strasburg and Speier, as well as on the Danube, but so far as I know, the grapes of these plants are useless. Respecting all these numerous habitats of the vine in its wild state, it

* Link's Urwelt und das Alterthum, &c. 2te Auflage. Berlin, 1834, i. p. 432.

may be with reason asked, whether it has not become wild in these places, and I think that this may be affirmed with certainty for France and Germany; perhaps less confidently for the more southern vine, the grapes of which, for example in Naples, are very good and yield a well-flavoured wine. With greater certainty the native country of the vine is placed in the East, in the old Cyrenaica, and in general in the countries lying between the Black and Caspian Seas.* In the forests of Mingreli and Imereti the vine is the queen of the trees;† it there attains the diameter of from three to six inches, and mounts to the tops of the highest trees, completely entwining them and binding them together. There is no true culture of the vine in those countries, and yet there is such an abundance of excellent grapes, that even the poor peasant does not gather all that grow in his little patch of land, but leaves them over the winter, and often a short time before Easter the grapes of the previous year are beaten off the trees. It may well be thought, that a plant is indigenous, where it produces beautiful and delicious fruit without the aid of the fostering hand of man. But besides this, it is known that the growing wild of an agricultural plant is of extremely rare occurrence, and when it does happen, the improved fruit at least disappears. Very probably the native country of the vine extends far beyond the Caspian Sea, towards India and even to the north of China, for in several countries there, for example in Cashmere, and the Deccan, the vine is cultivated, though the use of wine is unknown there as well as in China. It is still a question, whether the vine, which is at present cultivated over almost the whole globe, has descended from one species only; a distinguished botanist who has travelled much in the wine countries, viz., Link, seems to be of opinion that our vine is the offspring of several

* See Bieberstein, Flora Tauro-Caucasia, i. p. 174.

† See Parrot's Reise nach dem Ararat, i. p. 247.

wild species, to which conclusion he has been led principally by the shape and pubescence of the leaves.

The number of the varieties of the vine is extraordinarily great, about 200 might be enumerated ; but the most remarkable fact here is, that the same variety of vine produces perfectly different kinds of wine, in two places often lying close to each other. The difference between Johannisberg and Rüdesheim wine is sufficiently well known ; nay, the wines of the same mountain differ, according as the vine has grown on the upper part or at the base of the mountain. How extraordinarily different is the Leistenwein at Wurzburg from the Wurzburg and Steinwein which are produced quite near it. The true Leistenwein contains as much alcohol as madeira, although our northern wines are constantly reproached for sourness and weakness. These differences are, it is true, inexplicable, and nothing is more certain than that they are caused solely by the locality, although we do not know how. A vine, which grows on swine-stone, receives the peculiar odour of the stone, and it may be important to attend to this.

The fruit of the vine is not only consumed in making wine and brandy, but it frequently serves as a pleasant, and, on account of the sugar it contains, very nutritious food. Amongst the Mahommedans, and especially amongst the Turks, a pleasant kind of confection is made of grapes. Fresh grapes are eaten in quantities wherever the vine is cultivated ; but in many countries the grapes are dried and made into raisins, in order to preserve the longer so delightful an article of food. We find this the custom particularly in Northern Chili, in the province Coquimbo, where immense quantities of raisins are made.* In many countries, for example, in Persia,† Crete, Mingrelia, &c., the attempt is made to keep the grapes fresh on the vines throughout the

* Meyen's Reise um die Erde. I. p. 420, &c.

† See Chardin's Voyage en Perse. Tom. i. p. 53.

greater part of the year, which is perhaps possible in these countries on account of the very dry atmosphere of the winter. At Catania, at the foot of Etna, where the beautiful vine grows which yields the lachrymæ Christi, flowers and fruit are found on the same vine at the same time, a circumstance which was known to Pliny.

The distribution of the culture of the vine is regulated much less by the mean temperature of a place, than by the greater summer heat; but it is chiefly the length of the summer which influences the ripening of the fruit. At Moscow the grape ripens only in forcing-houses, although the summer heat there is as great as at Paris and on the banks of the Loire generally. But only June and July at Moscow show this high temperature; in August it sinks to 14° Cels., and in September, when the grape should ripen, the mean temperature there is only 9.9° Cels. and severe night-frosts destroy the crop.

The vine thrives exceedingly well with a mean temperature of 15° and 16° Cels.; for example, in the south of Italy and Sicily. The wine is not so sweet and contains less alcohol with a lower temperature, for example, of 9° or 8.7° Cels. mean annual heat, and along with the summer temperature must be 19° or 20° Cels. otherwise the grape does not become fully ripe, of which London gives an instance. The mean temperature at London is = 9.12° Cels., almost equal to that at Geneva, where the temperature of June and July is equal to 17°; but in Geneva the September and October are still so warm, that the vine can ripen, which is not the case in London. As to the maxima of heat, under which the vine will succeed, I believe that it will do so under every tropical heat, provided it be not united with too high a degree of humidity. Even with us the vine must not be in too damp a situation, but it delights in dry places, such as the slopes of mountains. On the western coast of South America, where, at least above Guyaquil, the climate is very dry, the vine is cultivated often close by the coast, as high as 6° south latitude.* The wine of Pisco, in 14° south latitude, is excellent; one sort made there, at a certain age resembles Cyprus; but the grapes of Pisco are generally used in the distillation of the celebrated brandy of that name, the quantity of which annually produced is worth about half a million of piastres. This brandy, which is so highly esteemed in Peru and Chili, is carried about in large earthen vessels, which are two, three, and four feet high, almost cylindrical, and pointed at the bottom. These pitchers are transported on the backs of mules, one being hung on each side of the animal in a creel woven of twigs. Droves of hundreds of mules, laden in this manner, are seen going from one place to another.

At Moquegua and Tacna also, between 16° and 18° south latitude, the culture of the vine is not inconsiderable; but in Northern Chili, in the provinces Copiapó, Huasco and Coquimbo,† an immense quantity of vines are grown, and the raisins serve as part of the subsistence of the miners. The vine-yards of Quillota supply the market of Valparaiso with grapes, and the quantity of wine made in that district must, according to Poeppig's account, be far from inconsiderable. At Conception the culture of the vine is of importance; the best wine in Chili is made there, and much is sent about the country. The culture of the vine extends southwards to Valdivia, thus to the 40th degree of south latitude.

On the other side of South America, the vine is grown at Buenos Ayres, and probably in many parts of Brazil, but notices of this are not numerous. Further northwards the vine was, even at the time of Humboldt's journey, planted at Cumana and bore excellent grapes; and since that time, during which great changes have been carried out there, it must have been planted at several other places. The vine is, besides, cultivated

* Poeppig's Reise nach Chile und Peru, u. s. w. Bd. i. p. 330.

† Meyen's Reise, &c. i. p. 420, &c.

at more or less considerable altitudes in South America, but especially in Mexico and Guatemala. It thrives exceedingly well in the valley of San Jago in Chile (33° south latitude); the grapes, which are gathered in the valley of Arequipa, at the height of 7797 feet, are also beautiful. Perhaps the whole plateau of Mexico is adapted to the culture of the vine, which in the interior provinces extends to the Passo del Norte (32.9° north latitude). This is according to Humboldt's account, but since that time, a great number of foreigners have settled in that country, and therefore the culture of the vine must have been much extended. The vine is grown on both sides of North America, but data are wanting to enable us to give the latitude to which it extends. According to earlier statements, it goes as high as 37° north latitude on the Ohio, while on the north-west coast it is found at St. Francesco in latitude 38°. These, however, are certainly not its extreme points. According to the general survey of climate which was given in an early part of this work, the isothermals rise higher on the west coast of this continent, than on the eastern; the vine will, therefore, be cultivated at a higher latitude on the west than on the east coast.

In the Old World, whence the vine has wandered to the New, the area of its cultivation is considerably more extensive; but it has here been cultivated from remote antiquity, and, at least in Northern Europe, has been introduced on account of the religious worship. In the interior of Europe, the culture of the vine extends furthest north; on the western side, where the isotheral lines, which, as we have already seen, principally determine its range, it goes up to 47°, 48°, and 49° north latitude, viz., on the western bank of the Seine to Noyon and Laon. On the Rhine the culture of the vine extends above Cologne, nay, even above Dusseldorf. In England, in latitude 52°, the vine ripens only in a warm summer, like that of 1834. Further eastwards, in the interior of the continent, where the isotheral line bends to the north, the culture of the vine rises with it, so that it is found at Berlin in the 53d degree. Our wine, indeed, is sour, but Berlin lies in the isothermal line of 7.9° Cels., and in the isochimenal of — 1.1° Cels. This, however, is not yet the polar limit of the vine. In the 14th century it was brought to Prussia by the German knights, and has long been cultivated there. Dantzic, lying in 54° 21′ north latitude, has a very high temperature, evidently in consequence of the proximity of the Baltic and the immense body of water in the Vistula ; it lies in the isothermal line of 7° 79′ Cels., while the summer heat is 16° 86′, and the winter temperature equal to — 0.73° Cels. Hence it follows that the climate of Dantzic is on the whole better than that of Berlin, only the summer heat is somewhat depressed by the influence of the coast climate. In this district, therefore, the vine can grow as well as at Berlin and Potsdam, and in earlier times, it was much cultivated at Elbing, Thorn, and other places. The culture of the vine has even been carried above Konigsberg (54° 42′), for at Tilsit, where the winter's cold is very severe, the vine is planted in gardens, and there is even a vineyard, which is situated on a high hill on the banks of the Memel. Indeed I have rarely missed the vine in the gardens of the rich at Memel, and at a country estate, about a mile from the Russian boundary, on the road to Polangen, I have seen a great number of vines.

It is true, that the mean temperatures of these high latitudes are yet unknown, but at Konigsberg (54° 42′) the mean temperature is equal to 6.49° Cels., with a winter temperature of — 3.26° Cels. and a summer heat of 15.87° Cels., a climate which cannot produce wine fit to drink.

We learn from history, that at an earlier period there were vineyards in these parts, which probably were again deserted 300 years ago. The question has been put, why the culture of the vine has disappeared in these districts, and there have not been wanting learned men who wished to demonstrate from this a gradual diminution of the heat of the atmosphere since that time ; yet it seems to us that the fact may be readily explained. We know how bad and sour the wine of our country is ; for example, the Potsdam country wine, nay, even the noble Grüneberger. This wine could only find a sale in the neighbourhood of such populous towns as Berlin and Potsdam, where it is used to adulterate other wines.

With the severer climate of East Prussia, the grape is still less perfectly ripened, and the wine pressed from it is excessively sour. At Tilsit, the grape always remains small and wrinkled together. It is easy to understand that after the sweeter wines had reached these countries by the more ready communication both by land and water, the inhabitants gave up using their sour liquor, and thus, with the disappearance of the Order of German Knights in East Prussia, the culture of the vine, which they had introduced, also ceased. It also must not be overlooked that the early night-frosts in autumn would destroy the crop ; at least, this must often have been the case in East Prussia ; and, therefore, the farmers of these countries would the sooner be inclined to give up this uncertain branch of culture for a certain one, viz., the cultivation of the cereals. Besides, as there was no export, the value of that wine must have been very small.

Still further east, towards Asia, the vine is cultivated above the latitude of the northern extremity of the Caspian Sea, for there are vineyards round Astracan. Between the Black and Caspian Seas, the culture of the vine is very extensive, and the vine is even met with beyond the Caspian. This plant is said to be altogether wanting in Siberia (?), yet this is not the case in China, as has hitherto been generally believed. The missionary Gutzlaff speaks of the vine, which he has seen in Northern China, and I myself have eaten, in Canton, delicious grapes grown in the north of China.

In the southern provinces of China, I have never seen the vine grown; yet it is evident that its culture has been there suppressed by that of the tea. Even Loureiro* says that the vine is found in China, and that it is cultivated, though rarely, in Cochin China also. It, therefore, extends far south here. The vine is said to succeed well in Sumatra, thus directly under the Line, but it is not cultivated by the natives.† Even at Pondicherry, the French are said to have grown the vine with great success, although the heat there is so excessive.‡ In the flat parts of the East Indies, on the Philippine and Sunda Isles, the culture of the vine is not general; the damp climate is opposed to its success. Yet the vine grows on Java, where the grapes are said to be so large and beautiful, that they equal the best kinds from Portugal.§ In the work referred to, it is said that the vine is found in all parts of Java, but chiefly at Batavia. Even on the island Moa, one of the Banda Isles, the vine grows very well when it is planted. It has also been grown on the island of Lethy; in these parts it bears fruit twice a year.‖ On the other hand, the culture of the vine is widely spread over the table-lands of India; for example, in the Deccan, and in the plains of Cashmere,¶ which lie at the height of 5400 or 5500 feet, and where the vine climbs to the top of the poplar.

In the elevated land of Kunawar, between latitudes 31° and 32°, rich vineyards are to be found at a height of 9000 or 10,000 feet, where the grapes are pressed in September, and are also used as raisins. In Bokhara, also, the vine is cultivated.

* Flora Cochinchin, i. p. 155.

† See Marsden, Hist. of Sumatra, 3d ed. p. 103.

‡ See Ainslie, Mat. Med. of Hind. i. p. 156, cited by Royle.

§ See Description of Batavia, &c., in the Dutch of J. J. Ebert. Leipzig, 1786, iv. p. 112.

‖ See Barchwitz, Ostindische Reisebeschreibung. Erfurt, 1751, p. 239, 2te Auflage.

¶ Royle's Illustrations. London, Fasc. i. 1833.

In Persia, on the banks of the Euphrates, in Syria, Lower Egypt, Abyssinia, and throughout Barbary, the culture of the vine is to be found, although in these countries the grapes are not pressed.

The wine made on the western side of Africa, the Canary Isles, and the Azores is very celebrated, yet the vine is also found on the Cape de Verd Islands, and even on St. Thomas and the coasts of Guinea, consequently also in the hottest zone. Schouw, indeed, supposes* that the mountains here favoured this branch of culture, but this supposition is not very well founded. The vine thrives on St. Helena also, and at the Cape of Good Hope the best kinds of wine are made.

The vine has been introduced to the English colonies in the south of New Holland, and it has wandered from America to the Sandwich Islands, where it succeeds well.

Thus we have shown how the culture of the vine has spread from its insurmountable polar limit between 49° and 55° north latitude, through every zone to the equator. In the southern hemisphere its polar limit lies in 40° south latitude, evidently on account of the peculiar configuration of the land and the prevalent coast climate.

Earlier views, according to which the equatorial limit of the culture of the vine stretched at farthest only to the tropic, have been exceedingly arbitrary, and could only have been adopted by a reliance on the defective account of travellers. A very hot and damp climate alone opposes the distribution of the vine.

### 7. *The Maguey-plant (species of Agave).*

Probably there are but few nations on the earth, who have not, in some way or other, prepared intoxicating liquors. A great number of examples might here be given.

* Geographie der Pflanzen, p. 209.

We shall take the maguey-plant, which is the vine of the Aztec nations, and forms a branch of trade of very great importance.

The maguey-plant, unfortunately, is not systematically determined so exactly as it deserves to be. According to Humboldt,* to whom we owe almost all our information respecting the growth of this Aztec wine, the pulque of the Spaniards (octli of the Aztecs), it is made from several varieties of the Agave americana, but not from the Agave cubensis *Jacq.* (A. odorata *Pers.*, and A. mexicana *Lam.* synon.), as has been affirmed by different writers.

On the plateau of Mexico, which lies above the height of 7000 feet, the maguey plantations are found chiefly in the provinces Puebla and Mexico. In them there are large tracts of land covered with maguey plantations, which must give the Mexican landscape an extremely singular character. Our countryman Deppe, has given us a view of such a maguey plantation, although of small extent, in the neighbourhood of Mexico (eight leagues to the north-east); we have but to compare such a landscape with the waving surface of a green corn field in our own country.

The most beautiful maguey plantations are those of Toluca, and in the plain of Cholula, where the plants are set in straight rows. The maguey plant here flowers after eight years, and at this period the juice is collected, which afterwards by fermentation yields the pulque. The countryman perceives the beginning of the development of the flower-shaft by the sudden rising of the root-leaves, which previously lie rather horizontally on the ground; he goes through the plantations every day that he may not miss the time when the plant begins to develope its flower-shaft. When he perceives this he cuts off the tuft of central leaves, widens the wound a little, and covers it with the upright side leaves, which he ties

* Neu-Spanien, &c. Bd. iii. p 95, &c.

together. The sap which ought to be employed in developing the flower-shaft, pours into this wound; this vigorous secretion of sap continues for two or three months, and it is collected thrice a day. The plant in general yields daily about 200 cubic inches of sap, nearly three Prussian quarts, of which three-eighths are obtained in the morning, two-eighths at mid-day, and three-eighths at six o'clock in the evening. A very vigorous plant is even said to yield 375 cubic inches a day, thus more than seven gallons, and that almost without interruption for four or five months. It is at the same time very worthy of remark, that this plant grows on the very dryest ground, which is often scarcely covered with mould.

The produce of the maguey plant, like that of the vine, is very unequal; usually the quantity of pulque which the countryman obtains in a day, is worth ten or twelve sous, and he reckons on 150 bottles from one maguey. An immense capital is laid out in the maguey plantations of Mexico, but he who plants them must have both patience and enterprise, for on the whole they are not profitable until after fifteen years. The maguey plant, which is exhausted by draining off its sap, dies, but hundreds of shoots spring from the root.

"The honey or Agave sap," says Baron von Humboldt,* "is pleasantly acid. On account of the sugar and mucilage it contains, it readily ferments, and to expedite this, a little old, sour pulque is added to it. The operation is thus completed in three or four days. The liquor then resembles cider, and has an extremely disagreeable smell like that of putrid flesh." However, when once the repugnance to this smell is overcome, foreigners prefer pulque to any other drink, for it is said to be very strengthening and nutritious. The pulque of Hocotitlan, north of Toluca, lying almost as high as the Nevado de Toluca, is said to be particularly excellent.

* L. c. p. 99.

We first recognise the high importance of the maguey culture to the economy of the state, when we learn that the import duties for pulque in the year 1793 amounted to 817,739 piastres for the towns of Mexico, Puebla, and Toluca alone, a sum which is equal to 3,800,000 francs.

From the sap of the maguey is also made a strong brandy, mexical or Aqua ardiente de maguey; but another species, viz. Agave potatorum *Zucc.*,* is that which is used for this purpose.

Besides yielding the Mexican wine, the leaves of the maguey-plant furnish the strongest hemp that is known, and also a substance which, like the papyrus of the ancients, is a substitute for paper, and was used by the old Mexicans for the delineation of hieroglyphical figures. In the manufacture of this paper the leaves of the agave are left to rot until the cellular tissue has disappeared, and all the layers of fibres stick to each other, a process quite similar to that by which stuffs are made of the bark of the Broussonetia on the South Sea Islands.

The elevation in which the maguey-plant occurs, is between 1168 and 1379 toises, and even still higher. The plateau of New Spain possesses a climate like that under the bright sky of Rome, viz., a mean temperature of 17° Cels.; even in January and February the mean daily heat at Mexico amounts to 13° or 14° Cels., while in summer the temperature is not above 24° Cels. All the mountain plains there, which lie higher than the plateau of Mexico, have a rude and unpleasant climate; even on the plain of Toluca, where there are the finest maguey plantations, the temperature scarcely ever exceeds 6° or 8° Cels.

It is evident that this sudden decrease of temperature can only be explained in this way, that the plateau of Mexico, in consequence of the reflection of the sun's rays from so extensive a surface, possesses a much higher

* Nova Acta Acad. C. L. C. Tom. xvi. P. ii. p. 675.

temperature than would properly belong to it in proportion to the altitude. The maguey-plant, therefore, properly thrives only in a climate which comes near to that of Southern Europe, where the various species of agave, in company with the prickly cacti, have become wild or rather naturalised. On St. Helena the Agave lurida has been planted on the sides of some roads, and it looks very stately when in flower.

It is to be expected from the nature of things that few will be inclined to promote the distribution of the maguey-plant, particularly with a view to the making of wine; nay, it will even diminish in extent in Mexico, as our vine becomes more frequent there.

### 8. *The Sugar Cane.*

The sugar cane is a plant of the Old World, which was cultivated in China and the South Sea Islands before the historical era. The Spaniards of the Canary Isles introduced it into America, and in the year 1520, Piedro de Atienza raised the first sugar cane on St. Domingo, whence it passed to Cuba and the continent of America. The Mexicans used the honey from the stalk of the maize, before their acquaintance with the sugar cane.

The culture of the sugar cane requires the same degree of heat as that of the cotton-tree, viz., a mean heat of 24° or 25° Cels. with which it succeeds best, though we still find large plantations of this plant in far colder parts, where the temperature amounts to 19° or 20° Cels. only. Accordingly, the tract which is capable of the sugar culture stretches far beyond the tropics; and this plant is even cultivated with advantage at some points in the south of Europe, for example, in Sicily, and in Spain also, formerly more frequently than now. Since, as we have already seen, the temperature diminishes with the increasing height in such a ratio that 1° Cels. corresponds with a height of 100 toises, the average temperature of 20° Cels., which the culture of the sugar cane requires, would be found at the height of 3000 toises; however, on extensive plateaux, the heat is so considerably increased by the reflection of the sunbeams, that on the mountains of Mexico and Columbia, the sugar cane is grown at a height of 4000, 5000, or even 6000 feet. Nay, the plain of the town of Mexico, whose climate should answer to the isothermal line, or 13.7°, has a mean heat of 17°, and sugar plantations were formed in this valley at the height of 6600 feet by Cortes himself. On the table lands of the Himalaya also, for instance, on the plateau of Nepaul, sugar and cotton are grown at the height of 4500 feet.*

The sugar cane is raised from shoots two or three feet in length, which are prepared from the shaft of the full-grown plant, and are set either horizontally or perpendicularly, or even in pairs converging towards each other. In fourteen days, the shoots spring from the joints, and in the space of a year the shaft is so far grown that it may be cut down. On land freshly taken in, which is not exposed to long inundations and is well planted, the sugar cane yields from twenty to thirty crops, since new shoots spring every year from the perennial root; nay, Humboldt saw a sugar plantation in Cuba which had been in existence for 45 years.†

In the various countries of the Old World, where the sugar cane is indigenous, there are very different varieties of it, one of which possesses more or less advantage over the others, so that it has on that account been transplanted to the most distant countries. It is known that the sugar cane of Otaheite has been brought to the West Indies, where its richer yield gives great satisfaction to the planters; in the same time and on an equal extent of ground, it yields one quarter more juice, and at the same time produces a larger and more solid mass

* Royle, Illustrations. London, 1833.

† Reise in die Æquatorial Gegenden. Bd. VI. p. 163.

of wood, which is used for fuel. But the sugar cane of the South Sea Islands is of remarkable excellence and strength; on the Sandwich Islands, as Cook mentions,* it attains the circumference of eleven inches.

It seems, however, that the East Indian sugar cane is far more productive than that from the South Sea Islands, for in Bengal the produce is twice as great as in Havanna,† while the daily wage of the Indian is nearly three times less than the cost of maintaining a slave. On the Philippines, the inhabitants, on account of its smaller yield, are quite against the culture of the cane from Otaheite, for the purpose of making sugar, but use it for eating.

I may suppose that the process of making raw sugar is already known; it consists in crushing the stem after the leaves have been stripped off by machinery, in boiling the mass, clearing the liquor, boiling it down and bringing it to crystallization. It is quite certain that we have learned the art of boiling and clearing sugar from India and China, for in that country, where foreign customs are admitted only in cases of absolute necessity, the same process has been followed since the earliest times, and in China it is even carried to greater perfection than with us.

Although with us sugar is only an almost indispensable article of luxury, in tropical countries it is a general article of food, partly as raw sugar, but also, and this is more commonly the case, in its unprepared state, viz., the ripe shaft of the plant, which is chewed and sucked after having been made soft by boring. It is incredible what enormous quantities of raw sugar cane are consumed in this manner; large ship loads are daily brought to the market of Manilla, and in Rio Janeiro. On the Sandwich Islands and at other places, every child has a piece of sugar cane in his hand.

* Dritte Reise um die Welt. Berlin, 1788, p. 294, ii.

† See Alexander von Humboldt. Ueber Neu-Mexico, iii. p. 116.

The profit of the culture of sugar is quite enormous; the quantity of spirituous liquor also which is made from the sugar cane, partly from the molasses, partly directly from the crushed canes, is quite incredible; and yet the molasses, at least in the Spanish colonies, are generally thrown away, because old privileges are said to prevent the making of brandy.

This is not the place to compare the profit of the different colonial articles with that of our cereals, but although it is true that the culture of sugar cane, on an equal space and in the same time, is considerably greater than that of corn, yet it is right to remark that in order to cultivate the sugar cane on a large scale, a large capital is required, which brings in no higher interest than the profit of agriculture with us. We must therefore not imagine that all the proprietors of plantations in the tropics are rich men, nor envy them as our farmers generally do. Besides the crop of these plantations is often, at least in the East Indies, destroyed by insects, viz., clouds of locusts, while a failure of the crops with us is very rare. Whenever the sugar-cane is near its full growth, the locusts, that is if they are not too numerous, are no longer dreaded, for then the loss of the leaves does not injure it, but the young plants are much hurt by losing them, and their growth checked. When the plantations are not very large, the Hacendado, with his numerous servants, tries to prevent the settling of the locusts.

## 9. *The Culture of Coffee (Coffea arabica, Lin.)*.

The coffee-tree grows in the hottest parts of the torrid zone; its artificial area, however, is so great that it passes far beyond the tropics, even above the 36th degree of north latitude, where it finds a mean temperature of 19½° or 20° Cels. The range of the coffee-tree therefore pretty nearly coincides with that of the cotton-tree. As the coffee seems to prefer a damp and shady situation, it grows best within the tropics at some height, for example from 1200 to 3000 feet, but it seldom succeeds above the height of 6000 feet.

The Old World is the native country of the coffee-plant, which is now one of the most important products of the New World; nay, in Brazil, not far from Rio Janeiro, in the woods of Corcovado, at a height of 1000 feet, I found several little coffee-trees growing wild. At present coffee is extensively cultivated in the East Indian islands, as on Java and even Luçon, and it grows exceedingly well on the South Sea Islands, yet its extension towards the East is prevented by the universal culture of tea. The coffee-plant was brought in the beginning of the last century from Persia to Java, where it has found its second native country.

Coffee is cultivated in the following manner:* the fresh beans are sown generally in the shade of coffee trees, and the little plants are taken up with the earth about them, so soon as they have reached the height of twelve inches. They are planted in a quincunx and so that the stems are from four to six feet apart. By cutting off the rankest shoots the coffee-trees in plantations are prevented becoming more than twelve feet in height, that the fruit which ripens in the 20th, or even not till the 32d month, after the transplantation, may be the more easily plucked. After four or five years the crop is very good, and then one servant is kept for every 1000 plants. The coffee-tree yields three crops annually, which furnish employment for almost the whole year. At Rio Janeiro the first crop commences in April, when only the perfectly ripe and red berries are gathered, the seeds of which are separated without difficulty from the pulp; the whole seeds are dried and then split by the aid of a machine.

* See Von Martius, Reise in Brasilien. I. p. 146.

## 10. *The Chinese Tea.*

The consumption of tea is so universal amongst a large portion of the inhabitants of the globe, and its culture is of such importance to the vast Chinese empire, that a full exposition of the culture, preparation, and uses of this plant will be here in its right place.

The plant which produces the common kinds of tea in commerce, is the Thea chinensis, of which there are a number of more or less constant varieties, so that several botanists have made not only two, but three, distinct species, viz., Thea viridis, Th. bohea, and Th. stricta. I shall afterwards return to the reasons which determine me to receive one species only as furnishing the Chinese tea.

China is the native country of the tea-plant; it is found there as high as the 40th degree of north latitude, as well as in the mountainous districts in the southern part of the country, particularly on the mountains which separate China from the Birman empire. That the culture of the Chinese tea is also carried on in Ava, the Birman empire, and on the eastern borders of Thibet, Ritter has proved from accurate sources.* But the tea-plant has recently been found wild in Assam, and, indeed, in the territory which belongs to the English, where the mountains are not above 6000 or 8000 feet high;† and, therefore, great hopes are entertained that the culture of tea on a large scale will soon be so successful there, that the trade in this article will soon be snatched from China. A large quantity of ordinary tea is produced in Cochin China and in Tonquin, yet here this branch of agriculture is much neglected. Whether the plant is wild here, or whether it has been introduced, we do not yet know; the latter might almost

* Ueber die Verbreitung der Theecultur.—Geogr. v. Asien, ii.

† See Wallich, Discovery of the genuine Tea Plant in Upper Assam; in the Journal of the Asiatic Soc. January, 1835.

be supposed, for the tea-plant thrives best in the subtropical zone, consequently it will be indigenous in it, as well as on the heights of mountains corresponding to this zone.

The use of the warm infusion of the leaves of this plant, which is known by the name of tea, stretches back into the earliest times of Chinese history, and it is at present so universal throughout the empire, that the consumption of tea-leaves can scarcely become greater, that is if the population does not increase.

Much has already been written on the origin and on the native country of the tea, and old Chinese writings are continually brought forward as authorities. Siebold has very lately tried to disseminate the opinion that the tea has been introduced into China from Kaorai,* which, however, Klaproth has shown to be erroneous.‡ He has shown that the oldest notices of the use of tea stretch back to the years from 265 to 419. In the Chinese writing which bears the title Schi schue, we find that in the middle of the fourth century, a minister of the public buildings, Wang-mung by name, used the tea-plant, which in Chinese is called Ming. In the year 600 the plant was recommended by a priest to an emperor of China, who suffered much from headache, and as the illness was soon cured by using tea, the use of it was speedily adopted everywhere. Tschha is a synonyme for the plant ming, and by that name the dried plant was first brought into notice by the Portuguese and Spaniards; the word Tschha is also common in all the northern provinces of China. Klaproth considers the word Thea to be the Malay Teh, which is derived from the Chinese word Thee.

The culture of tea in China must have been very considerable as early as the 8th century, for in 763, at a taxing of the empire, a duty of 10 per cent. was laid on

* See Nippon. Heft ii.

‡ Haude-und Spenersche Zeitung. Berlin, 1834. 11ten Dec.

tea, and since that time the government has always derived a large revenue from this useful plant. At the present time the tax on tea is collected in the following manner; none of the country people are allowed to sell tea without permission; but these permits are obtained at the custom-house of the province, and for the precise quantity which is to be sold a double receipt is issued, one of which the seller keeps, while the buyer receives the other, in order at any time to prove the legality of the sale.

Tea has been known in Japan since 810, and it has been cultivated in the Corea since 828. Its culture has also been attempted in Bengal, and great success is expected from it, nay, this question has been very recently discussed by Royle, yet, as it seems, with great partiality to India.* I shall subsequently attempt to show that even though the tea plant can be grown in all the cooler parts of the tropical zone, over the whole sub-tropical, and even far within the temperate zone, to about the 40th parallel, yet there are other circumstances necessary in order to a profitable culture of the plant. The chief one is a low price of labour, which is, indeed, very low in Bengal, as well as in China; but that it is in India only one-fourth or one-half as high as in China, which Reeves has asserted in Royle's work, may well be doubted; if it were so, then Bengal would soon furnish cheaper tea than China. The day's wage may indeed be eightpence at Canton, where Mr. Reeves lived, but in the interior of China it is only a fourth of this sum. Besides in these countries, tea has been planted in Ceylon and Java, whence some thousands of chests are annually exported. The tea of Java has lately reached the market of Amsterdam, and excited much attention there, for above 1,400,000 lbs. have already been obtained, so that the Dutch will probably

* Illustrations of the Botany of the Himalaya Mountains. Fasc. iv. London, 1834.

within twenty years, draw the whole quantity of tea they require from Java. According to Marsden, few tea bushes were cultivated in Sumatra in the last century. The tea has also been planted at the Cape of Good Hope, in St. Helena, and at Rio Janeiro, where there are still large plantations of it in the botanic garden, but these are in a miserable condition.*

Plantations of tea are formed by sowing the seeds, which are set more or less regularly. In the first year the middle shoot is stopped, that it may not grow tall and slender, but may become bushy, and be covered with a greater quantity of leaves. The crop of leaves begins in the fourth or fifth year. I have visited such tea plantations, and found them in hilly situations, which is said to be the case throughout the country. The plants in these plantations were in general only two and a half or three feet high, and they stood about three feet apart; a few branches only shot beyond the others and reached a height of five feet. I have found women sitting beside these bushes and plucking off the leaves by hand in the ordinary way. From the different statements respecting the time of gathering them, it seems to vary much in different parts of the Chinese and Japanese empires; however, the principal crops end in May and June, for in September and October fresh supplies of tea from the interior of the country reach Canton.

The mode of manuring these plantations differs much in different parts; in China, however, the most usual manure is a compost of human excrement and calcareous clay. We every where see in the Chinese fields close to the cultivated land, large walled-in pits or enclosures sunk in the ground, and filled with this compost. In Japan, according to Siebold's statement, other very strong manures are used for tea, viz., the expressed juice of the Japanese mustard, dried anchovies, also the oil-cake of Brassica orientalis and other plants.

* See Meyen's Reise um die Erde. Bd. i. p. 102.

The leaves of the tea-plant, when freshly plucked, have nothing of the odour and flavour of the dried leaves; they have neither a sharp, nor an aromatic, nor a bitter taste. The properties which they afterwards show as prepared tea, and for which they are so highly prized, viz., the pleasant taste and delightful odour, are the effects of the roasting by which the leaves are dried. We need wonder the less at this as it is the same with the coffee; every one knows that unroasted coffee possesses nothing of the pleasant aroma and ethereal odour which are proper to it after being roasted. The tea leaves are dried upon great iron plates, which are excessively heated, and in large flat iron pans, the sides of which are somewhat slanted. The leaves are first made to shrivel up in these pans by being constantly stirred with a gentle heat, and are then gradually dried by keeping up the heat. After this the hot leaves are turned out upon mats and rubbed with the palms of the hands; after having cooled they are again put in the pans and again roasted until the tea is perfectly dried, which is done by repeating this operation from four to six times. In drying the leaves lose three-fifths of their whole weight, so that 3 lbs. of fresh leaves produce only 1 lb. of dry tea.

The differences in the colour, shape, and pubescence of the dried tea leaves at first induced botanists to think that the green and black teas were prepared from different species; this, however, is not the case, but both kinds of tea can be made from the leaves of the same plant, as Abel learned during the journey of Lord Amherst. But when once prepared, the tea, as Mr. Reeves also mentions, cannot well be changed; at least, black tea cannot be made into green tea, though the green may be changed, imperfectly at least, into black.

It is a singular circumstance that the dispute amongst botanists whether tea is made from one species of the genus Thea, or from two different species of this genus, has never yet been terminated. In Japan, where black as well as green tea is made, the tea shrubs, according to the observations of Kaempfer, Thunberg, and Siebold, belong to the same species, of which F. Nees v. Esenbeck also has convinced himself from the specimens brought home by Siebold; consequently the opinion so sharply expressed by Mr. Reeves, formerly tea-taster at Canton to the English East India Company, that the black and green teas are obtained from two perfectly different plants, is shown to be incorrect.

*Observation.*—I do not think that more importance can be attached to the opinion of Mr. Reeves, even though he has lived so long at Canton and Macao, than to the botanists by profession, who must know better what are to be considered as species and what as varieties. Besides, Mr. Reeves has never been in the provinces of China, where the culture of tea is universally carried on; nay, he does not even seem to have once visited the tea-plantations in the neighbourhood of Canton.*

Mr. Reeves wonders how any one, who has been in China, who has only seen the different infusions of green and black tea, can consider both kinds of tea the leaves of one and the same plant,† and this expression, which we must consider very extravagant, seems to find great approval. But had Mr. Reeves known how the different kinds of tea are prepared, he would no longer have wondered. He refers us to the figures of the two sorts of tea which are given in Loddige's Bot. Cab. Tab. 226 and 227, and remarks that the two species, which furnish the black and green teas, are there very well characterized. But that this is not so very extraordinarily clear as Mr. Reeves thinks, will perhaps be confirmed by the unprejudiced comparison of most botanists. Even in our agricultural plants of the kind, which have but small areas, in comparison with the tea-plant, greater

* Meyen's Reise, &c. ii. p. 375, &c.

† Loudon's Gard. Mag. ix. p. 713.

differences between the varieties than we find here may be pointed out. It is true that Hooker* also has admitted the existence of two species of tea, but his characters are founded on plants which had been grown in England.

If we take a number of leaves of all the various sorts of tea, which come to us in trade, soften them in hot water, and lay them out side by side, we shall soon be convinced that there are no characters which distinguish the different kinds of black tea from the green teas; provided a great number of leaves be observed. Accum, who is at present at Berlin, has lately executed, and laid before the Society for the advancement of Industry in Prussia, a work, well worthy of attention, by which the transition of the leaves of all the different sorts of tea is shown.

The correctness of Mr. Reeves's opinion on this subject may, therefore, at least, be doubted; I think that it may even be refuted.

The green tea is prepared in the manner I have already stated; the black, on the contrary, is made in what is called the moist way. The fresh leaves are laid on large sieves and these are placed over boiling water, so that the leaves are permeated and strongly infused by the hot steam. After this the leaves are dried on iron frames in the manner previously described. By this infusion with hot steam the fresh tea is deprived of its astringent principles, viz., the gallic acid and tannin; the leaves also in consequence contain fewer of the delightful aromatic particles which are present in green tea in such quantities. Thus, according to the known analyses of chemists, black tea contains less gallic acid and tannin than green tea; nay, the latter alone contains theine, an alkaloid, which the black tea is probably deprived of solely by the infusion with the hot steam.

Although it is now decided that all kinds of tea are

* Bot. Mag. Tab. 3148.

prepared from the same species of Thea, it must not be thought that all the sorts can be made in one and the same district and from one bush. In one place the black tea chiefly is grown, in another the green tea; here the tea is but little curled up, there very much, so that it becomes quite globular; this, however, is by no means a mark of very fine tea. I do not think that one can be surprised at this, for the same holds good with other agricultural plants with us, of which there are likewise hundreds of varieties. I here call to mind the making of our wine; the vine is almost everywhere the same species, and yet how differently do the wines taste and smell. The tea plantations, the leaves of which are of a particular flavour, are just as limited as the place in which this or that vine of a peculiar flavour occurs; and it is not the case, that the pleasant odour of particular kinds of tea is given by other fragrant substances. However, I here remark, that I have seen large quantities of the flower-buds of Olea fragrans, which are a real article of trade in China, and are used by connoisseurs to improve the flavour of green tea; but every one mixes this substance according to his own taste.

Inferior sorts of tea, which generally do not come to us in commerce, but are kept for home consumption, are prepared by taking entire branches and shoots from common plants and drying the leaves partly with the stalks, partly only stripped off by the hand. The brick-tea is made of this sort of tea. It comes into trade in hard cakes, very like thin bricks, but is chiefly consumed in Northern China, and in the interior of Asia, for example, by the nomades in the desert of Cobi; it consists of bad and dirty leaves, mixed with stalks, which are glued together by some clammy substance, pressed in the form of cakes and dried in ovens.

In using this brick-tea some pieces are broken off, and after having been reduced to powder, it is boiled with water or with milk, meal and fat.* The Chinese sol-

* See Timkowsky's Journey to China, i. p. 46.

diers on the northern frontiers receive this brick-tea as pay, and what they do not need themselves, is sold at Kiachta. Indeed this tea is bartered in all parts of Mongolia and in Dauria.* Large caravans of camels laden with this tea journey through the desert of Cobi. At an earlier period it was quite usual both in China and Japan to make tea from the powdered leaves.

The oldest work yet known, in which tea is mentioned by an European, is the Historia indica of Maffei, which appeared at Leyden in 1580 ; yet tea was first brought to Europe by Dutch merchants in 1610. As early as the year 1660 a tax was laid on the sale of tea by an act of parliament. In the year 1638 ambassadors from Moscow brought tea as a present to the Czar.

Now that we have become acquainted with the culture, the preparation, and the distribution of tea, we shall proceed to the consideration of the immense quantities which are annually produced from this useful plant, and consumed. We know that at present so large a quantity is consumed in England, that more than 1½ lbs. must be reckoned for each inhabitant ; but the consumption of tea in China must certainly be much greater, for there every one who can, drinks tea the whole day. However, if we reckon only a pound and a half for each person, as the population of the Chinese Empire is at least 200 millions of souls, we obtain the enormous amount of 300 millions of lbs. If we also take into consideration the consumption of tea in Japan, Cochin China, and the neighbouring states, 450 millions of lbs. of this dried herb is probably not too great a sum for the east of Asia. Now think of the mass of fresh leaves and the number of hands which are necessary to prepare this immense quantity of tea! Of such importance is agriculture in China and Japan for this object only.

We do not indeed know with sufficient exactness the

* See C. Ritter's Historico-Geographical Inquiry into the Distribution of the Tea Culture, in his Geographie von Asien, ii.

quantity of tea which is annually exported from China, but we know what is imported to Europe and the European colonies. From accurate sources I have calculated the whole quantity of tea exported by Europeans from the port of Canton, at 45,000,000 lbs. for the year 1830;* to this must be added the tea brought to Russia by caravans, which is said to have amounted, in the year 1830, to no more than 5,405,990 Prussian lbs.

The large quantity of tea, with which China supplies the Indian Empire by land carriage, is unfortunately not known, and data also are wanting to enable us to estimate its approximate amount; yet to judge from various accounts, the consumption of tea there must be very great. Over all Thibet and in Nepaul tea is the common beverage at every meal. But disregarding entirely the quantity of tea which is brought by land carriage to India, yet 50,000,000 lbs. are exported to Europe and its colonies, for which a sum of about eighteen millions of Prussian dollars flows into the Chinese empire, for on an average the teas are worth one-fourth of a piastre the lb. at Canton. We have, however, seen in what an exceedingly singular way this immense sum of money is again drawn from the Chinese empire; but unfortunately it flows into other hands, so that Europe by the use of tea always suffers a considerable loss of gold.

I may mention, that of the 50,000,000 lbs. of tea which are exported by sea from Canton, and sent by land through Kiachta, not more than about 200,000 lbs. are used in the Prussian States, while England consumes 26,000,000 or 27,000,000 lbs.;† accordingly England, in proportion to the population, consumes one hundred times more than Prussia.

* See my Reise um die Erde. Bd. ii. p. 382, &c.

† In the course of the years 1834–35, after the privilege of trading to China was taken from the English East India Company, and the high duty on tea was lowered, about 36,000,000 lbs. of tea, according to the newspapers, were consumed.

The culture of tea is now of great importance to the eastern part of Asia, and yet scarcely 100 years have passed since the use of tea became general in Europe ; in the meantime the taste of the people for tea is daily increasing, and therefore it may be foreseen that this branch of agriculture in half a century will be a new and important source of prosperity to certain tropical countries.

The chief circumstance to be regarded in introducing the culture of tea into other countries, is the value of labour in them. The preparation of tea demands much labour, and as the value of tea of itself is very small, of course a very low wage can be given for the labour of its preparation ; therefore in a country where labourers are few and the day's wage high, tea can never be grown with advantage ; this is the case with the culture of tea in Brazil, where slaves are so enormously dear. As the average price of a pound of tea at Canton is about 8½d., the cultivator must sell it at 5d. or 6d., for the merchant who disposes of the tea to foreigners must have about 30 per cent. profit after all expenses are deducted, for the Chinese merchant borrows money at Canton at from 20 to 26 per cent., with which he goes to the tea plantations in the interior and buys the crop on the bush for ready money, just as the wine makers buy the grapes with us.

## 11. *The Pepper.*

The culture of pepper, the most common spice, which is used in every country of the globe that has reached any degree of civilization, is likewise of great interest. The pepper-plant (Piper nigrum, *L.*) is a perennial climber, which can be cultivated in tropical countries only. The East Indies is the native country of the pepper, and indeed, almost Malabar alone, where Buchanan found the pepper-plant growing wild in the woods. It has not been found wild in Sumatra and the other Sunda Isles, on which it is at present cultivated, and it has certainly been introduced into them from the West, from whence it has been attempted to bring it to every country within the tropics where the speculative spirit of man has seen the prospect of deriving pecuniary profit from it. The pepper of Malabar is also much stronger than that of Sumatra, and is on that account more highly valued; so that from this also we may conclude on the original country of this plant. The culture of the pepper is noticed in several books, but we have received an ample description of it from Marsden,* who, as an officer of the East India Company, has observed for years this branch of agriculture so important to commerce.

For the pepper plantations moderately high ground overgrown by trees is chosen; those plantings where mountain rice has once been cultivated, can also be made use of. The ground is cleared of the wood upon it in the usual manner, viz., by felling, drying, and burning, and is then divided into regular beds, six feet square. In each pepper plantation there may generally be counted a thousand of these square fields, and in each square stands the support on which the pepper-plant winds.

The supports are first put in the ground; they are very various according to the customs of the different countries. Sometimes a slender tree is made use of, sometimes common poles, yet the living tree is generally preferred in order to obtain at the same time some shade for the pepper-plant. On Sumatra, where much pepper is grown, the supports employed are the cut branches of Erythrina corallodendron, which are planted some months earlier than the pepper-plant, that they may send out roots to enable them to support the plant which twines upon them. The upper branches of the Erythrina corallodendron are peculiarly fit for this pur-

* The History of Sumatra, &c. 3d Edition. London, 1811, p. 130–148.

pose, because they grow rapidly and are set with little thorns, by means of which the pepper-plant clings the more firmly. So soon as the Erythrina begins to shoot, all the lateral twigs are broken off, and only the centre and straightest branch allowed to shoot until it reaches the height of about fifteen feet, when the top of it is cut off to prevent it becoming higher. In other parts the Morinda citrifolia and Erythrina indica are made use of. As the pepper-plant lasts several years, these rooted supports are particularly advantageous, for common poles in so damp a climate rot in a short time, and the pepper-plant is injured by the insertion of fresh poles. The branches and leaves of the rooted poles are carefully cleared away from the trunk, and the top is cut into the shape of a fan, that it may give the proper shade to the pepper-plant. The pepper plantations require much attention; they must constantly be kept free from weeds, which in that hot and damp climate would soon take the upper hand and choke the plants. But during the hot summer months, when the atmosphere is very dry, a long grass is allowed to grow in the plantations, which is said to induce greater moisture of the soil.

The pepper is planted by slips, which are cut from the out-runners of an old plant; a single joint is sufficient. In general two plants are set beside a pole; in three years they reach the height of eight or twelve feet, and begin to bear fruit. About this time, when the pepper has become ripe, the whole plant is cut down to the length of three feet; it is carefully separated from the support and laid horizontally in a circle on the ground, so that the ends again take root. The plant now shoots out afresh and bears every year a large quantity of fruit, while without this previous laying down it would exhaust itself chiefly in leaves. The middle shoot is generally left on the support, and the side shoots only cut down. If the new pepper plantations were formed of these long shoots they would bear fruit in the second year.

Usually the plant sends up too many shoots from the branches laid down, and in that case most of them are cut away, and only one or two allowed to climb up. After the pepper-plant begins to bear, it increases in fruitfulness until the seventh or eighth year, when it remains stationary for a few years, till it gradually becomes too old for bearing.

It is true that the pepper plantations are decried by the inhabitants of hot countries on account of the great labour they require, yet it is said that 1000 pepper-plants can be taken care of by two persons, who can at the same time attend to their rice.

As in most agricultural plants, there are a number of varieties of the pepper-plant, one of which is esteemed for one reason, another for another quality. The time in which this plant flowers and bears berries is very variable ; yet it yields two crops in the year ; sometimes it bears flowers and fruit throughout the whole year, but in some countries it bears them only once a-year. In Java the crop is sometimes so great that the leaves of the plant cannot be seen for the immense quantity of berries. In Sumatra the produce of 1000 plants in full bearing is usually equal to 400 or 450 lbs.

The berries of the pepper require four or five months to ripen ; they are at first green, but when ripe they are a beautiful red, and they then fall if not plucked at the right time. Whenever a part of the berries are ripe, the whole are taken off and spread on mats or on the bare ground, when they become dry and assume a black colour with a wrinkled surface. The riper the berries were when plucked, the more wrinkled they become in drying.

White pepper is made from the black by taking off the black outer coating by allowing it to rot. For this purpose the black pepper is laid in ditches or pools, where after soaking for fourteen days, the pod bursts, and the white seeds which were contained in it are dried in the sun, after the pods have been completely removed by rubbing in the hands. It is, indeed, alleged that the white pepper has lost much of its strength by this soaking, but on account of its smaller pungency it is more agreeable to many persons.

An inferior sort of white pepper is made from the over-ripe berries which had fallen, the black pods of which burst, and it seems to follow from this, that the pungency of the pepper diminishes as it ripens.

We find it also stated that a pepper which bears white berries grows in India, but this perhaps needs to be confirmed.

In the fourth volume of Ritter's description of the earth, which has just appeared, the learned author has given an estimate of the whole quantity of pepper produced. From the calculations of Crawford, a concise view of which I give here, the whole production of pepper at present amounts to 50,000,000 lbs., one-third of which comes to Europe ; the largest quantity is consumed by the Chinese. If the whole quantity of pepper be divided amongst a thousand millions of persons, the daily consumption of each is 1.05 gr. As the use of pepper, as the commonest spice, is increasing from year to year, and is gradually extending amongst the rudest peoples, it may be foreseen that the production of pepper will become far greater.

## V. ON SOME OF THE PRINCIPAL PLANTS, THE FIBRES AND WOOL OF WHICH ARE USED IN THE MANUFACTURE OF STUFFS AND OTHER MATERIALS INDISPENSABLE TO MAN.

It is evident that the subject before us would furnish material for a very extensive work, if it were treated of in detail. It is known how variously the stuffs, which serve for the ordinary dress of the South Sea Islanders, are prepared. I may here pass over the manufacture of these cloths from the bark of different trees, as it has been so often described in various works ; the account in the classic voyages of Cook must be recommended to any one, who wishes more exact information concerning it. Here I confine myself to the mention of those plants which are used in the preparation of cloth by the inhabitants of different countries.

The best known to us of all these plants are the flax (Linum usitatissimum) and the hemp (Cannabis sativa), the various uses of which are so universally known. The Old World is the native country of the flax, but in North America is found another species which can be used for the same purposes. Towards the east and south the culture of flax is superseded by that of the cotton plant, which yields a much more valuable crop. On the cold table lands of India, where the cotton does not succeed, flax is cultivated, but chiefly for extracting the oil. Whether our flax is grown in China is still unascertained ; for the fabric, which comes to us under the name of Chinese linen, is made from another plant, still unknown to us.

### 1. *The Cotton-Plant.*

The cotton-plant is one of the most useful to mankind of all that are cultivated on the globe ; it is very probable that a greater number of persons are clothed by cotton than by any other stuff.

Not only is the cotton-plant cultivated in the tropical parts of every land of the Old and New Worlds, but it extends far beyond the tropics even to countries whose mean temperature is between 13° and 16° R., which is the case in the most southern parts of Europe. In particular cases, for example in the Crimea, where other causes moderate the winter's cold, the culture of the cotton tree in Europe ascends to 45° north latitude ; in Asia even to Astracan. In the south of Spain and Italy, in Sicily and Greece, as well as all round the Mediterranean Sea, in Syria, Egypt, and Asia Minor, the Gossypium herbaceum is cultivated for its cotton. It is the common plant with white cotton, but in Sicily and Greece, a bush with yellow cotton, probably Gossypium religiosum, is grown ; in these warm countries, however, where the olive and orange thrive so well, but little cotton is cultivated. This branch of culture, on the contrary, is more considerable in Asia Minor, Egypt, in the neighbouring lands of Asia, in China, and in Japan, where it almost always extends to latitudes 40° and 41°. In the account of the distribution of the cotton tree by Royle,* we find that Gossypium vitifolium is cultivated round Cairo, in the West Indies, and in the neighbourhood of Rio de Janeiro, where G. herbaceum and G. barbadense are also found. On Porto Rico Gossypium racemosum is grown, and generally G. hirsutum in the French West Indian possessions.

The number of species of this genus, which furnish cotton, sometimes white, sometimes yellow in colour, is very great, and is still far from having been ascertained. The culture of the cotton is, it is true, more widely diffused than of any other useful plant, yet the single species of Gossypium which are cultivated in different places occupy a less extensive surface. However, we have yet too few data, to permit us to enter upon investigations of this kind. The Gossypium herbaceum is that which in Europe extends furthest north ; in North America the culture of cotton is carried on to the 40th parallel. In South America, according to Aug. St. Hilaire, the culture of this plant extends to the 30th parallel on the eastern coast, and on the west coast of this continent I have seen a few cotton trees in 30° and 33° south latitude, though true plantations are not to be found here, but further north, for example in the province of Copiapó. The most southern stations of the cotton tree are at the Cape of Good Hope, and in New Holland, at the English colonies there.

Von Martius gives an ample account of the mode of

* Illustrations. Fasc. iii.

cultivating the cotton plant in Brazil.* After the ground is prepared, which is generally done by rooting up and burning the trees and bushes, the seeds are sown in the month of January; five, six, or even twelve of them are put in holes three or four inches deep and five or six feet apart. The cotton tree now grows with incredible rapidity, and lasts 12, 15, or 20 years, but it flowers, and ripens seed twice a year; the first crop commences in the 9th or 10th month after sowing. In the province of Pernambuco the principal crop falls in July and August, and the first produce of a cotton tree is the best; the strongest trees then yield 2½ lbs. of clean cotton, the weakest only about five ounces.

When the seed is once sown, the cultivator does not usually trouble himself about the plantation until the time of gathering the crop; however, this negligence is often severely punished, for the mass of weeds amongst the plants becomes so great that they cannot grow. Von Martius† names these weeds; they are Ipomœa quamoclit, *L.*, I. hederacea, *R. Br.*, Momordica macropetala, *Mart.* &c. The industrious countryman roots out the weeds twice a year. At the same time he tries to keep the trees of these plantations about five or six feet high, by breaking off the upright branches.

Gathering the cotton-capsules in large plantations is a severe labour, and a great number of negroes are required for it, as a slave cannot collect more than between one and two arrobas full in a day. But the labour of separating the cotton from the seeds is still worse; at present this, as well as the packing of the great sacks of cotton, is done by machinery.

In China and Japan the cotton culture is in its most flourishing condition; yet the demand for it is far from being supplied, so that immense quantities of this article are exported from the East Indies to China; not to

* Reise nach Brasilien, ii. p. 815.

† L. c. p. 816.

mention the large importations of woven stuffs. In the year 1828 raw cotton, worth 1,322,361 piastres, was imported into China alone.*

The Gossypium which yields the well-known yellow nankin cotton, was formerly considered to be a variety of Gossypium religiosum, to which plant China and Japan were assigned as a native country. A more special investigation, however, showed me that the true nankin cotton is produced by a distinct species, which I have named Gossypium nanking.† It is very interesting to learn, that that Gossypium which Forster brought from the South Sea Islands is not the Gossypium religiosum, but is identical with the true nankin plant.‡

We find ample accounts of the culture of cotton in India in Royle's work, already mentioned. In the northern provinces it is sown in the middle of March and April, and the crop is gathered in October and November when the rains have ceased. In the beginning of February the plant puts forth new leaves and flowers, and during March and April the cotton is gathered a second time. In central India also, to the height of 4000 feet, there are two crops, one after the rains, another in February and March. In Georgia and Guiana the harvest is in September.

It is a well known complaint, that the Indian cotton is short and therefore less fit for using than other sorts, but by Roxburgh's advice the cotton plant of Bourbon has been transplanted to India, and the cotton from it is said to answer all expectations. England annually consumes 300 million lbs. of cotton; of this quantity India furnishes one-twentieth only.

* Meyen's Reise, ii. p. 397.

† Meyen's Reise, ii. p. 397, und Verhandlungen des Vereins zur Beförderung des Gartenbaues in Preussen. Berlin, 1836, xi. 2 Hft.

‡ See Royle's Illustr. Fasc. iii.

## 2. *The Hemp.*

The culture of hemp in Northern Europe, in Asia, and in North America, is of extraordinary importance ; nay, it is incredible what masses of this article are produced in the Russian possessions alone. The plant in these countries is of yet greater importance, as its seeds yield an excellent oil, which is there universally used in cooking. It is true that even at the present time the artificial area of the hemp is becoming more and more enlarged, yet the manufacture of a stronger hemp from various other plants, which I shall presently name, will, perhaps, be at last destructive to our hemp culture.

When speaking of the culture of the banana (page 327), I mentioned the hemp of Manilla, the avaça, and gave an account of its preparation from the Musa textilis, to which I now refer. From the stems of other species of banana also may be obtained very strong fibres, which are chiefly suitable for ropes and fishing lines.

## 3. *The New Zealand Hemp.*

The flax or hemp of New Zealand was made known on Cook's first voyage round the world ; it is prepared from the fibres of the leaves of Phormium tenax, a large and handsome reed. The natives of New Zealand prepared from the leaves not only their whole clothing, but also all kinds of line and ropes ; nay, even fine thread, like silk, may be prepared from this plant.

Cook himself* perceived of what importance it would be to England, if this plant could be introduced and cultivated there, for which the climate seemed to him favourable. It is known that for a long time attempts have been made to introduce the New Zealand flax into England, as well as to the continent, but unfortunately

* Reise um die Welt. Berlin, 1774, ii. p. 34.

all meteorological observations in its native country were wanting, from which it is necessary to set out. Even now I am acquainted with no observations of the kind from New Zealand, but we have the mean temperatures of two places on Van Diemen's Land, an island which lies in the same latitude and under similar conditions, so that we can substitute the temperatures of Van Diemen's Land for the same latitudes in New Zealand.

The temperature of Macquarie Harbour and Hobart Town in Van Diemen's Land is known;* the former possesses a mean annual temperature of 13° Cels., and is situated on the western side of the island, while the latter, on the eastern side, possesses a mean temperature of 11.3° Cels. This great difference is occasioned chiefly by the higher winter temperature, which prevails at Macquarie Harbour, while the temperature of the summer months is nearly the same. We do not indeed know the first causes of this great difference, but in all probability they rest on the prevailing winds. In order to show how exactly the climate of Van Diemen's Land agrees with that in some parts of England, so that there can remain no doubt, that the flax of New Zealand would succeed in England, particularly on the western side, and above all in Ireland, I have compared the variations of temperature at Macquarie Harbour and at Hobart Town with those at London (I am not acquainted with good observations from Ireland), and drawn them on the prefixed Table. From this graphical representation of the mean temperature, it will at once be seen that the New Zealand flax cannot be cultivated in our climate, as the winter cold is much too severe; this plant is, however, now cultivated in the south of France and in Dalmatia.

Phormium tenax has been introduced into New Holland, and around Sydney there are large plantations of

* Edinb. Journ. of Science, 1825, p. 75.

it, which at present produce so much hemp that it has been exported to England.* In the colony excellent whaling-line, for harpooning the whale, is made of it, and in all probability the New Zealand hemp will also make excellent ropes, when once the prejudice against it has been overcome. The leaves of this plant attain a length of six and seven feet, and therefore yield longer fibres than those of our European hemp.

Another kind of hemp is made from the fibres of the agave leaves, which, as we have already seen (page 377), are taken off the maguey plants from which pulque is obtained, and which die down after this operation. The fibres of the agaves are considered the strongest, and have long been an article of commerce. In the East Indies the Agave vivipara is planted as the inclosures of gardens and fields, and from its leaves is obtained a hemp extremely well adapted for making ropes.

We have likewise already mentioned the manufacture of the thick cables which serve to fasten anchors, from the fibrous husk of the cocoa nut. This manufacture is of considerable importance in those parts of the East Indies where the largest plantations of cocoa-palms are to be found. In this case also the fibres are separated from the hard shell by rotting, and from each other by beating, and are then twisted together.

Lastly, I name those plants which in various countries furnish materials for clothing. The paper mulberry tree (Broussonetia papyrifera) is the best known of these; it is chiefly indigenous on the South Sea Islands and in China, and stuffs are made from the fibres of the bark. On the Sandwich Islands I have seen Bœhmeria albida, *Hook*, and Neraudia melastomæfolia, *Gaud*. used for this purpose; we find there large plantations of the former plant. Similar stuffs are also made from the inner bark of the breadfruit-tree, and from the bark of

* See Bennett's Wanderings in New South Wales. London, 1834, i. p. 72, &c.

Aletris nervosa and Celtis orientalis. In Eastern Asia it is chiefly species of Corchorus which are reduced to fibres like the hemp-plant, but furnish much finer cloths. Corchorus olitorius is cultivated in Bengal, C. capsularis principally in China, but also in India, C. japonicus in Japan. There are besides various species of Sida, Hibiscus, and Malva, from which are made cloths in India and on the South Sea Islands.

### 4. *The Culture of the Indigo-plant.*

Although all the species of the genus Indigofera produce indigo, the Indigofera tinctoria is the one chiefly cultivated. Throughout the East Indies, where by far the greatest quantity of this article is produced, is cultivated this species only, which is said to yield a greater proportion of the dye than any other. On the Philippines another Indigofera, perhaps still undetermined, is cultivated with great success; and in America, along with Indigofera tinctoria, several varieties of Indigofera anil are grown.

We have learned the use of the indigo from India. Pliny and Strabo both speak of the beautiful blue dye which is furnished by the Indicum, from which the word indigo has been derived. But yet many other peoples of the Old World were acquainted with the dyeing properties of the species of Indigofera without having ever been in communication with India. Before the discovery of the passage round the Cape of Good Hope, the whole trade in indigo to Europe passed through Aleppo;* and after the discovery of America, this branch of agriculture spread to that continent, and is now carried on at an immense number of points throughout the torrid and sub-tropical zones. The best indigo is made on the west coast of Mexico from the Indigofera argentea.

* J. Phipps, a Series of Treatises on the Principal Products of Bengal. Nro. i. Indigo. Calcutta, 1832.

The dye stuff of Nerium tinctorium, Isatis tinctoria, Galega tinctoria, Spilanthus tinctoria, Amorpha fruticosa, and several other plants, is employed in the adulteration of good indigo.

The indigo-plant requires a very rich and light soil; at least the produce of dye-stuff is greater, the more these demands are answered, even by artificial means. In the East Indies, and, indeed, in all the Indian countries of the Old World, the plant is sown from March to May, and the crop then follows from July to September. These periods occur earlier or later in different districts, but they are regulated so that the crop may be gathered in before the commencement of the rainy season, and, therefore, in many parts the seed time begins as early as November and December.

The whole of the plant is used in the preparation of indigo, and the dye is separated from the surrounding tissue by gentle fermentation. To effect this the indigo-plant is mowed at the time of flowering, and put into large vessels of water where it passes into fermentation; after being several times stirred, all the dye is extracted by the water, which is then poured off into other vessels, where the dye settles as a precipitate at the bottom. The indigo is by no means a product of the fermentation, but it existed already perfectly developed in the plant, in union with mucilaginous, resinous, and different woody substances, the separation of which is the true art of the indigo-maker. It is at first yellow, but becomes blue by contact with the air. Afterwards the moisture is steamed off from the sediment by boiling, which is continued until the mass no longer froths. The mass obtained in this way is put into wooden moulds and pressed into pieces such as come into trade; these pieces are then allowed to become perfectly dry, and lastly are packed.

One may form some idea of the enormous extent of this so important branch of culture from the fact that the importation of indigo into England from the English colonies amounts to six and a half million lbs., of which more than 2,000,000 lbs. are used in England, while the remaining 4,000,000 lbs. go to the Continent. The average prices of indigo are from 1 to 3½ dollars the lb., according to its quality.

Besides this the export of indigo from the earlier Spanish dominions in America, as well as from the southern provinces of North America, and from the possessions of the Dutch, Spanish, and Portuguese in India, is very considerable. From Manilla alone, in the last few years, on an average more than a million lbs. have been exported,* so that if sufficient data had been collected, we should probably find that more than nine or ten millions of lbs. of this article annually come into trade.

Adanson gives us a very interesting account of the preparation of dye stuff from the indigo-plant by the negroes of Senegal. The negroes pluck the leaves of the plant and pound them in mortars to a paste, which they press into lumps and allow to dry. When about to use the dye, some of this paste is dissolved in a ley of the ashes of Sesuvium portulacastrum *L*, and the solution immediately assumes its blue colour.

* See Meyen's Reise. Theil ii. p. 276.

THE END.

# LIST OF ERRATA.

*Page* 7, *line* 1, *for* allgemeiner *read* allgemeinen.
... 16, ... 17, *for* that *read* than.
... 18, ... 22, *for* minimum *read* summer.
... 48, ... 13, *for* Jungermanniæ *read* Jungermanniaceæ.
... 54, ... 33, *for* Nymphæ *read* Nymphæaceæ.
... 55, ... 10, *for* Ocillatorieæ *read* Oscillatorieæ.
... 63, ... 5, *for* Gysophila *read* Gypsophila.
... 70, ... 31, *for* Cryptogamiæ *read* Cryptogamia.
... 84, ... 33, *for* Eucalyptæ *read* Eucalypti.
... 92, ... 21, *for* Crytogamiæ *read* Cryptogamia.
... 103, ... 18 *and* 28, *for* Eucalyptæ *read* Eucalypti.
... 131, ... 21, *for* Vacciniæ *read* Vaccinia.
... 243, ... 4 *from bottom*, *for* Rheaiæ *read* Rhexiæ.

# INDEX.


Abies, 130.
Abietinæ, 104, 130.
    region of, 243.
Acacia, 103, 128.
Acaciæ, 103, 128, 170.
Acorus calamus, 315.
Acrocomia sclerocarpa, 165.
Actiniæ, 52.
Adansonia digitata, 162.
Ærides, 155.
Æschinomene grandiflora, 354.
Africa, northern, vine in, 367.
    north of, 336.
    southern, extremity of, 185.
    south of, 150.
Agave, 376.
    form, 118.
    sap, 378.
    vivipara, 406.
Agriculture, history of, 291.
Agrostis arundinacea, 213.
Aira cæspitosa, 213.
Albania, forest of oaks, 347.
Alder, 138.
Aletris neroosa, 407.
Alexandria, soda of, 49.
Alfalfa, 59.
Alfonsia, 331.
Algæ, 51.
    as articles of food, 214.
Alnus, 203.
    viridis, 40.
Aloe plants, 120.
    vulgaris, 120.
    macra, 120.
    dichotoma, 121.
Alpinia, 113.
Alpine plants, region of, 250, 252.
    vegetation, 85.
Alps, Swiss vegetation of, 247.
Ambauba, 170.
Amomum, 113.
Amorpha fruticosa, 408.
Amsterdam, 388.
Anabasis, 59.
Anacardiæ, 162.
Ananas, form, 116.
Anemones, sea, 52.
Andiræ, 170.
Andromeda polifolia, 63.
Angelica, 203.
Anonæ, 162.
Anona squamosa, 351.
Anonas, 344.
Antipates, 52.
Anlochthones, 266.
Apennines, vegetation of the, 246.
Apple of Paradise, 325.
Aquatic plants, 50.
Arabia, 120, 337.
Arabis Halleri, 41.
Aracacha, 320.
Aralia, 152.
Araras, 170.
Araucaria, 104, 132, 348.
Areca catechu, 351.
    palm, 351.
Arequipa, 2, 366.
    valley of, 319.
    volcano of, 31, 101.
Arica, 366.
Aristolochiæ, 152, 165, 166.
Aron, 314,
Aroideæ, 153.
Aroyclea, 146.
Arrack, 302.
Arrudææ, 165.
Artocarpi, 162.
Artocarpus incisa, 141, 321.
Arum, 154, 314.


2 B

Arum macrorrhujon, 314.
Arundo arenaria, 62, 214.
 Donat, 107.
 phragmites, 57.
 tibialis, 354.
Ash, 210.
Asia, 404.
 eastern, 301.
 interior of, 310.
Aster subulatus, 60.
Astragali, 129.
Atobapo, 167.
Atmosphere, moisture of, 38.
Atyee wine, 377.
Avaca, 327, 404.
Averrhoa Belimbi, 351.
Avicenniæ, 168.
Axes, 320.
Ayalea, 203.
Azores, number of species in the, 262.

Baeckiæ, 136.
Bambusaceæ, 108, 109.
Banana, 9, 111, 324, 404.
Bananas, region of, 229.
Banaria, 325.
Banisteria, 152, 165.
Baobab, 162.
Barley, 296.
Barringtoniæ, 59, 138.
Bartsia, 203.
Batata, 318.
Batavia, 336.
Bauhimæ, 152, 165.
Begaria ledifolia, 248.
Behring's Straits, 22.
Bengal, 387, 407.
Bentang, island of, 355.
Berlin, temperature of, 19.
Bertholletiæ, 162.
Bertholletia excelsa, 350.
Betel nut, 351.
 pepper, 353.
 rolls, preparation of, 352.
Betula alba, 82.
Bignonia, 152.
Birch, 138.
Bœhmeria albida, 406.
Bokhara, vine in, 375.
Bolat, 146.
Bolivia, climate of, 11.
Borago officinalis, 75, 77.
Borneo, 45, 340.
Borassus flabelliformis, 342.
Botanical geography, works on, 6.
Brandy of Pisco, 371.
Brassica, 329.
 orientalis, 388.
Braza alpina, 93.
Brazil, 155, 165, 298-300, 318, 331, 384, 394, 402.
 forests of, 165, 169.
 vegetation of, 61.
Brazilian nut, 349.
Bread fruit, 321.
 tree, 141.
Broihan, 305.
Bromelia ananas, 116.
 pinguia, 117.
Broussonetia, 379.
 papyrifera, 406.
Bryoniæ, 151.
Buckwheat, 309.
Buenos Ayres, 371.
Butomus umbellatus, 57.

Cactaceæ, 5, 47, 99.
Cacti singular, 86.
Cactus ficus indica, 147.
 form, 146.
Cadiz, 337.
Cæsalpiniæ, 162.
Cairo, 401.
Caladium, 314.
 arborescens, 154.
 esculentum, 299.
Calami, 71.
Calamus, 121, 152.
Calandrina, 41, 254.
Calceolaria, 41, 94.
Calcutta, climate of, 18.
California, 302.
 New, 330.
Calla, 203.
 æthiopica, 154.
 palustris, 154.
Callitriche, 213.
Caltha palustris, 60.
Camburi, 324.
Camelliæ, 94.
Camola, 318.
Campalius, 121.
Campanula, 203.
Canna, 113.
Cannabis sativa, 400.
Canada, climate of, 21.
Canary Isles, 103, 149, 238, 300, 329, 376.

Canary Isles, vegetation of, 176.
number of species in, 261.
Canen, 305.
Canneæ, 111.
Canta, 305.
Canton, 9, 10, 18, 91, 131, 344, 394.
price of tea at, 395.
sale of opium, 360.
Cape of Good Hope wine, 94, 135, 376, 388.
Cape Horn, 4.
heat of, 12.
Carabatos, 170.
Caraccas, tobacco of, 362.
Carex, 108.
arenaria, 62.
Caryotæ, 162.
Caschu, 354.
Cashmere, 344.
valley of, 2.
Caspian Sea, 51.
Cassava, 317.
Castanea vesca, 346.
Castor oil, 329.
Casuarina, 133.
Catania, vine in, 370.
Catappa fruit, 348.
Catechu, 356.
Catingas, 169.
Caucasus, 346.
Cazavi, 317.
Cecropia palmata, 364.
pellata, 170.
Cedrela odorata, 43.
Celtis orientalis, 407.
Cenomyce rangiferina, 217.
Centaurea cyanus, 76.
Cereals, 91, 107.
culture of, 290.
Cerei, 143.
Cereum, 340.
Ceropegia, 148.
Cerros de Guayanna, 350.
Cevennes, 329, 346.
Ceylon, 332, 387.
forests of, 325.
Chamæcrops, 121, 123.
humilis, 123.
Chancay, 312.
Chara vulgaris, 213.
hispida, 213.
Charæ, 49, 57.
Charpentiera obovata, 173.
Chenopodium quinoa, 3, 308.
vulgare, 75.
Chesta de Zapata, 239.
Chesnut, 138, 346.
Chicha, 305.
Chili, 84, 94, 109, 122, 128, 143, 198, 302, 309.
vegetation of, 189.
climate of, 11.
flora of southern part, 198,
vines in, 372.
use of grape in, 369.
Chilian palm, 338.
China, 9, 298, 303, 314, 320, 329, 334, 344, 346, 351, 382, 385, 387, 400.
Cochin, 326, 393.
cotton in, 402.
grapes grown in, 374.
salt of, 110, 319.
southern, 344.
temperature of, 10.
tobacco, consumption in, 361.
vine in, 368.
Chinese tea, 385.
use of opium, 357.
Chorisia ventricosa, 170.
Cholula, 377.
Chuquito, 363.
plain of, 96.
plateau of, 249.
Cichorium Intybus, 77.
Cinchona, 83.
lancifolia, 243.
Circæa alpina, 40.
Cissus, 152.
Cistus plants, 104.
ladaniferus, 193.
Climbing plants, 151.
Clove pinks, 91.
Clusiæ, 166.
Clusia alba, 165.
Cobi, 392.
Cochlearia danica, 203.
Cochineal insect, 147.
Cocoa, 331.
palm, 123, 331.
plant, 362.
nut oil, 333.
Cocos butyracea, 342.

Cocos chilensis, 123.
nucifera, 331, 341.
Coffea arabica, 383.
Coffee, cultivation of, 384.
culture of, 303.
Colletiæ, 239.
Compositæ, 340.
Conception, 361.
vines at, 371.
Conferva rivularis, 58.
glomerata, 50.
fenestrata, 75.
dendrita, 75.
Conium maculatum, 320.
Convallaria majalis, 35.
Convolvulus arvensis, 76
Batatas, 318.
sepium, 157.
Copaiferæ, 170.
Copiapo, 109.
valley of, 85.
Corallinea, 52.
Corchorus olatorius, 407.
japonicus, 407.
capsularis, 407.
Corcovado, woods of, 384.
Cordilleras, 2, 10, 48, 101, 305, 339.
of Mexico, 320.
South American, 311.
Cordon, 178.
Corea, 387.
Cornus succica, 203.
Corylus avellana, 347.
Corypha umbraculifera, 340.
Cotton capsules, 402.
plant, 400.
trees, 162.
yellow, 401.
Cratægus, 203.
Crete, use of grapes in, 309.
Crimea, 329.
Crinum, 150.
Cryptogamia, 47, 92, 264.
Cuba, 332, 380.
culture of tobacco in, 362.
Cumana, 371.
Cycas circinalis, 340.
revoluta, 340.
Cymbidium, 155, 203.
Cyperaceæ, 5, 108, 284.
Cypripedium calceolus, 50, 62, 155.
Cyperoideæ, 108.
Cyperus, 108.
polystachys, 93.
Cyrenaica, 368.

Dahue, 309.
Date, 336.
palm, 123.
white variety, 336.
Dauria, 393.
De Candolle, 158.
Dendrobium, 155.
Denmark, 3.
Diatomeæ, 50.
Dicranum glaucum, 15, 80, 156.
Dicotyledonous trees, 138.
Dioscorea alata, 320.
Dodoneæ, 168.
Dodecatheon, 203.
Dominico, 324.
Draba verna, 213.
Dracaena, 113, 173, 197.
australis, 198.
draco, 114.
Dracontium polyphyllum, 321.
perfusum, 154.
Dragon tree, 114.
Drimys chilensis, 239.

East Indies, 3, 306.
East coast of North America, climate of, 23.
East and west coasts, difference in climate of, 24.
Ebenaceæ, 162.
Echinocactus, 145.
Egypt, 337.
deserts of, 60.
Elais guineensis, 341.
Elm, 138.
Elymus, 203.
England, 272.
Enontekis, temperature of, 17.
Ephedra americana, 239.
Epidendrum, 155.
Epilobium, 81.
Equatorial zone, plants of, 162.
Ericeæ, 94.
Erica vulgaris, 95.
Erigeron canadensis, 40.
Eriocaulon, 110.
septangulare, 43.
Eriophora, 110.

Erythrina indica, 354, 397.
corallodendron, 396.
Erythroxylum coca, 364.
Escalloniæ, 239.
Etna, 337.
Eucalyptus, 137.
Eugenia, 59, 175.
Euphorbia, 148.
balsamifera, 177.
canariensis, 178.
Euphorbiæ, 100.
Europe Southern, 328.
Exanthema of plants, 72.

Fagus sylvatica, 96.
Falkland Islands, flora of the, 204.
Fan palms, 122, 338.
Faroe Islands, vegetation of, 212, 311.
Ferns, the form, 82, 125, 126, 275.
Field plants, 77.
Ficoideæ, 149.
Fig trees, 175.
Figs, region of, 231.
Flax, 400.
Fragosa, 146.
France, 272-3.
mountain plants of, 286.
vegetation of, 94.
vine in, 368.
Friendly Isles, 222.
Fritillaria, 150, 203.
Fucus benjamina, 230.
natans, 53.
pyriferus, 51.
Sargasso, 53.
Fuci, 51, 87.
Fungi social, 87.
Fungus singular, 101.
Furcrœa, 118.
longæva, 119.
gigantea, 119.

Galega tinctoria, 408.
Galinsogea parviflora, 40.
Galium boreale, 203.
Gambir extract, 355.
Gerardia maritima, 60.
Germany, 272, 313, 346.
climate of, 10.
vine in, 368.
vegetation of, 201.
Gentianæ, 5.
Gentiana uliginosa, 26.
Georgia, 403.
Geum intermedium, 203.
Gladiolus, 150.
Glaux maritima, 50, 59.
Gleditschia, 128.
triacanthos, 24.
Glyceria fluitans, 307.
Glychirrhiza hirsuta, 61.
Grapes, uses of, 369.
Grasses, 106, 107.
Greece, 104, 120, 346.
Grenada, 39.
Grusia, western, 346.
Gorgonia, 52.
Gossypium barbadense, 401.
herbaceum, 400, 401.
racemosum, 401.
religiosum, 401.
vitæfolium, 401.
Guaivrare, 167.
Guarauni, 339.
Guavas, 138.
Guiana, 403.
Guilandina moringa, 354.
Guinea oil palm, 341.
Gum cistus, 193.
Gynerium Neesii, 109.
speciosum, 109.
Gyrophora, 156.

Hæmodorum, 321.
Havannah, 18.
tobacco, 361.
Hawaii, 18.
Harz mountains, 40.
Havettiæ, 165.
Hayti, 361.
Hazel nut, 347.
Heaths, vegetation of, 78.
Hedera helix, 74.
Helena St., 388.
Hemerocallis, 150.
Hemp, 400, 404.
New Zealand, 404.
Heracleum, 203.
Herniaria glabra, 62.
Hibiscus, 140.
Himalayas, 30, 90, 309.
vegetation of, 90.
climate of, 31.
Hindostan plains of, 41.
Hobart Town, 405.
Hop, 151.

Huallaga river, 365.
Hydrocharis morsus ranæ, 58.
Hymenophyllum, 163.

Iceland, 213, 311.
  climate of, 33.
  moss, 214.
  vegetation of, 212.
Igname, 320.
Imeriti forests of, 368.
India, 108, 298, 303, 311, 326, 329, 334, 344, 355, 382, 400, 403, 407.
  damp coasts of, 332.
  vegetation of, 110.
Indian empire, 394.
  archipelago, 298, 321.
  East, isles, 127.
  islands. 340.
  West, isles, 331.
  Ocean, 332.
Indies, East, 91, 306, 314, 319, 329, 340, 343, 351, 395, 406.
Indigo, 408.
  plant, 407.
Indigofera tinctoria, 407.
Indicum, 407.
Ipomæa quamoclit, 402.
  hederacea, 402.
  tuberosa, 318.
Iris, 150.
Isaria, 73.
Isariæ, growth of, 268.
Isatis tinctoria, 408.
Isis, 52.
Italy, 346.
Ixia, 150.

Jagua palms, 122.
Jalapa, vegetation of, 236.
Jambosa malaccensis, 173.
Japan, 303, 393.
  flora of, 194.
  preparation of tea in, 389.
  cotton in, 402.
  tea in, 387.
Japanese mustard, 388.
Jatropha manihot, 317.
Java, 45, 91, 114, 299, 300, 320, 326, 340, 342, 355, 384, 387.
  vine in, 375.
  Isle of, 91.
Juan Fernandez, 338.
Juca dulce, 317.
Juca amarga, 317,
Juncus bufonius, 60.
Juncoideæ, 110.
Jungermanniæ, 72, 173.
Juvia tree, 349.
Juvias, 348.

Kamschatka, 134.
  flora of, 215.
Kaorai, 385.
Kiachta, 393.
Kunawar, grapes in, 375.

Lachrymæ Christi, 370.
Ladanz, 300.
Laguna de Baz, 56, 301.
La Maire, straits of, 51.
Laminariæ, 51.
La Paz, 363.
Lapland, 2, 3, 14, 72, 311.
Larix, 130.
Lathræa stelleri, 203.
Laurel, 138.
Laurelia serrata, 239.
Laurineæ, 103, 162.
  region of, 234.
Laurus aromatica, 239.
  permio, 239.
Lavatera, 140.
Lecanora muralis, 74.
Lecythidiæ, 162.
Leistenwein, 369.
Lemna, 55, 92.
  minor, 56, 92.
  trisulca, 56.
Leptospermum, 136.
Lianas, 151.
Lichens, 48, 92, 156.
Lily form, 149.
  of the Nile, 154.
Lilium, 150.
Lime, 138.
Limosella aquatica, 60.
Limousin, 346.
Linnæa alpina, 40.
Liniræ borealis, 80, 93, 203.
Linum usitatissimum, 400.
Liquidambar, 235.
Lithuania, plants of, 26.
Littoral plants, 49.
Llanura de Ranegua, 143.
Lobelia dortmanna, 94.
Lolium temulentum, 76.

Lonicera xylosteum, 151,
Loranthus, 69, 164.
aphyllus, 143.
Los Attos de Toledo, 31.
Luçon, 56, 300, 326, 334.
Island of, 165.
Lycopodia, 146.
Lygodiun, 167.

Macao, 9, 18, 39, 131,. 344.
Macassar, 340.
Macauba palm, 165.
Macquarie Harbour, 405.
Madagascar, 112.
Madeira, flora of, 179.
Madrepores, 52.
Magdalena, 83.
Maglia, 312.
Magnolias, 195.
Magellan, Straits of, 51.
climate of, 33.
flora of, 208.
Maguey plant, 376.
Maipu, volcano of, 101.
Maize, 95, 302.
stalk of the, 306.
fermented liquors from, 305.
Malabar, 342, 395.
Malacca, 356.
Malays, 360.
Malpighiaceæ, 162.
Mammillaria, 145.
Mango, 343.
Mandiocca, 316.
Manioc, 316.
Manioca, 318.
Manilla, 302, 320, 351, 404, 409.
cigars, 362,
Mangrove, 61, 168.
Manicaria saccifera, 122.
Manihot utilissima, 317.
cirpi, 317.
Maranhaõ, 298.
Marian Isles, 303.
Marcgraavia, 165.
Marine plants, 51.
Mauritius palm, 338.
Maypu, volcano of, 66.
Meadow plants, 77.
Medicago sativa, 95.
Mediterranean, vegetation of the coasts of the, 192.
Melaleuca, 136.
leucodendrum, 175.
Meliaceæ, 162.
Melocactus, 145.
Melville's Island, 220.
Mesembryanthema, 103.
Metrosideros polymorpha, 173.
Mexico, 119, 147, 318, 330, 377, 381.
vines in, 372.
flora of, 238.
maize in, 304.
Mexican wine, 379.
Millet, 306.
Turkey, 306
Mimosa form, the, 127.
Mingreli, 368.
Minz, 386.
Missouri, forests of, 196.
Moisture of atmosphere, 38.
Molinæa microcoocos, 123, 338.
Moluccas, 321, 348.
Momordica macropetala, 402.
Mongolia, 393.
Monkey bread tree, 162.
Montia fontana, 58.
Moquegua, 371.
Morina citrifolia, 397.
Moscow, 393.
climate of, 24.
vine at, 370.
Mosses, 48, 92.
Moss reindeer, 217.
form, 156.
Mulberry tree, 406.
Musæ, 324.
Musaceæ, 111.
Musa paradisiaca, 324.
sapientum, 324.
Myrtaceæ, region of, 234.
Myrtle, 9, 13.
Myrtles, 239.
trees, 136.
Myrtus, 136.
communis, 136.

Nasturtium, 60.
palustre, 60.
Nauclea gambia, 355.
aculeata, 355.
New Holland, extent of, 83, 84, 103, 128, 131, 405.
forests of, 133, 183.

New Zealand, 121, 196, 319.
flora of, 196.
hemp, 405.
Negro corn, 306.
Nelumbium speciosum, 9.
Nepaul, 394.
Neraudia melastomæfolia, 406.
Nerium tinctorum, 408.
Nicotiana, 361.
Nigra, 121.
palm, 123, 342.
Norantea, 165.
North America, 60, 404.
climate of, 23.
sub-tropical zone, 182.
southern states of, 298.
North Pacific Ocean, coasts of, 203.
Norway, flora of, 212.
Nostoc, 57.
Nymphesæ, 54, 57.

Oahu, 109, 316.
Oats, 296.
Oak, 138.
of Jupiter, 347.
Oaks, 347.
fruit of, 347.
Oca, 320.
Ochroma, 140.
Œnothera biennis, 40.
Olea fragrans, 392.
Europea, 328.
Olive, 138, 328.
Olives, Mount of, 331
Ophrys, 155.
Opium, culture of, 357.
Turkish, 357.
Indian, 358.
Opuntia tunas, 147.
Orange, 9, 343.
Orchideæ form, the, 155.
Origanum vulgare, 79, 90.
Orinoco, mouths of the, 338
Orizaba, volcano of, 312.
peak of, plants at the base, 244.
Oryza sativa, 297.
Oryzeæ, 108.
Oscillatoria, 54.
flos aquæ, 57.
Otaheite sugar-cane, 381.
Oxalis tuberosa, 320.

Paddee, 301.
Pāddih, 301.
Palms, 121.
cane, 121.
jagua, 122.
fan, 122, 338.
date, 123.
cocoa, 123.
macauba, 165.
of Thebes, 124.
region of, 229.
Chilian, 338.
Guinea oil, 341.
wine, 342.
areca, 351,
Mauritius, 338,
Panax horridum, 203.
Pancreatium, 150.
Pandanus, 55, 59.
forms of, 113.
odoratissimus, 115.
fruit of, 115.
Pandani, 114.
Paplionaceæ, 129.
Parasites, 67.
true, 69.
false, 69.
Paraquitos, 170.
Parmelia parietina, 74.
Passiflora, 152, 165.
Paullinia, 152, 165.
Pedicularis, 203.
Peplis portula, 60.
Pepper, 395.
white, 398.
Pernambuco, 402.
Persia, 336.
grapes in, 369 376.
Peru, 304, 364.
southern, 95, 101, 308.
climate of, 11.
western coast of, 330.
Peruvian coca, 362.
Petro Paul's Haven, 202.
Phanerogamiæ, 92.
Phellandrium aquaticum, 58.
Philippines 2, 114, 163, 301, 324, 342, 351, 357, 407.
culture of tobacco in, 362.
Phœnix dactylifera, 336.
sylvestris, 342.
Phormium tenax, 405.

Phytelephas 113.
Pigafetta, 351.
Pihring, 299.
Pinus, 130.
cembra, 347.
pinea, 347.
Pine, 210.
apple, 116, 344.
seeds, 347.
Pinguicula alpina, 64, 209.
Piper nigrum, 395.
Piperaceæ, 173.
Pisco, wine of, 371.
Pistia, 56.
stratiotes, 93.
Pisum maritimum, 203, 214.
Pitcarniæ, 164.
Plants, aquatic, 50.
marine, 51.
fresh-water, 54.
floating, 56.
fountain, 58.
shore, 60.
land, 62.
sand, 62.
chalk, 62.
bog, 64.
rock, 65.
gravel, 66.
cultivated, 75, 290.
field, 75, 77.
mountain, 79.
bush, 79.
forest, 79.
wall, 73.
social, 80.
board, 74.
fallow, 76.
pasture, 75,
rubbish, 74.
meadow, 77.
statistics of, 259.
Plantago, 203.
Platano, 324, 325.
Harton, 324.
Plantain, 324.
Poa amnia, 82.
thalassica, 61.
pratensis, 213.
Poe, 316.
Pogni, 312.
Polypodiaceæ, 163.
Polygonum aviculare, 82.
fagopyrum, 309.
tartaricum, 310.
Pomegranate, 9.
Pondicherry, 375.
Pontederia, 58.
Poplar, 138, 183.
Poppy oil, 360.
Portugal, 104, 306, 347.
climate of, 13.
vine in, 367.
Potato, 3, 310.
Potamagetons, 54, 57.
Pothos plants, 70, 153.
Prosopis, 128.
Prunella vulgaris, 90.
officinalis, 213.
Prussia, 311.
east, climate of, 374.
Pteris aquilina, 126.
Pulque, 377.
de Mahio, 306.
Pyrolæ, 80.

Quercus ægilops, 347.
ballote, 347.
Quillota, 371.
Quinoa, 3, 308.

Racodium cellare, 75.
Radishes, 91.
Ranunculus acris, 203.
arvensis, 90.
fluviatilis, 58.
Raphanus, 329.
Rasamala woods, 235.
Rattan canes, 165.
Restiaceæ, 110.
Rhine, 303.
cultivation of the vine on the, 372.
Rhio, 356.
Rhipsalis, 148.
Rhizophora, 61.
Rhododendrons, 131.
Rhododendron ferrugineum, 40.
region of, 247.
Rhone, valley of the, 329.
Ribes, 203.
Rice, 297.
marsh, 299.
mountain, 300.

Ricinus, 329.
Rinz, 312.
Rio Janeiro, 91, 171, 297, 388, 401.
Madeira, 297.
Rome, 39.
Rubus odoratus, 203.
spectabilis, 203.
Rumex acetosella, 76.
Ruyschia, 165.
Rye, 296.

Saccharineæ, 108, 109.
Sagittaria sagittata, 320.
Sagittariæ, 57.
Sago palm, 340.
Sagus Rumphii, 340.
Salicornia, 59.
Salix Babylonica, 139.
Salsola, 59.
glomerata, 61.
kali, 49.
Samdschu, 302.
Samolus valerandi, 49.
Sandwich Isles, 113, 114, 172, 173, 314, 319, 382, 406.
forests of, 172, 325.
San Fernando, 92.
Sapindæ, 162.
Sarcostemma, 148.
Sargasso Sea, 53.
Sargassum vulgare, 53.
Savannahs, 77.
Swauhr, 299.
Saxifraga rivularis, 66.
Saxony, 311.
Schinus, 169.
Schow, division of vegetation, 158.
Scirpus, 110.
lacustris, 57.
palustris, 57.
Scitamineæ, 111.
Sea anemones, 52.
Sedum acre, 62.
rupestre, 65.
telephium, 74.
Semperviveæ, 149.
Sempervivum, 103.
Senecio viscosus, 75.
Senegal, 409.
Serratula arvensis, 213.
Sesamum orientale, 329.
Sesuvium portalacastrum, 409.
Siberia, forests of, 215.
southern, 310.
vegetation of, 16.
vine in, 374.
Siberian stone pine, 347.
Sicily, 149, 337, 380, 401.
climate of, 245.
Sida, 66.
Sitka, forests of, 203.
Skye, Isle of, 43.
Smegdadermos, 239.
Snow limit, height of, 31.
Society Islands, 321, 351.
Solanum tuberosum, 310.
Söndmor, temperature of, 19.
Sophoræ, 127.
Sorbus, 203.
South America, 94, 306, 339, 350.
climate of, 11.
difference of climate of east and west coasts, 25.
plains of, 78.
west coast of, 87.
South Sea Islands, 2, 55, 82, 123, 127, 140, 316, 325, 332, 379.
sugar-cane of the, 382.
Spain, 104, 337, 347, 400.
Spartina glabra, 60.
Spergula arvensis, 76.
Sphagna, 63.
Sphagnum, 156.
palustris, 80.
Spilanthus tinctoria, 408.
Stapeliæ, 148.
Statice tartarica, 61.
Statistics of, 259.
Stelis, 155.
Sterculia, 140.
Sterculiæ, 162.
Stockholm, climate of, 16.
Stone pine, 347.
Stratiotes, 55.
aloides, 58.
Struthiopteris germanica, 126.
St. Bernard, 28.
St. Gothard, climate of, 17.
St. Helena, 263, 376, 380.
climate of, 70.
Sub-tropical zone, vegetation of, 176.
Succulent plants, 148.
Succus catechu, 354.

Sugar-cane, 333, 380.
  of the East Indies, 382.
Sumatra, 114, 299, 300, 320, 340, 351, 355, 388, 396.
  vine in, 375.
Sunda Isles, 396.
Surinam, 91.
Sweden, 272.
Swieteniæ, 152.
Switzerland, 26, 272, 346, 347.
  climate of, 33.
Sydney, 184.
Syngenesia, 40, 252.

Tacca pinnatifida, 321.
Tacna, 366.
Tapioca, 317.
Taro, 315, 316.
Tarro, 299.
Taxus, 130.
Tea, Chinese, 385.
  preparation of, 389.
  sorts of, 392.
Temperate zone, colder, 199.
  warmer, 190.
  forests of the, 191.
Teneriffe, peak of, 245.
  plants on, 239, 245.
Terminalia catappa, 348.
Terra del Fuego, 209.
  vegetation of, 209.
Terra japonica, 354.
Thea chinensis, 385.
  viridis, 385.
  bohea, 385.
  stricta, 385.
  Ava, 385.
Thibet, 294.
  tea in, 385.
Thlaspi brusa pastoris, 213.
Tillandsiæ, 117, 164.
Tillandsia, usneoides, 117.
Titicaca, Isle of, 304.
  Lake of, 27, 55, 96, 146, 309, 312.
Tobacco, 361.
  Havannah, 361.
  South American, 361.
Toluca, maguey plantations at, 377.
Tonquin, 385.
Tournefortiæ, 168.
Trapæ, 57.
Trapa bispinosa, 344.
Trapa bicornis, 344.
Tree ferns, 126.
  region of, 231.
Trichomanes, 163.
Trifolium arvensis, 213.
Triglochin, 203.
Triticum sativum, 292.
Tropical zone, plants of, 171.
Tschha, 386.
Tulipa, 150.
Tunas, 147.
Turkey millet, 306.
Turks, 357.
Turritis hirsuta, 203.
  glabra, 203.
Tyrcua, 150.
Tyrol, 346, 347.

Ubi, Island of, 326
Urania, 111.
  amazonia, 112.
Urtica dioica, 77.
  urens, 76.
Usneæ, 70.
Utriculariæ, 55, 57.

Vaccinium oxycoccus, 63.
Valparaiso, 38.
Van Diemen's Land, 405.
  flora of, 198.
Vanilla, 155.
Vegetation, divisions of, 161.
  regions of, 223.
Venezuela, 298, 332, 362.
Verbena, 146.
Veronica anagallis, 58.
  beccabunga, 58.
Vine, 152, 367.
  locality of the, 91.
  native country of the, 194.
  temperature for the, 370.
Virginia, 313.
Viscum, 69.
Vitex 152.
Vitis vinifera, 367.

Walnut tree, 345.
Water nut, 344.
Weeds, garden, 76.
West coast of North America, climate of, 23.
West Indies, 298.
Wheat, 292.

Willows, 183.
Willow, 139.
weeping, 139.
Willdenow, division of vegetation, 157.
Winds on the ocean, 42.
Wine palm, 342.
Wines, difference in, 369.
Wolga, 201.
Works on botanical geography, 6.

Yam, 320.
Yucca, 149.
Yuccæ, 119.

Zambitas, 363.
Zamiæ, 125.
Zea Mays, 302.
Zizania aquatica, 297.
Zone, arctic flora of, 216.
polar flora of, 221.
sub-arctic, 209.
sub-tropical, 176.
Zosteræ, 51.

THE END.

# RAY SOCIETY.

## LIST OF SUBSCRIBERS.

| | |
|---|---|
| ABERDEEN . *Local Sec.* | DICKIE, GEORGE, M.D. |
| | Dallachay, Mr. John |
| | Dyce, Dr. |
| | Gordon, P. L. esq. Craig Myle |
| | Lee, W. S. esq. Morningside |
| | Leslie, James, esq. Chanonry |
| | Lizars, Professor A. J. M.D. |
| | Mitchell, Alexander, esq. |
| | Ogilvie, J.F. esq. Morningside Lunatic Asylum |
| | Rattray, W. esq. |
| ABINGDON . . . . | Tomkins, C. M.D. |
| ALNWICK . . . . | Embleton, R. esq. |
| | Tate, George, esq. F.G.S. |
| ALTON, *Hants.* . . | Curtis, W. jun. esq. |
| ARMAGH, *Ireland* . . | Robinson, Rev. George |
| AYLESBURY, *Bucks* . . | Lee, Dr. F.R.S. F.R.A.S. &c. Hartwell |
| | Reade, Rev. J. B. Stone Vicarage |
| BARTON-UNDER-NEEDWOOD, *Lichfield* . . . | Conway, P. L. esq. |
| BATH . . *Local Sec.* | GEORGE, RICHARD F. esq. |
| | Broderick, the Honorable Miss Charlotte |
| | Pearce, J. C. esq. Bradford |
| | Fox, G. T. esq. F.R.S. G.S. |
| BECCLES, *Suffolk*, *Local Sec.* | DAVEY, H. esq. |
| BELFAST . . *Local Sec.* | THOMPSON, W. esq. Pres. Nat. Hist. Soc. |
| | Hyndmam, George C. esq. |
| | Natural History Society Library |
| | Patterson, Robert, esq. |
| | Whitla, F. esq. |
| BELFORD . . . . | Broderick, W. esq. |
| | Clark, Rev. J. D. |
| BERNE . . . . | Shuttleworth, Robert James, esq. |
| BERWICK-UPON-TWEED, HON. SEC. . . | JOHNSTON, GEORGE, M.D. LL.D. F.R.C.S.E. |
| | Clarke, H. G. M.D. |
| | Clay, Mrs. P. Newwater Haugh |
| | Fyfe, W. W. esq. |
| | Tancred, Sir Thomas, Bart. Shotley hill. |
| | Weddell, Robert, esq. |
| BEVERLEY, *Yorks.* *Local Sec.* | MORRIS, BEVERLEY, R. A.B. M.D. |
| | Boulton, R. G. esq. M.R.C.S. |
| | Sandwith, T. M.D. |
| BEXLEY, *Kent* . . . | Cottingham, E. esq. |
| BIRMINGHAM . *Local Sec.* | PERCY, J. M.D. |
| | Bindley, S. A. esq. |
| | Blount, J. H. esq. |

| | |
|---|---|
| BIRMINGHAM . . . | Buckman, J. esq. |
| | Dawes, J. S. esq. F.G.S. West Bromwich |
| | Fletcher, Bell, M.D. |
| | Hodson, Joseph, esq. |
| | Knowles, Professor C. B. |
| | Lee, Rev. James P. |
| | Palmer, Dr. S. |
| | Russell, James, M.D. |
| | Wickenden, Joseph, esq. |
| BLACKHEATH . . . | Carr, W. esq. |
| BOSTON, *Lincolnshire* | Rawson, J. esq. Skerbeck |
| | Reckitt, Wm. esq. |
| | Shaw, Dr. Hophouse |
| BOLTON-LE-MOORS, *Lancash.* | Sharp, Henry, esq. |
| BOULTIBROOKE, *Presteign, Radnorshire* . . | Brydges, J. H. esq. F.L.S. |
| BRADFORD, *Yorkshire* . | Birkbeck, Morris, esq. |
| | Kay, David, M.D. |
| | Hailstone, S. esq. F.L.S. Horton Hall |
| BRIDLINGTON QUAY, *Yorks.* | Strickland, Arthur, esq. |
| BRIGHTON . . . | Winter, T. B. esq. |
| | Weeks, R. esq. F.L.S. Hurst Pierpont |
| BRISTOL . . *Local Sec.* | W. BUDD, M.D. |
| | Braikenridge, G.W. esq. F.S.A.G.S. Brislington. |
| | Bompas, C. S. esq. |
| | Broome, C. E. esq. M.A. Clifton |
| | Fry, J. S. esq. |
| | Howell, R. M.D. F.R.S.E. Clifton |
| | Hetling, G. H. esq. |
| | Knapp, A J. esq. Clifton |
| | Library Society |
| | Microscopical Society |
| | Mordaunt, John, esq. Ashton Water |
| | Ormerod, W. esq. |
| | Reading and Conversational Society |
| | Russel, F. esq. Brislington |
| | Smith, A. esq. M.R.C.S. |
| BROMSGROVE, *Worcestersh.* | Maund, B. esq. |
| BUNGAY . . . . | Garneys, W. C. esq. |
| | Stock, D. esq. |
| | Webb, J. E. esq. |
| BURFORD, *Oxon.* . . | Cooke, W. R esq. |
| BURRISOKANE, *Ireland* . | Waller, Edward, esq. |
| BURTON-ON-TRENT . . | Brown, Edwin, esq. |
| | Thornwell, R. esq. |
| BURY ST. EDMUNDS, *Loc. Sec.* | SMITH, C. C. esq. M.R.C.S. |
| | Dunnett, J. esq. |
| | Image, Rev. T. A.M. F.G.S. Whepsted. |
| | Oakes, J. esq. |
| BUXTON, *Derbyshire* . | Robertson, W. H. M.D. |
| CAMBRIDGE . *Local Sec.* | BABINGTON, C. C. M.A. F.L.S.G.S. St.John's Col. |
| | Anthony J. esq. Caius College |
| | Babington, C. esq. B.A. St. John's College |
| | Carter, J. esq. |
| | Clark, Rev. Professor, M.D. F.R.S |
| | Collings, W. esq. Trinity College |

| | |
|---|---|
| CAMBRIDGE . . . . | Douglas, R. C. esq. B.A. Corpus Christi College |
| | Fitton, W. J. esq. B.A. Trinity College |
| | Frere, H. T. esq. Corpus Christi College |
| | Gandy, J. H. esq. Trinity College |
| | Garnons, Rev. P. L. B.D Sidney College |
| | Henslow, Professor, M.A. F.L.S. |
| | Hewetson, J. esq. Corpus Christi College |
| | Newnham, W. O. esq. St. John's College |
| | Ollivant, Rev. Professor, D.D. |
| | Power, Rev. Joseph. M.A. |
| | Sedgwick, Rev. Professor, M.A. F.R.S. |
| | Smith, Rev. J. J. M.A. Caius College |
| | Townsend, F. esq. Trinity College. |
| | Walton, R. esq. |
| | Wollaston, T. V. esq. B.A. Jesus College |
| | Wooley, J. esq. B.A Trinity College |
| CANTERBURY . . . | Bartlett, J. P. esq. Kingston Rectory |
| | Conyngham, Lord A. K.C.H. F.S.A. |
| | Harrison, Rev. J. B. Barham |
| | Kenrick, Miss, Stonehouse |
| CASTLE DOUGLAS . . | Maxwell, W. esq. Munshes |
| CHELMSFORD, *Essex, L. Sec.* | MEGGY, G. esq. |
| | Greenwood, A. esq. |
| | Barlow, J. N. esq. Writtle |
| CHELSEA . . . | Warburton, H. esq. M.P. F.R.S. PRES G.S. |
| CHELTENHAM . *Local Sec.* | CARY, W. esq. |
| | Allardyce, Dr. |
| | Anthony, Dr. M. |
| | Bodley, T. esq. F.G.S. |
| | Wright, T. esq. |
| CHESTER . . . | Fox, Rev. Darwin, M.A. Delamere Rectory |
| CHESTERFIELD . . | Walker, J. esq. |
| CHICHESTER, *Sussex* . | Bayton, Rev. W. S. |
| | Gruggen, J. B. esq. |
| CHIPPENHAM, *Wilts.* . | Alexander, Dr. R. C. Corsham |
| CHIPPING NORTON, *Oxon.* | Goatley, Thos. esq. |
| CHORLEY, *Lancashire* . | Williams, Rev. J. Euxton Glebe |
| CLAPHAM COMMON . . | Deane, Henry, esq. |
| | Horne, C. esq. |
| | Howells, J. esq. |
| | Hudson, R. esq. F.R.S. G.S. |
| | Potts, W. J. esq. |
| | Wollaston, Miss. |
| CLAPHAM RISE . . | Pollexfen, Rev. J. H. B.A. M.D. |
| CLAPTON . . . | Kennedy, B. esq. |
| COLCHESTER, *Essex* . | Walker, D. esq. M.A. |
| COLDSTREAM, *Berwickshire* | Melrose, J. esq. |
| CONGLETON . . . | Hall, J. esq. |
| CORK . . *Local Sec.* | HARVEY, J. R. M.D. |
| | Clear, W. esq. |
| | Haines, Dr. |
| | Harrison, Dr. |
| | Meredith, W. L. M.D. |
| | Popham, J. M.D. |
| | Power, T. M.D. |
| | Wood, Sam. A.M. M.D. Bandon |
| CORTLING STOCK, *Notts.* . | Harley, F. esq. |

| | | |
|---|---|---|
| Darlington . . . | | Backhouse, W. J. esq. |
| | | Pease, Mrs. Feetham |
| Derby . | *Local Sec.* | Bell, R. J. esq. |
| | | Hewgill, A. esq. Ripton |
| | | Ray, J. esq. Hennor Hall |
| | | Robertson, J. esq. |
| Devonport . | *Local Sec.* | Hore, Rev. W. S. M.A. F.L.S. |
| | | Devon and Cornwall Nat. Hist. Soc. |
| | | Head, Edw. esq. |
| | | Shepheard, Jas. M.D. Stonehouse |
| | | Morrison, Dr. |
| | | Mackey, Dr. G. |
| Dorking . . . | | Brown, Isaac, esq. |
| Doncaster . . . | | Hindle, J. esq. Norton |
| Down, *Kent* . . . | | Darwin, C. esq. M.A. |
| Downham, *Norfolk* . . | | Salmon, R. esq. |
| Dublin . | *Local Sec.* | Ball, R. esq. M.R.I.A. |
| | | Allman, Professor |
| | | Ball, J. esq. B.A. M.R.I.A. |
| | | Banks, Dr. |
| | | Beatty, Prof. T. E. M.D. M.R.I.A. |
| | | Bergin, T. esq. M.R.I.A. |
| | | Callwell, R. esq. M.R.I.A. |
| | | Crampton, Sir Philip |
| | | Crocker, Dr. C. P. M.R.I.A. |
| | | Farren, C. M.D. |
| | | Griffiths, Richard, Esq. M.R.I.A. |
| | | Harrison, Professor |
| | | Harvey, W. H. M.D. Trinity College |
| | | Hutton, Thos. esq. F.G.S. M.R.I.A. |
| | | Irvine, Hans, esq. M.B. |
| | | James, Capt. R.E. |
| | | Litton, Prof. M.D. R.D.S. |
| | | Lyle, Mrs. Acheson |
| | | Mackenzie, S. esq. LL.D. |
| | | Mackay, J. T. esq. M.R.I.A. |
| | | Oldham, Professor |
| | | Royal Society |
| | | Royal Irish Academy |
| | | Staunton, T. B. esq. |
| Edinburgh . | *Local Sec.* | Trinity College |
| | | Warren, T. W. esq. |
| | | Maclagan, Douglas, M.D. F.R.S.E. |
| | | Alison, Prof. W. P. M.D. |
| | | Balfour, Prof. M.D. F.L.S. |
| | | Chambers, R. esq. |
| | | Dalyell, Sir J. G. Bart. |
| | | Dundas, Mrs. Murrayfield |
| | | Goodsir, Professor John, F.R.S. |
| | | Goodsir, H. D. S. esq. M.W.S. R.N. |
| | | Greville, R. K. LL.D. F.R.S.E. |
| | | Grut, N. esq. F.R.S.E. |
| | | Knapp, Dr. |
| | | Lizars, W. H. esq. |
| | | Macdonald, A. esq. |
| | | Melville, A. esq. |
| | | Mercer, J. M.D. F.R.C.S.E. |

| | | |
|---|---|---|
| EDINBURGH . . . | | Neill, P. LL.D. F R.S.E. L.S. |
| | | Parnell, R. M.D. F.R.S.E. |
| | | Sellers, Wm. M.D. |
| | | Simpson, Prof. J. Y. M.D. |
| | | Stark, R. M. esq. |
| | | Taylor, Dr. Robert |
| | | Thompson, Prof. A. |
| | | Torrie, T J. esq. F.G.S. |
| ELGIN . . . . . | | Gordon, Rev. J. Manse of Birnie |
| EMSWORTH, *Hants* . . | | Miller, G. esq. M.R.C.S.E. |
| ESHER, *Surrey* . . . | | Spicer, Rev. W. W. |
| | | Spicer, J. W. G. esq. |
| EXETER . . | *Local Sec.* | SCOTT, W. R. esq. PH.D. |
| | | Shapter, Thos. M D. |
| | | Hinckes, Rev. T. |
| EXMOUTH . . . . . | | Kane, W. esq. B.A. T.C.D. |
| FALMOUTH . . . | | Fox, A. L. esq. |
| | | Squire, L. esq. |
| | | Cocks, W. P. esq. |
| FARNHAM, *Surrey* . . | | Newnham, W. O. esq. |
| FAVERSHAM, *Kent* . . | | Munn, W. A. esq. |
| FOWEY, *Cornwall* . . | | Holman, J. R. esq. |
| FRITHAM, *Stoneycross, Hants*. | | Munro, Miss |
| GARGRAVE, *Yorks*. . . | | Currer, Miss, Eshton Hall |
| GATESHEAD . . . | | Hardy, J. esq. |
| GLASGOW . | *Local Sec.* | THOMPSON, R. D. M.D. |
| | | Ballock, R. esq. |
| | | Blackie, W. G. esq. |
| | | Cuthbertson, D. LL.B. |
| | | Gourlie, W. J. esq. |
| | | Gray, T. esq. |
| | | Mackintosh, G. esq. Campsie |
| | | Philosophical Society |
| | | Rogers, Mrs. |
| | | Watson, J. M.D. |
| GLOUCESTER . | *Local Sec.* | HITCH, J. W. M.D. |
| | | Cockin, J. esq. R.N. |
| | | Huxley. J. E. M.D. |
| | | Wilton, J. W esq. |
| | | Wintle, G. S. esq. |
| GODALMING . | *Local Sec.* | SALMON, J. D. esq. |
| | | Bull, Henry, esq. |
| | | Chandler, A. T. esq. |
| GOSPORT . . . | | Richardson, Sir John, M.D. F.R.S.L.S. &c. Haslar Hospital |
| GREENOCK . . . | | Gray, J. esq. |
| GREENWICH . | *Local Sec.* | BUSK, G. esq. |
| | | Burton, J. esq. |
| | | Clapp, W. esq. |
| | | Jefferson, J. esq. |
| | | Leeson, Dr. |
| | | Liddell, Dr. |
| | | Purvis, P. M.D. |
| | | Society for the Diffusion of Useful Knowledge |
| | | Stratton, Miss Dora |
| GUILDFORD, *Surrey* . . | | Austin, R.A.C. esq. SEC. O.S. Merrow |
| | | Chandler, Mrs. Bramley |

| | |
|---|---|
| HACKNEY . . . | Mackintyre, Dr. F.L.S. |
| HALIFAX, *Yorks. Local Sec.* | INGLIS, J. M D. |
| | Adam, T. esq. |
| | Bowman, J, esq. |
| | Bramley, Lawrence, esq. |
| | Holland, W. esq. Light Cliff |
| | Jubb, Abraham, esq. |
| | Leyland, E. esq. |
| | Mac Taggart, D. esq. |
| | Rawson, T. W. esq |
| | Rudd, E. J. esq. |
| | Sutherland, Capt. G. M. |
| | Waterhouse, J. esq. F.R.S. |
| HAMPSTEAD . . . | Hinckes, Rev. W. F.L.S. |
| | Edwards, F. C. esq. Downshire Hill |
| HARMSTON, *Lincolns.* . | Clark, Hamlet, esq. |
| HARROW . . . . | Curtis, — esq. |
| | Hewlet, G. esq. |
| HARROWGATE . . . | Brown, C. F. esq. |
| HASTINGS, *Sussex* . . | Gabb, G. H. esq. |
| | Hodson, Miss |
| | Literary and Scientific Society |
| | Ranking, Robert, esq. |
| HATFIELD. *Herts. Local Sec.* | THOMAS, W. Lloyd, esq. |
| | Branton, J. esq. |
| | Church, J. esq. |
| HENFIELD, *Sussex* . . | Borrer, W. esq. F.R.S.L.S. |
| HEREFORD . *Local Sec.* | PRICE, T. Tucker, esq. |
| | Bodenham, F. L. esq. |
| | Boniville, A. C. de, esq. |
| | Lingin, C. M D. |
| | Lingwood, R. M. esq. M.A. F.L.S. Ross |
| | Lye, J. Bleeck, M.D. |
| | Wright, E. G. esq. |
| HERTFORD . *Local Sec.* | REID, F. G M.D. |
| HITCHIN, *Herts.* . . . | Curling, W. esq. |
| HOLMES CHAPEL, *Cheshire* | Hall, C. R. M.D. |
| HOMERTON . . . | Smith, Dr. Pye |
| HORNSEY, *Middlesex* . | Buckton, G. B. esq. |
| | Hands, B. esq. |
| HORTON, *near Slough* . | Stevens, W. V. esq. |
| HOUNSLOW . . . | Statham. Rev. S. F. |
| HUDDERSFIELD . . | Welch, R. esq. |
| HULL . . *Local Sec.* | NORMAN, G. esq. |
| | Huntingdon, F. esq. |
| | Kitching, A. esq. |
| | Sandwith, H. M.D. |
| | Thompson, Thomas, esq. |
| | Young, J. esq. |
| HUNMANBY, *Yorks.* . . | Mitford, Capt. |
| HUNTINGDON . . | Isaacson, W. esq. |
| IPSWICH, *Suffolk, Local Sec.* | KING, J. esq. |
| | Johnson, J. W esq. |
| | Philosophical Society |
| | Ransome, Geo. esq. |
| | Ransome, Robert, esq. |
| ISLE OF SKYE . . . | Martin, Dr. Glendall |

| | |
|---|---|
| ISLE OF WIGHT *Local Sec.* | SALTER, T. B. M.D. F.L.S. Ryde |
| | Barnes, Rev. H. F. |
| | Broomfield, W. A. M.D. F.L.S. |
| | Ward, Geo. esq. Cowes |
| KEITH, *Banffshire* . . | Christie, Dr. J. |
| KELSO . . . . | Brisbane, Sir T. M. Bart. K.C.B., &c., Mackerstown |
| | Douglas, F. M.D. |
| | Wilson, Charles, M.D. |
| KELVEDON, *Essex* . . | Varenne, E. G. esq. |
| KUNMERGHAME . . | Bonar, Miss |
| KINGSBRIDGE . . . | Elliot, J. esq. |
| KINGSLAND . . . . | Humphries, Ed. esq. |
| KIRKPATRICK, *Moffat* . | Little, Rev. W. |
| LANCASTER . *Local Sec.* | SIMPSON, S. esq. |
| | Amicable Library |
| | Giles, Jas. esq. |
| | Howitt, Thos. esq. |
| | Natural History Society |
| LEE, *Kent* . . . | Carr, W. esq. M.R.C.S. |
| LEEDS . . *Local Sec.* | TEALE, T. P. esq. |
| | Allanson, J. esq. |
| | Bearpark, G. E. esq. |
| | Chadwick, C. M.D |
| | Denny, H. esq. |
| | Gott, J. esq. Wither |
| | Gott, W. esq. |
| | Hall, M. esq. Wortley |
| | Hey, Samuel, esq. |
| | Morley, G. esq. |
| | Nunnerly, Thos. esq. |
| | Price, W. esq. |
| | Teale, Joseph |
| LEICESTER . *Local Sec.* | HARLEY, JAS. esq. |
| | Bates, H. W. esq. |
| | Hudson, E. D. Esq. Little Thorpe |
| | Kirby, T. B. esq. |
| | Moore, J. esq. |
| | Nedham, J. esq. F.L.S. |
| | Plant, J. esq. |
| | Stallard, J. P. esq. |
| LENHAM LODGE, *Kent* . | Fielding, G. H. M D F.R.S., &c. |
| LEWISHAM, *Kent* . . | Steel, C. W. esq. |
| LISCARD, *Cheshire* . . | Boyd, J. C. esq. |
| | Lowe, Jas. esq. |
| LIVERPOOL . *Local Sec.* | DICKINSON, J. JOSEPH, M.D. |
| | Archer, F. esq. |
| | Davies, W. M. esq. |
| | Ellison, K. esq. |
| | Fleming, — esq. Bootle |
| | Henderson, T. esq. |
| | Holt, G. esq. |
| | Inman, Dr. Thos. |
| | Lyceum Library |
| | Mc Andrew, R. esq. |
| | Medical Institution |

LIVERPOOL . . . Robson, H. E. esq.
Royal Institution
Sweetlove, J esq.
Thompson, G. esq.
Tudor, R. esq. Bootle
Wells, W. esq.

LLANBEDR HALL, *Ruthin* . Ablett, J. esq.

LOCKERBIE, *Dumfries* Jardine, Sir W. Bart. F.R.S.E. L.S., &c.
Wilson, T. J. esq.

## LONDON LIST.

Adlard, C. and J. . . . . . Bartholomew close
Allchin, W. H. esq. M.B. . . . .
Ansted, Prof. D. T. M.A. F.R.S . King's College
Athenæum Club . . . . Pall Mall
Bailey, — esq. . . . . . Museum of Economic Geology
Baird, W. M D. . . . . . British Museum
Barrow, Mrs. . . . . . . 65, Old Broad street
Bell, Professor, F.R S. L.S. G.S. &c. . New Broad street, City
Bell, J. T. esq. . . . . . 10, Northwick terrace
Benson, John, esq.
Bennett, James Risdon, M D. . . 24, Finsbury Place
Bennett, J. I. esq. F.R S. F.Z.S SEC. L.S. British Museum
Botfield, J. B. esq. M.P. F.R.S L.S. &c. 9, Stratton street
Bowerbank, J. S. esq. F.R.S. G.S. &c. 3, Highbury Grove
Bowerbank, E S. esq. . . . . Sun street, Bishopsgate
Bowman, William, esq. F.R.S. . . Golden square
Brodhurst, B. esq.
Browell, E. esq. . . . . . Ambassadors' Court, St. James's Palace
Brown, Robert, D C L. F.R S. V.P.L.S. &c. British Museum
Browne, W. M. esq. . . . . Westminster Fire Office, King st., Covent Garden
Buchanan, Walter, esq. F.L.S. . . 2, Sussex Place, Hyde-park gardens
Budd, Dr. Geo. F.R S. . . . Dover street, Piccadilly
Burlington, Lord, F R S. . . . 10, Belgrave square
Britton, —, esq. . . . . . Holborn Bars
Carpenter, W. B. M.D. F.R.S. . . Fullerian Prof. Royal Inst.
Children, J. G esq. F R.S. S.A. L S. &c Torrington square
Cumming, H. esq. F.L S. . . . 80, Gower street
Dalrymple, John, esq . . . 56, Grosvenor street, Bond street
Day, Dr. . . . . . . . 3, Southwick street
De la Beche, Sir H. F.R S. G.S. &c. Craig's court
Dennes, G. E. esq. F.L.S. &c. . . Vine street, Golden square
Dickson, Dr. F.L.S. . . . . Curzon street, May Fair
Dilke, C. Wentworth, esq. . . 76, Sloane street
Doubleday, Edward, esq. F.L.S. &c. British Museum
Dunn, Robert, esq. . . . . 15, Norfolk street, Strand
Edgeworth, M. P. esq. . . . 1, North Audley street
Edwards, Thomas, esq.
Elsey, J. R. esq. . . . . Bank of England
Evans, Mrs. . . . . . . 72, Oxford terrace, Edgeware road.
Falconer, Dr. H. F R S. L S. &c . 23, Norfolk street, Strand
Farre, Dr. Frederic, F.L.S. . . Bridge street, Blackfriars

| | |
|---|---|
| Fellows, Sir C. . . . . . | 30, Russell square |
| Fennell, James H. esq. . . . | Devereux court |
| Ferguson, Wm. esq . . . . | 18, Wharton street, Lloyd square |
| Finsbury Medical Book Society . | 2, St. Mary Axe |
| Fitton, E. B. esq. . . . . . | 53, Upper Harley street |
| Fitzwilliam, Earl . . . . | |
| Flower, J. W. esq. . . . . . | 61, Bread street, City |
| Forbes, Professor Edward, F.R.S. &c. | King's College |
| Forbes, J. M.D. F.R.S. F.L.S. &c. . | Old Burlington street |
| Forster, Edward, esq. F.R.S. V.P.L.S. &c. | Mansion House street |
| Frampton, Dr. A. . . . . . | New Broad street, City |
| Gough, G. S. esq. M.R.I.A. Gren. Gds. | 45, Park street, Grosvenor square, or Athenæum |
| Gould, John, esq. F.R.S. L.S. Z.S. &c. | 21, Broad street, Golden square |
| Gratton, James, esq. . . . . | Shoreditch |
| Greening, William, esq. . . | Marlborough place, Kent road |
| Gull, Dr. . . . . . . | Guy's Hospital |
| Gutch, J. G. esq. . . . . . | Great Portland street |
| Hall, Marshall, esq. . . . . | 14. Manchester square |
| Hamilton, Rev. James . . . | 7, Lansdown place, Brunswick sq. |
| Hanson, Samuel, esq . . . | 15, Trinity square, Tower hill |
| Hardwicke, William, esq. . . | Calthorpe street, Gray's Inn lane |
| Haywood, J. W. esq . . . . | |
| Henderson, W. T. esq. . . . . | London & Westminster Bank, Lothbury |
| Henfrey, Arthur, esq. F.L.S. &c. . | Middlesex Hospital |
| Henry, J. H. esq. F.R.S. . . . | Brick lane Brewery |
| Hodgkin, Dr. F.R.S. . . . | Lower Brook street |
| Hogg, J. esq. M.A. F.R.S. L.S. . . | 12, King's Bench walk |
| Holman, J. R. esq. . . . . . | London Hospital |
| Hope, A. J. B esq. M.P. . . . | 1, Connaught place |
| Horner, Leonard, esq. F.R.S. . . | 2, Bedford place |
| Horsfield, Dr. F.R.S. L.S. G.S &c. . | India House, London |
| Hoyer, J. esq. . . . . . | 8, Great St. Helen's, Bishopsgate |
| Ibbotson, S. L. B. esq. . . . | Museum of Economic Geology |
| Janson, Joseph, esq. F.L.S. . . . | 32, Abchurch lane |
| Jones, Capt. M.P. F.G.S. &c. . . | 30, Charles street, St. James's |
| Jones, — esq. . . . . . | 23, Soho square |
| Joy, C. A. esq. . . . . . | King's College, London |
| Kelaart, Dr. F.L.S. | |
| Keman, J. esq. | |
| Kitching, W. V. esq. . . . | 46, Conduit street |
| Langmore, Dr. . . . . . | 40, Finsbury square |
| Langmore, W. B. esq. . . . | 4, Christopher street, Finsbury sq. |
| Langstaff, J. esq. . . . . . | 9, Cambridge square, Hyde park |
| Lankester, E. M.D. F.R.S. L.S. &c. . | 22, Old Burlington street |
| Laycock, Dr. C. R. | |
| Lindley, Professor, D. PH. F.R.S. &c. | 21, Regent street |
| Lloyd, W. Horton, esq. F.L.S. Z.S. &c. | Park square, Regent's park |
| London Institution . . . . . | Finsbury circus |
| London Library . . . . | St. James's square |
| Lovell, G. esq. . . . . . | 12, Ely place, Holborn |
| Lyell, Charles, M.D. F.R.S. L.S. G.S. &c. | 16, Hart street, Bloomsbury |
| Macmeikan, John, esq. . . . | London Hospital |
| Marshall, Matthew, esq. . . . | Bank of England |
| Mitchell, D. W. esq. . . . | 28, Great Russell st. Bloomsbury |

| | |
|---|---|
| Moore, J. C. esq. | 37, Hertford street, May Fair |
| Mullinix, — esq. | 7, Adelphi terrace |
| Murchison, Sir R. J. F.R.S. &c. | Belgrave square |
| Murdoch, W. M.D. | 320, Rotherhithe street |
| Museum of Economic Geology | Craig's court |
| Nasmyth, Alex. esq. F L.S. G.S. &c. | 13, George street, Hanover square |
| Newport, George, esq. F.R.S. | Southwick street, Cambridge terrace |
| Northampton, Marquis of, F.R.S. &c. | 145, Piccadilly |
| Ogilby, W. esq. | Upper Gower street |
| Otter, Captain | Hydrographic Survey |
| Owen, Professor, F.R.S. L.S. &c. | Royal College of Surgeons |
| Parkinson, esq. F.L.S. | Cambridge terrace, Hyde park |
| Partridge, Richard, esq. F.R.S. | King's College, London |
| Percival, W. esq. 1st Life Guards | Hyde park Barracks |
| Pereira, J. M.D. F.R.S. &c. | Finsbury square |
| Playfair, Lyon, Dr. | Museum Economic Geology |
| Powell, Hugh, esq. | Clarendon street, Somers town |
| Pratt, S. P. esq. | 53, Lincoln's inn fields |
| Prevost, J. L. Swiss Consul | 3, Suffolk place |
| Prestwich, Joseph, jun. esq. | Mark lane |
| Pritchard, A. esq. | 3, Canonbury lane, Islington |
| Quain, Richard, M.D. | University College Hospital |
| Quekett, E. J. esq. F.L.S. | 50, Wellclose square |
| Ramsay, A. esq. | Museum Economic Geology |
| Reeve, Lovell, esq. A.L.S. &c. | King William street, Strand |
| Reynolds, Henry, esq. | 42, Moorgate street |
| Ross, Andrew, esq. | 21, Featherstone buildings, Holborn |
| Rothery, H. C. esq. M.A. | 10, Stratford place |
| Royle, Prof. J. F. M.D. F.R.S. L.S. G.S. | 4, Bulstrode street, St. Marylebone |
| Russell Medical Book Society | |
| Rutherford, W. esq. | |
| Ryder, Charles, esq. | Stamp and Taxes Office, Somerset House |
| Salter, H. H. esq. | King's College |
| Saull, W. D. esq. F.G.S, A.S. &c. | 15, Aldersgate street |
| Seager, J. L. esq. | Millbank, Westminster |
| Searle, George, esq. | Cumming street, Pentonville |
| Semple, Robert H. esq. | 3, Han's buildings, Islington |
| Sion College Library | London Wall |
| Solly, E. esq. F.R.S. | Bedford row |
| Solly, R. H. esq. F.R.S. L.S. &c. | Great Ormond street, Queen square |
| Sowerby, J. de C. esq. F.L.S. Z.S. &c. | Royal Botanic Garden, Regent's park |
| Stokes, Charles, esq. F.R.S. G.S. &c. | 4, Verulam buildings, Gray's Inn |
| Strickland, J. H. esq. | 22, North Audley street |
| Spence, W. esq. F.R.S. L.S. &c. | 18, Lower Seymour st. Portman sq. |
| Taylor, Richard, esq. F.L.S. S.A. &c. | Red Lion court, Fleet street |
| Tennant, James, esq. F.G.S. &c. | 149, Strand |
| Tosswill, C. S. esq. | 8, Torrington place |
| Twining, William, M.D. | 13, Bedford place, Russell square |
| Walton, John, esq. F.L.S. &c. | 9, Barnsbury square, Islington |
| Warington, Robert, esq. | Apothecaries' Hall |
| Ward, N. B. esq. F.L.S. Z.S. &c. | Wellclose square |
| Waterhouse, G. esq. | British Museum |
| Wheatstone, Professor | King's College |
| White, H. H. esq. | 13, Old square, Lincoln's Inn |
| White, Adam, esq. | British Museum |

| | |
|---|---|
| White, Alfred, esq. F.L.S. &c. . . | Cloudesley square, Islington |
| Wilcock, W. esq. . . . . . | 6, Stone buildings, Lincoln's Inn |
| Wilson, Dr. . . . . . . | 3, Devonport street, Sussex square |
| Wix, Rev. S. F.R.S. A.S. . . . | St. Bartholomew's Hospital |
| Wood, S. V. esq. F.G.S. &c. | Bernard street |
| Wood, Basil, T. esq. F.L.S. &c. . | 14, Cumberland street |
| Wordsworth, — esq. | London Hospital |
| Wright, G. P. esq. . . . . | Auburn Cottage, Hawley road Kentish town |
| Yarrell, William, esq. F.L.S. Z.S. &c. | Ryder street, St. James's street |
| Young, Dr. J. F. F.Z.S. &c. . . | Upper Kennington lane |

---

| | |
|---|---|
| LONDONDERRY, *Ireland* . | Neville, Henry, esq. |
| LOUGHBOROUGH . . | Kennedy, J. M.D. Woodhouse |
| LOUTH . . . . | Smyth, W. H. esq. South Elkington |
| LUDLOW . . . . | Natural History Society |
| LYDD, *Kent* . . . | Plomley, F. M D. PH.D. F.S.A. &c. |
| LYNN, *Norfolk* . *Local Sec.* | SAYLE, G. esq. |
| | Bransby, Rev. John |
| | Lynn Book Society |
| | Mumford, Rev. Geo. East Winch Vicarage |
| | Rolfe, Rev. S.C.E.N. |
| | Whiting, J. B. esq. |
| MACHYNLLETH, *Montgomery*. | Bevan, E. esq. |
| MAIDSTONE . . . | Hepburn, J. esq. M.A. F.L.S. |
| MANCHESTER - *Local Sec.* | BARROW, P. esq. |
| | Fleming, Dr. |
| | Goadsby, E. esq. |
| | Holme, Dr. E. F.L.S. |
| | Jepson, W. M.D. Salford |
| | Literary and Philosophical Society |
| | Natural History Society |
| | Satterfield, J. esq. |
| | Wilkinson, E. M.A. M.D. |
| | Windsor, J. esq. F.L.S. F.R.C.S.E. |
| MARKET RASEN, *Lincolns.* | Cooper, W. W. esq, |
| MARLBOROUGH, *Wilts* . | Kemm, T. esq. |
| MAUCHLINE, *Ayrshire* . | Mitchell, Dr. |
| MELKSHAM, *Wilts* . . | Cooke, J. esq. |
| MORPETH, *Northumberland* | Bell, Miss, East Shafto |
| | Sidney, M. J. T. esq. Cowper |
| NANTWICH, *Cheshire* . | Broughton, Rev. D. |
| | Gretton, Rev. H. |
| NEATH, *Glamorganshire* . | Wallace, A. R. esq. |
| NEWBURY, *Berks* . . | Winterbottom, J. E. esq. M.A. B.M. F.L.S. |
| NEWCASTLE-ON-TYNE, *Loc.Sec* | THORNHILL, J. esq. |
| | Alder, J. esq. |
| | Bold, T. J. esq. |
| | Currie, R. esq. |
| | Embleton, Dennis, M.D. |
| | Hancock, J. esq. |
| | Hutton; — esq. |

| | |
|---|---|
| Newcastle-on-Tyne . | Literary and Philosophical Society |
| | Loftus, W. K. esq. |
| | Pigg, T. jun. esq. |
| | Robertson, W. esq. F.Z.S. |
| Northampton . . | Hartshorne, Rev. C. |
| Northumberland . | Fryer, J. H. esq. |
| Norwich . *Local Sec.* | Brightwell, T. esq. F.L.S. |
| | Birkbeck, H. esq. Keswick |
| | Copeman, G. esq. Collishall |
| | Crosse, J. G. esq. |
| | Dalrymple, D. esq. |
| | Gurney, Mrs. R. |
| | Gurney, J. H. esq. Earlham |
| | Hudson, A. esq. |
| | Norfolk and Norwich Literary Institution |
| | The Lord Bishop of |
| | Wells, R. esq. |
| Norwood . . . . | Grainger, R. D. esq. F.R.S. Anerley |
| Nottingham . *Local Sec.* | Eddison, B. esq. |
| | Fletcher, Rev. C. Southwell |
| | Higginbottom, J. esq. F.R.C.S. |
| | Percy, E. esq. |
| | Subscription Library |
| | Taylor, H. M.D. |
| Oakland, *near Llanrwst, Denbighs.* . . . . | Blackwall, J. esq. F.L.S. |
| Orkney . . . . . | Haddle, Robert, esq. Melseter |
| Oulton Park, *Cheshire* . | Egerton, Sir P. G. Bart. M.P. F.R.S. G.S. &c. |
| Oxford . *Local Sec.* | Daubeny, Prof. M.D. F.R.S. &c. |
| | Ackland, — esq. |
| | Ashmolean Society |
| | Baxter, W. esq. |
| | Duncan, P. esq. M.A. F.G.S. |
| | Freeborn, J. J. S. esq. |
| | Greenhill, Dr. |
| | Jackson, Dr. |
| | Parker, C. L. esq. |
| | Radcliffe Library |
| | Strickland, H. E. esq. M.A. F.G.S. &c. |
| | Twiss, Prof. T. |
| Paisley . . . . . | Kay, J. esq. |
| | Steuart, Sir J. Bart. |
| | Stewart, A. M.D. |
| Panshanger, *Herts* . | Beck, Rev. J. E. Cowper |
| Petworth, *Sussex* . . | Peachey, W. esq. Fittleworth |
| Plymouth . . . | Budd, Dr. |
| | Fuge, J. H. esq. F.R.C.S.E. |
| | Lane, E. esq. |
| | Miller, O. T. esq. |
| | Rohloff, — esq. |
| | Tripe, Dr. |
| Pontefract . . . | Thorpe, Rev. W. Womersley |
| Pont-y-Pool . . . | Bladon, J. esq. |
| Poole, *Dorset* . . | Lacy, E. esq. M.R.C.S. |
| Poplar . . . . | White, C. G. esq. |
| Portsmouth . . . | Raper, W. A. M.D. |

| | |
|---|---|
| **Preston**, *Lancashire* . | Brown, R. esq. |
| | Heslop, R. C. M.D. |
| | Literary and Philosophical Society |
| **Preston Kirk**, N.B. | Hepburn, A. esq. Whittingham |
| | Hepburn, Sir T. B. M.P. Smeaton |
| **Rammerscales**, *Lochmaben* | Macdonald, W. B. esq. |
| **Reading** . . . . | Harrison, J. esq. |
| **Reigate**, *Surrey* . . | Hepburn, T. B. esq. Chipstead |
| | Hunson, W. esq. |
| | Reigate Institution |
| **Repton** . . . . | Hewgill, Arthur, M.D. |
| **Rochdale** . . . | Buckley, N. esq. |
| | Coates, J. esq. |
| | Fenton, J. esq. Lynn Hall |
| **Rochester**, *Kent* . . | Martin, D. A. |
| **Rockhill**, *Letterkenny* . | Stewart, J. V. esq. |
| **Rugby** . . . . | Paxton, J. M.D. |
| **Rye**, *Sussex* . . . | Plomley, J. T. esq. |
| **Saffron Walden**, *Essex* . | Gibson, G. S. esq. |
| **Saltcoats**, *Ayrshire* . | Landsborough, Rev. D. |
| **Scarborough**, *York*. *Loc. Sec.* | Murray, P. M.D. |
| | Atkinson, Rev. J. C. B.A. |
| | Bean, W. esq. |
| | Leckenby, J. esq. |
| | Rowntree, J. esq. |
| | Theakstone, J. W. esq. |
| **Sedburgh**, *Yorks*. . . | Pinder, Rev. Geo. |
| **Settle**, *Yorks*. . . | Tatham, J. jun. esq. |
| **Sheffield** . *Local Sec.* | Jackson, H. esq. |
| | Gardner, S. J. esq. |
| | Heppenstall, J. esq. Upper Thorpe |
| | Kitching, — esq. |
| | Literary and Philosophical Society |
| | Newbold, Rev. W. M. |
| | Thomas, H. esq. |
| **Sherborne** . . . | Dale, J. C. esq. M.A. F.L.S. C.P.S. |
| **North Shields** . . | Clarke, W. B. M.D. |
| **South Shields** . . | Winterbottom, P. M. M.D. Westoe |
| | Ingham, R. esq. M.A. F.G.S. Westoe |
| **Shrewsbury** . . . | Eyton, T. C. esq. F.G.S. Z.S. |
| | Leighton, Rev. W. A. B.A. F.B.S.E. Luciefield |
| | Slaney, W. H. esq. |
| | Stokes, W. R. esq. |
| **Sidmouth** . *Local Sec.* | Cullen, W. H. M.D. |
| | Creswell, Rev. R. |
| | Cutler, Miss, Budleigh-Salterton |
| **Sissinghurst** . . . | Cleaver, Mrs. H. |
| **Sittingbourne**, *Kent* . | Ray, Geo esq. Milton |
| **South Petherton** . . | Norris, H. esq. |
| **Southampton** . . | Wheeler, Mrs. |
| | Toovey, —, esq. Ordnance Survey |
| **Spilsby**, *Lincolnshire* . | Spence, Rev. John, East Keal |
| **Spofforth**, *Yorks*. . . | Herbert, the Very Rev. Dean of Manchester |
| **St. Alban's** . . . | Blagg, T. W. esq. |
| | Lewis, T. esq. |
| **Stepney** . . . | Goldsworthy, J. H. esq. |

| | |
|---|---|
| Stockbridge, *Hants* . | Elwes, J. esq. |
| Stockport . . . | Hill, W. T. esq. |
| Stockton-on-Tees . | Keenlyside, R. H. m.d. |
| Stoke Newington . | Miller, C. M. esq. |
| Stoneport, *Worcestershire* | Clifton, J. esq. |
| Stowmarket, *Suffolk* . | Bree, R. C. esq. |
| | Thurlow, Lord |
| Stratford, *Essex* . . | Allcard, J. esq. f.l.s. |
| | Lister, J. J. esq. f.r.s. Upton |
| Sunderland . . . | Backhouse, T. J. esq. |
| Swaffam Bulbeck . | Jenyns, Rev. L. m.a. f.l.s. |
| Swaffham, *Norfolk* . | Dowell, E. W. esq. |
| Swansea . . . | Dillwyn, L. L. esq. |
| | Jeffreys, J. G. esq. f.r.s. |
| | Llewellyn, — esq. |
| | Moggridge, M. esq. |
| | Wood, Mrs E. |
| Swindon, *Wilts* . . | Prower, Rev. J. M. Purton |
| Taunton . *Local Sec.* | Higgins, C. H. esq. |
| | Alford, H. esq. |
| | Bendon, W. jun. esq. |
| | Cathett, Dr. |
| | Crotch, Rev. — m.a. |
| | Gillet, W. E. esq. |
| | Langton, W. G. esq. |
| | Raban, Capt. R. B. r.n. |
| | Woodford, Dr. |
| Tenby, *South Wales*, *Local Sec.* | Falconer, Dr. R. W. |
| | Player, Mrs. |
| | Wilson, Edw. esq. |
| Thorncoffin, *Somerset* . | Sabine, Rev. W. b.a. |
| Tollerton, *Yorks.* . | Bird, G. esq. |
| Torquay . *Local Sec.* | Battersby, Dr. |
| | Fyshe, Rev. F. |
| | Griffith, Mrs. |
| | Natural History Society |
| | Spragge, W. K. esq. Tor |
| | Spragge, F. esq. Tor |
| Tottenham . . . | Fowler, W. esq. |
| | Lewis, J. esq. |
| | Moon, W. esq. |
| Truro, *Cornwall* . . | Boase, — esq. |
| | Cornwall Library |
| | King, R. L. esq. |
| | Tweedy, M. esq. Alverton |
| Tuam, *Ireland* . . | Clarke, Rev. B. |
| Twizel, *Northumberland* . | Selby, P. J. esq. f.l.s. g.s. &c. |
| Uckfield, *Sussex* . . | Hough, W. esq. Maresfield |
| Upton, *Cheshire* . . | Webster, W. jun. esq. |
| Upper Woodside, *Cheshire* | Byerly, J. esq. |
| Wansford . . . | Berkeley, Rev. M. J. m.a. f.l.s. &c. King'sCl |
| Wakefield, *Yorks.* *Loc Sec.* | Naylor, G. F. esq. |
| | Corsellis, C. C. m.d. |
| | Statter, W. esq. |
| | Winn, C. esq. Nostell |
| Ware, *Herts* . . . | Say, Rev. F. H. S. |

| | |
|---|---|
| Warrington . . . | Natural History Society |
| Warwickshire . . | Natural History Society |
| Wells, *Somerset* . . | Walker, J. jun. esq. Axbridge |
| | Whish, H. esq. B A. |
| Welster, *Orkney* . . | Weddle, R. esq. |
| West Malling, *Kent* . | Phelps, Rev. H. D. |
| Welwyn, *Herts* . . | Clifton, A. C. esq. |
| Whittingham, *Northumberl.* | Collingwood, F. J. W. esq. |
| Whitehaven . *Local Sec.* | Dickenson, W. esq. |
| | Stanger, — esq. |
| | Whitehaven Library |
| Wigan, *Lancashire* . . | Walmsley, V. O. esq. |
| Winchester . *Local Sec.* | White, A.D. M.D. |
| | Wickham, W. J. esq. |
| | The Very Rev. the Dean of Winchester |
| Witham, *Essex* . *Local Sec.* | Pattisson, Jacob H. esq. LL.B. |
| | Baker, R. esq. Writtle |
| | Blood, J. H. esq. |
| | Butler, W. esq. |
| | Ducane, Capt. R.N. |
| | Luard, W. W. esq. |
| | Pattisson, W. H. esq. |
| | Witham Literary Institution |
| Wolverhampton . . | Bidwell, H. esq. Albrighton |
| Woodbridge, *Suffolk* . | Kirkman, Dr. W. Melton |
| Woodhill, *by Ripley, Surrey* | Ainslie, W. esq. |
| Woolwich . . . | Sabine, Col. |
| Worcester . . . | Streeten, R. J. N. M.D. |
| Yarmouth . . . | Nelson, J. G. esq. |
| | Turner, D. esq. F.R.S. &c. |
| Yetholm, *Roxburghshire* . | Baird, Rev. John |
| York . *Local Sec.* | Tuke, W. H. esq. |
| | Allis, T. H. esq. |
| | Barker, T. esq. |
| | Davies, Mrs. |
| | Gray, W. esq. |
| | Hodson, H. B. esq. |
| | Husband, W. D. esq. |
| | King, H. esq. |
| | Matterson, Wm. jun. esq |
| | Meynell, T. esq. |
| | Phillips, Prof. John, F.R.S. |
| | Philosophical Society. |
| | Swineard, T. esq. |
| | Travis, T. H. esq. |
| | Tuke, J. H. esq. |
| | Wellbeloved, Rev. C. |

## FOREIGN.

| | |
|---|---|
| AUSTRALIA . . . | Cotton, John, esq. Doogullock Station, River Goulburn, Port Philip |
| | Ewing, Rev. Thos. |
| BENGAL . . . . | Maclagan, Lieut. R. Bengal Engineers |
| CANADA . . . . | Maclagan, P. W. M.D. Assistant Surgeon in the Army |
| CANINO . . . . | The Prince of |
| CAPE OF GOOD HOPE . | Stanger, W. M.D. |
| FRANKFORT-ON-MAINE . | Rüppell, Herr |
| JAMAICA . . . | Wharton, Rev. T. Bath |
| LIEGE . . . . | Köninck, L. de, esq. |
| | Longchamps, M. de, Selys |
| LEYDEN . . . . | Höven, Van der |
| MADRAS . . . . | Jerdon, T. C. esq. Literary Society |
| NEUFCHATEL . . . | Rougement, M. de |
| NEW BRUNSWICK . . | Robb. J. M.D. King's College, Fredericton |
| PARIS . . . . | Verneuil, M. E. de |
| UMBALA, *India* . . | Edlin, Dr. E. |